LOW POWER DIGITAL CMOS DESIGN

LOW POWER DIGITAL CMOS DESIGN

Anantha P. Chandrakasan
Massachusetts Institute of Technology

Robert W. Brodersen
University of California/Berkeley

KLUWER ACADEMIC PUBLISHERS
Boston / Dordrecht / London

Distributors for North America:
Kluwer Academic Publishers
101 Philip Drive
Assinippi Park
Norwell, Massachusetts 02061 USA

Distributors for all other countries:
Kluwer Academic Publishers Group
Distribution Centre
Post Office Box 322
3300 AH Dordrecht, THE NETHERLANDS

Consulting Editor: Jonathan Allen, Massachusetts Institute of Technology

Library of Congress Cataloging-in-Publication Data

A C.I.P. Catalogue record for this book is available
from the Library of Congress.

Printed on acid-free paper.

Printed in the United States of America

PREFACE

The genesis of low power microelectronics can be traced to the invention of the transistor in 1947. The elimination of the crushing needs for several watts of heater power and several hundred volts of anode voltage in vacuum tubes in exchange for transistor operation in the tens of milliwatts range was a breakthrough of virtually unparalleled importance in electronics. The capability to fully utilize the low power assets of the transistor was provided by the invention of the integrated circuit in 1958. Historically, the motivation for low power electronics has stemmed from three reasonably distinct classes of need: 1) the earliest and most demanding of these is for portable battery operated equipment that is sufficiently small in size and weight and long in operating life to satisfy the user; 2) the most recent need is for ever increasing packing density in order to further enhance the speed of high performance systems which imposes severe restrictions on power dissipation density; and 3) the broadest need is for conservation of power in desk-top and desk-side systems where a competitive life cycle cost-to-performance ratio demands low power operation to reduce power supply and cooling costs. Viewed in toto, these three classes of need appear to encompass a substantial majority of current applications of electronic equipment. Low power electronics has become the mainstream of the effort to achieve GSI.

The earliest and still the most urgent demands for low power electronics originate from the stringent requirements for small size and weight, long operating life, utility and reliability of battery operated equipment such as wrist watches, pocket calculators and cellular phones, hearing aids, implantable cardiac pacemakers and a myriad of portable military equipments used by individual foot soldiers. Perhaps no segment of the electronics industry has a growth potential as explosive as that of the personal digital assistant (PDA) which has been characterized as a combined pocket cellular phone, pager, e-mail terminal, fax, computer, calendar, address directory, notebook, etc. To satisfy the needs of the PDA for low power electronics, comprehensive approaches are proposed in this book that include use of the lowest possible supply voltage coupled with architectural, logic style, circuit and CMOS technology optimizations.

The antecedents of these concepts are strikingly evident in publications from the 1960's, in which several critical principles of low power design were formulated and codified (J. D. Meindl, *Micropower Circuits*, J. Wiley and Sons,

1969). The first of these was simply to use the lowest possible supply voltage, preferably a single cell battery. The second guideline was to use analog techniques wherever possible particularly in order to avoid the large standby power drain of then available bipolar digital circuits. A third key principle of micropower design that was convincingly demonstrated quite early is the advantage of selecting the smallest geometry, highest frequency transistors available to implement a required circuit function, e.g. a wideband amplifier, and then scaling down the quiescent current until the transistor gain-bandwidth product f_T just satisfies the relevant system performance requirements. A fourth generic principle of low power design that was clearly articulated in antiquity is the advantage of using "extra" electronics to reduce total power drain. This trade-off of silicon hardware for battery hardware was demonstrated e.g. for a multi-stage wideband amplifier in which total current drain was reduced by more than an order of magnitude by doubling the number of stages from two to four while maintaining a constant overall gain-bandwidth product. A final overarching principle of low power design that was rigorously illustrated for a wide variety of circuit functions including audio and video is that micropower design begins with a judicious specification of the required system performance and proceeds to the optimal implementation that fulfills the required performance at minimum power drain.

The advent of CMOS digital technology removed quiescent power drain as an unacceptable penalty for broadscale utilization of digital techniques in portable battery operated equipment. During the 1970's a variety of new micropower techniques were introduced and by far the most widely used product exploiting these techniques was and is the electronic wristwatch (E. Vittoz, keynote speech at the 1994 ISSCC). In the 1980's, the increasing level of power dissipation in mainstream microprocessor, memory and a host of application specific integrated circuit chips prompted an industry wide shift from NMOS and NPN bipolar technologies to CMOS in order to alleviate heat removal problems. The greatly reduced average power drain of CMOS chips provided a relatively effortless interim solution to the problems of low power design. However, the relentless march of microelectronics to higher packing densities and larger clock frequencies has, during the early 1990's, brought low power design to the forefront as a primary requirement for mainstream microelectronics which is addressed in this book.

James D. Meindl
Georgia Institute of Technology

TABLE OF CONTENTS

ACKNOWLEDGEMENTS

The majority of work described in this book began in 1991 and includes contributions from a number of U.C. Berkeley researchers. We would particularly like to acknowledge the contribution of Sam Sheng who was involved in developing the key ideas in the early phases of our research and to Prof. Jan Rabaey and his students who have been continually involved in all aspects of this work. We would also like to acknowledge the invaluable feedback on our early research efforts from Professors Teresa Meng, Charles Sodini, and Mark Horowitz. We would like to thank Prof. Jim Meindl (Chapter 2 and preface), Tony Stratakos, Charles Sullivan and Prof. Seth Sanders (Chapter 5), Lars Svensson (Chapter 6), and Mani Srivastava (Chapter 10) for their contributions to this book. We would also like to thank Arthur Abnous, Rajeevan Amirthara, David Lidsky, Thomas Barber, Tom Miller, Tom Simon, and Carlin Vieri for reading this book and providing valuable feedback.

The design tool described in Chapter 8 (HYPER-LP) was a result of a collaborative effort with Miodrag Potkonjak and Prof. Jan Rabaey. We would also like to thank Sean Huang for coding the local transforms used in the CAD tool, Paul Landman for providing software to generate data with different statistics, Renu Mehra for generating the various controllers and building the controller model, Ingrid Verbauwhede for providing the wavelet example, Scarlett Wu for switch-level simulation of the wavelet filter, and the rest of the HYPER team for software support.

The design and demonstration of the InfoPad chipset and terminal (described in Chapter 9) was a joint effort of a number of people. The first generation radio and protocol group consisting of Bill Baringer, Kathy Lu, Trevor Pering, and Tom Truman was responsible for designing the radio module that interfaced to the low-power chipset. Tom Burd (who designed the radio receiver module), Andy Burstein (who designed the SRAM), and Shankar Narayanaswamy (responsible for the pen interface) directly contributed to the design of the chipset. Of critical importance was the low-power cell-library that was largely developed by Tom Burd and Andy Burstein. Brian Richards provided valuable suggestions concerning the chip designs and Roger Doering, Susan Mellers, and Ken

Lutz setup and supported the infrastructure for designing and testing boards. Fred Burghardt, Brian Richards, and Shankar Narayanaswamy designed the basestation software and applications. For the video chipset, Ian O'Donnell prototyped the logic to verify the modified NTSC timing suggested by Ken Nishimura. Cormac Conroy and Robert Neff provided valuable suggestions on the DAC design. Prof. Mark Horowitz of Stanford university provided us with a modified version of IRSIM which we used extensively for power estimation. Finally, Tom Lookabaugh and Prof. Eve Riskin provided the source code for VQ which was used for the InfoPad video demonstration.

Of key importance was the support obtained from our Government and Industrial sponsors. The primary funding for the work described here was from the Computer Information Systems Technology Office of ARPA. In particular, the Deputy Office Director, John Toole gave us the freedom to redirect our efforts into a major activity focussed on low power design and Program Manager, Bob Parker, has continued to give us the flexibility to follow up promising research areas and John Hoyt of the Justice department who has always been available to help in not only the administrative aspects of our project, but in the technical areas as well. Also, provided by ARPA support was access to the MOSIS fast turnaround implementation service of the Information Science Institute of the University of Southern California which was critical in allowing to try out many new ideas in circuit design with a minimum level of overhead.

LOW POWER DIGITAL CMOS DESIGN

1

Introduction

With much of the research efforts of the past ten years directed toward increasing the speed of digital systems, present-day technologies possess computing capabilities which make possible powerful personal workstations, sophisticated computer graphics, and multi-media capabilities such as real-time speech recognition and real-time video. A significant change in the attitude of users is the desire to have access to this computation at any location, without the need to be connected to the wired power source. A major factor in the weight and size of these portable devices is the amount of batteries which is directly impacted by the power dissipated by the electronic circuits.

Even when power is available as in non-portable applications, the issue of low power design is becoming critical. Up until now, this power consumption has not been of great concern, since large packages, cooling fins and fans have been capable of dissipating the generated heat. However, as the density and size of the chips and systems continues to increase, the difficulty in providing adequate cooling might either add significant cost to the system or provide a limit on the amount of functionality that can be provided.

Figure 1.1 shows a plot of power consumption for various microprocessors that have been reported at the International Solid-state Circuits Conference for the last 20 years. The obsession of increasing the clock rate without regard to power dissipation has resulted in single chip power levels in excess of 30W.

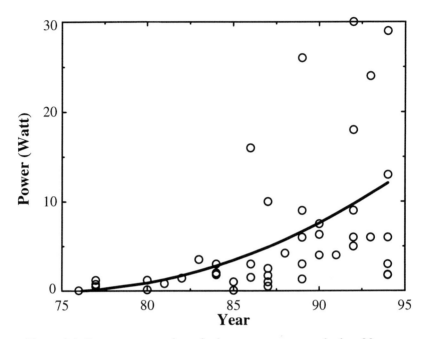

Figure 1.1: Power consumption of microprocessors over the last 20 years
[Rabaey94].

In addition to heat removal, there are also economic/environmental incentives for low-power general purpose computation. Studies have shown that personal computers in U.S. use $2 billion dollars of electricity, indirectly produce as much CO_2 as 5 million cars, and account for 5% of commercial electricity consumption (10% by the year 2000) [Nadel93]. These compelling economic and environmental reasons alone have resulted in the requirement for "green computers".

Although the traditional mainstay of portable digital applications has been in low-power, low-throughput uses such as wristwatches and pocket calculators, there are an ever-increasing number of portable applications requiring low-power and high-throughput. For example, notebook and laptop computers, representing the fastest growing segment of the computer industry, desire the same computation capabilities as found in desktop machines. Equally demanding are developments in portable communications, such as the current generation of digital cellular telephony networks which employ complex speech compression algorithms and sophisticated radio modems in a pocket sized device. Even more dramatic are the proposed personal communication service applications, with universal, portable multimedia access supporting full-motion digital video and control via speech recognition. In these applications, not only will voice be transmitted via wireless links, but data and video as well. It is apparent that portability can no longer be associated with limited computation rates; instead, vastly increased capabilities, actually in excess of that demanded of fixed workstations, must be placed in a low-power, portable environment to support these multimedia capabilities.

In spite of these concerns, until recently, there has not been a major focus on a design methodology of digital circuits which directly addresses power reduction, with the focus rather on ever faster clock rates and logic speeds. The approach which will be presented here takes another viewpoint, in which all possible aspects of a system design are investigated with the goal of reducing the power consumption. These considerations include the technology being used for the implementation, the circuit and logic topologies, the digital architectures and even the algorithms being implemented.

Maintaining a given level of computation or throughput is a common concept in signal processing and other dedicated applications, in which there is no advantage in performing the computation faster than some given rate, since the processor will simply have to wait until further processing is required. This is in contrast to general purpose computing, where the goal is often to provide the fastest possible computation without bound. One of the most important ramifications of only maintaining throughput is that it enables an architecture driven voltage scaling strategy, in which aggressive voltage reduction is used to reduce power, and the resulting reduction in logic speed is compensated through parallel architectures to maintain throughput. However, the techniques to be presented are also applicable to the general purpose environment, if the figure of merit is the amount of processing per unit of power dissipation (e.g. MIPS/Watt). Since in this case the efficiency in implementing the computation is considered and voltage scaling decreases the energy expended per operation.

The optimization to minimize area, has only been secondary to the fixation on increasing circuit speed and again this should be examined with respect to its effect on power consumption. Some of the techniques that will be presented will come at the expense of increased silicon area and thus the cost of the implementation will be increased. The desirability of this trade-off can only be determined with respect to a given market situation, but in many cases a moderate increase in area can have substantial impact on the power requirements. It is clear that if power reduction is more important than increasing circuit clock rate, then the area consumed by large clock buffers, power distribution busses and predictive circuit architectures would be better spent to reduce the power dissipation.

1.1 Overview of Book

1.1.1 Hierarchy of Limits of Power [Chapter 2]

Low power microelectronics was conceived through the invention of the transistor in 1947 and enabled by the invention of the integrated circuit in 1958. Throughout the following 36 years, microelectronics has advanced in productivity and performance at a pace unmatched in technological history. Therefore, it is imperative that we gain as deep an understanding as possible of where we have been and especially of where we may be headed with the world's most important technology. Chapter 2 treats the most important theoretical limits of power dissipation associated with each level of the design hierarchy, which can be codified as: 1) fundamental, 2) material, 3) device, 4) circuit and 5) system [Meindl83]. At each level there are two different kinds of limits to consider, theoretical and practical. Theoretical limits are constrained by the laws of physics and by technological invention. Practical limits, of course, must comply with these constraints but must also take account of manufacturing costs and markets.

1.1.2 Low-power Design Methodologies [Chapters 3-7]

Chapter 3 will provide the fundamentals of power consumption in CMOS circuits and will provide the background to understand the various power minimization techniques presented in Chapters 4 through 7. Four components of power consumption will be described: switching power, short-circuit power, leakage power and static power. For a properly designed CMOS circuit, it is usually the switching component of power which dominates, contributing to more than 90% of the total power. The central role of switching activity (the number of capacitive transitions per clock cycle) will be discussed and the various factors affecting switching activity will be described.

1.1.2.1 Voltage Scaling Approaches [Chapter 4]

It is evident that methodologies for the design of high-throughput, low-power digital systems are needed. Since power is proportional to the square of the supply voltage, it is clear that lowering the power supply voltage, V_{dd}, is the key to power reduction. Lowering V_{dd} however comes at the cost of increased circuit delays and therefore lower functional throughput. Chapter 4 presents various approaches to scaling the power supply voltage without loss in throughput.

Fortunately, there are clear technological trends that give us a new degree of freedom, so that it is possible to satisfy contradictory requirements of low-power and high throughput. Scaling of device feature sizes, along with the development of high density, low-parasitic packaging, such as multi-chip modules, will alleviate the overriding concern with the numbers of transistors being used. When MOS technology has scaled to 0.2µm minimum feature size it will be possible to place from 1-10×10^9 transistors in an area of 8" by 10" if a high-density packaging technology is used. The question then becomes how can this increased capability be used to meet a goal of low power operation. It will be shown here that for computationally intensive functions that the best use for power reduction is to provide additional circuitry to parallelize the computation based on an architecture driven voltage scaling strategy.

Figure 1.2 shows an overview of various approaches to power supply voltage scaling which range from optimizing the technology and circuits used to the architectures and algorithms. The various techniques currently used to scale the supply voltage will be reviewed which include optimizing the technology and the device reliability. These techniques dictate a limited amount of voltage scaling, typically going from 5V down to 3V. Our architecture driven voltage scaling strategy will be presented which trades area for lower power and allows for voltage scaling down to 1V without loss in system performance. Combining architecture optimization with threshold voltage reduction can result in scaling of the supply voltage to the sub-1V range. The key to architecture driven voltage scaling is the exploitation of concurrency. For this, algorithmic transformations will be used (as discussed in Chapter 8).

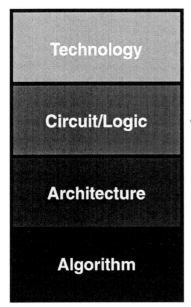

Feature Size Scaling, LDD Structures
Threshold Voltage Reduction

Transistor Sizing, Fast Logic Structures

Parallelism and Pipelining

Transformations to exploit concurrency

Figure 1.2: A system level approach to supply voltage scaling.

1.1.2.2 DC Power Supply Design in Portable Systems [Chapter 5]

Chapter 5 describes the high-efficiency DC-DC converter circuitry required for efficient low voltage operation. Various techniques will be presented to maximize the efficiency and minimize the size of DC-DC converters for low supply voltages and low current levels. This chapter will also focus on design issues at the power system level for applications in which a number of DC voltages must be supported from a single voltage source.

1.1.2.3 Adiabatic Switching [Chapter 6]

The voltage-scaling approach to power dissipation minimization assumes that the signal voltage swing can be freely chosen. In cases when the swing is determined by external constraints, conventional techniques offer little help once unnecessary transitions have been eliminated and the driven capacitances have been minimized. The emerging field of adiabatic switching addresses these situations by offering ways around the CV^2f limit. An introduction to the field and an overview of its current status is given in Chapter 6.

1.1.2.4 Minimizing Switched Capacitance [Chapter 7]

Since CMOS circuits do not dissipate power if they are not switching, a major focus of low power design is to reduce the switching activity to the minimal level required to perform the computation. This can range from simply powering down the complete circuit or portions of it, to more sophisticated schemes in which the clocks are gated or optimized circuit architectures are used which minimize the number of transitions. An important attribute which can be used in circuit and architectural optimization, is the correlation which can exist between values of a temporal sequence of data, since switching should decrease if the data is slowly changing (highly positively correlated). An example of the difference in the number of transitions which can be obtained for a highly correlated data stream (human speech) versus random data is shown in Figure 1.3. For circuits which do not destroy the data correlation, the speech data can switch up to 80% less capacitance compared to the random case.

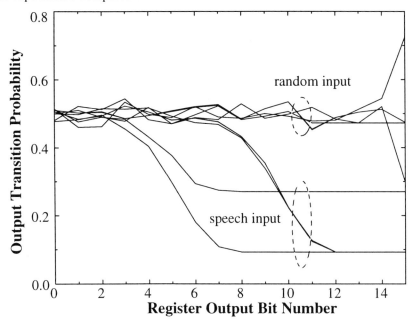

Figure 1.3: Dependence of activity on statistics: correlated vs. random input.

Figure 1.4 shows an overview of the techniques that can be used to reduce the effective switched capacitance. The knowledge about signal statistics can be exploited to reduce the number of transitions. Various techniques are described which once again span all levels of the system design ranging from the technology and circuits used to the architectures and algorithms. At the physical design level, the layout can be optimized for low-power. Logic minimization and logic level power down are key techniques to minimize the transition activity. Data dependent power down will be introduced which is a dynamic technique to reduce switched capacitance. At the architecture level, the trade-offs between time-multiplexed vs. parallel architectures, sign-magnitude vs. two's complement datapaths, ordering of operations, etc. will be studied for their influence on the switched capacitance. At the algorithmic level, the computational complexity (minimizing the number of operations) and data representation will be optimized for minimizing the switched capacitance.

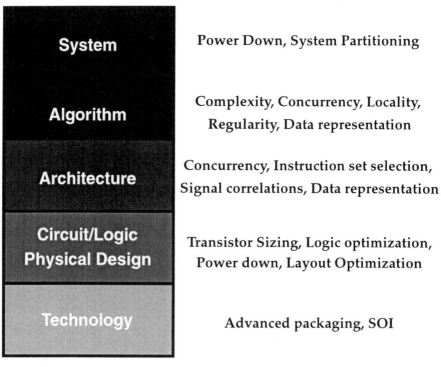

Figure 1.4: A system level approach to minimizing the switched capacitance.

1.1.2.5 Computer Aided Design Tools [Chapter 8]

The focus of Chapter 8 is on automatically finding computational structures that result in the lowest power consumption for DSP applications that have a specific throughput constraint given a high-level algorithmic specification. The basic approach is to scan the design space utilizing various algorithmic flowgraph transformations, high-level power estimation, and efficient heuristic/probabilistic search mechanisms. While algorithmic transformations have been successfully applied in high-level synthesis with the goal of optimizing speed and/or area, they have not addressed the problem of minimizing power.

There are two approaches taken to explore the algorithmic design space to minimize power consumption. First, is the exploitation of concurrency which enables circuits to operate at the lowest possible voltage without loss in functional throughput. Second, computational structures are used that minimize the effective capacitance that is switched at a fixed voltage: through reductions in the number of operations, the interconnect capacitance, the glitching activity, and internal bit widths and using operations that require less energy per computation.

1.1.2.6 System Design Example: A Portable Multimedia Terminal [Chapter 9]

Future terminals will allow users to have untethered access to multimedia information servers that are interconnected through high bandwidth network backbones. Chapter 9 describes a low-power chipset for such a terminal, which is designed to transmit audio and pen input from the user to the network on a wireless uplink and will receive audio, text/graphics and compressed video from the backbone on the downlink. The portability requirement resulted in the primary design focus to be on power reduction. Six chips provide the interface between a high speed digital radio modem and a commercial speech codec, pen input circuitry, and LCD panels for text/graphics and full-motion video display. This chipset demonstrates the various low-power techniques that are developed in the earlier chapters.

A system approach to power reduction will be described which involves optimizing physical design, circuit design, logic design, architectures, algorithms, and system partitioning. This includes minimizing the number of operations at the algorithmic level, using the architecture driven voltage scaling strategy described in Chapter 4 to operate at supply voltages as low as 1.1V, using gated clocks to minimize the effective clock load and the switched capacitance in logic circuits, optimizing circuits to minimize the voltage swings and switched capacitance, using self-timed memory circuits to eliminate glitching on the data busses, utilizing a low-power cell-library that features minimum sized devices, optimized layout, single-phase clocking, and using an activity driven place and route. The chips provide protocol conversion, synchronization, error correction, packetization, buffering, video decompression, and D/A conversion at a total power consumption of less than 5 mW - orders of magnitude lower than existing commercial solutions.

1.1.2.7 Low Power Programmable Computation [Chapter 10]

Chapter 10 describes approaches for low-power programmable computation. Two key architectural approaches for energy efficient programmable computation are described: predictive shutdown and concurrency driven supply voltage reduction. Many computations are "event-driven" in nature with intermittent computation activity triggered by external events and separated by periods of inactivity - examples include X server, communication interfaces, etc. An obvious way to reduce average power consumption in such computations would be to shut the system down during periods of inactivity. Shutting the system down - which would make power consumption zero or negligible - can be accomplished either by shutting off the clock (f=0) or in certain cases by shutting off the power supply (V_{dd} = 0). However, the shutdown mechanism has to be such that there is little or no degradation in speed - both latency and throughput are usually important in event-driven computations. There are two main problems in shutdown that are addressed- *how* to shutdown and *when* to shutdown. An aggressive shut down strategy based on a predictive technique will be presented which can reduce the power consumption by a large factor compared to the straightforward conventional schemes where the power-down decision is based solely on a predetermined idle time threshold.

REFERENCES:

[Meindl83] J.D. Meindl, "Theoretical, Practical and Analogical Limits in ULSI," *IEEE IEDM*, pp. 8-13, 1983.

[Nadel93] B. Nadel, "The Green Machine," *PC Magazine*, vol. 12, no. 10, pp. 110, May 25, 1993.

[Rabaey94] J. Rabaey, keynote talk at the Berkeley Industrial Liaison Program (ILP), 1994.

2

Hierarchy of Limits of Power

James D. Meindl
Joseph M. Petit Chair Professor of Microelectronics
Georgia Institute of Technology

2.1 Introduction

Low power microelectronics was conceived through the invention of the transistor in 1947 and enabled by the invention of the integrated circuit in 1958. Throughout the following 36 years, microelectronics has advanced in productivity and performance at a pace unmatched in technological history. Minimum feature size F has declined by about a factor of 1/50; die area D^2 has increased by approximately 170 times; packing efficiency PE, defined as the number of transistors per minimum feature area has multiplied by more than a factor of 100 so that the composite number of transistors per chip $N=F^{-2} D^2 PE$ has skyrocketed by a factor of about 50×10^6, while the price range of a chip has remained virtually unchanged and its reliability has increased manifold [Meindl93]. An inextricable concomitant advance of low power microelectronics has been a reduction in the switching energy dissipation E or power-delay product $Pt_d=E$ of a binary transition by approximately $1/10^5$ times. Consequently, as the principal driver of the modern information revolution, the ubiquitous microchip has had a profound and pervasive impact on our daily lives. Therefore, it is imperative that we gain as deep an understanding as possible of where we have been and especially of where we may be headed with the world's most important technology.

Almost two decades ago Gordon Moore of Intel Corporation observed that the number of transistors per chip had been doubling annually for a period of 15 years [Moore75]. This astute observation has become known as "Moore's Law." With a reduction of the rate of increase to about 1.5 times per year, or a quadrupling every three years, Moore's Law has remained through 1994 an accurate description of the course of microelectronics. This discussion defines a corollary of Moore's Law which asserts that "future opportunities to achieve multi-billion transistor chips or gigascale integration (GSI) in the 21st century will be governed by a hierarchy of limits." The levels of this hierarchy can be codified as: 1) fundamental, 2) material, 3) device, 4) circuit and 5) system [Meindl83]. At each level there are two different kinds of limits to consider, theoretical and practical. Theoretical limits are constrained by the laws of physics and by technological invention. Practical limits, of course, must comply with these constraints but must also take account of manufacturing costs and markets. Consequently, the path to GSI will be governed by a hierarchical matrix of limits as illustrated in Figure 2.1.

Hierarchical Matrix of Limits

	Theoretical	Practical
5. System		
4. Circuit		
3. Device		
2. Material		
1. Fundamental		

Figure 2.1: Hierarchical Matrix of Limits on GSI.

Following this introduction, Section 2.2 provides a brief retrospective view of low power microelectronics in which the antecedents of many current innovations are cited. Then, Section 2.3 treats the most important theoretical limits associated with each level of the hierarchy introduced in the preceding paragraph. In order to elucidate opportunities for low power microelectronics, many of these limits are represented by graphing the average power transfer P during a binary switching transition versus the transition time t_d. For logarithmic scales, diagonal lines in the Pt_d plane represent loci of constant switching energy. Limits imposed by interconnections are represented by graphing the square of the reciprocal length of an interconnect $(1/L)^2$ versus the response time of the corresponding circuit τ. For logarithmic scales, diagonal lines in the $(1/L)^2\tau$ plane represent loci of constant distributed resistance-capacitance product for an interconnect. The twin goals of low power microelectronics are to drive both the Pt_d and the $(1/L)^2\tau$ loci toward the lower left corners of their allowable zones of operation reflecting switching functions consuming minimal power and time, and communication functions covering maximal distance in minimal time.

Virtually all previous and contemporary microchips dissipate the entire amount of electrical energy transferred during a binary switching transition. This assumption is made in deriving the hierarchy of limits represented in the Pt_d plane in Section 2.3. However, in Section 2.4 this stipulation is removed in a brief discussion of a new hierarchy of limits on quasi-adiabatic switching operations that recycle, rather than dissipate, a fraction of the energy transferred during a binary switching transition [Landauer61][Landauer88].

In Section 2.5 practical limits are compactly summarized in a sequence of plots of minimum feature size, die edge, packing efficiency and number of transistors per chip versus calendar year. Then the results of the discussions of theoretical and practical limits are joined by defining the most important single metric that indicates the promise of a technology for low power microelectronics, and that is the chip performance index or CPI which equals the quotient of the number of transistors per chip and the associated switching energy or $CPI=N/Pt_d$. Section 2.6 concludes with a speculative comment on a paramount economic issue.

2.2 Background

The genesis of low power microelectronics can be traced to the invention of the transistor in 1947. The elimination of the crushing needs for several watts of heater power and several hundred volts of anode voltage in vacuum tubes in exchange for transistor operation in the tens of milliwatts range was a breakthrough of virtually unparalleled importance in electronics. The capability to fully utilize the low power assets of the transistor was provided by the invention of the integrated circuit in 1958. Historically, the motivation for low power electronics has stemmed from three reasonably distinct classes of need [Keonjian64] [Meindl69] [Degrauwe94] [Kohyama94] [Molhi94] [Chandrakasan92]: 1) the earliest and most demanding of these is for portable battery operated equipment that is sufficiently small in size and weight and long in operating life to satisfy the user;

2) the most recent need is for ever increasing packing density in order to further enhance the speed of high performance systems which imposes severe restrictions on power dissipation density; and 3) the broadest need is for conservation of power in desk-top and desk-side systems where a competitive life cycle cost-to-performance ratio demands low power operation to reduce power supply and cooling costs. Viewed in toto, these three classes of need appear to encompass a substantial majority of current applications of electronic equipment. Low power electronics has become the mainstream of the effort to achieve GSI.

The earliest and still the most urgent demands for low power electronics originate from the stringent requirements for small size and weight, long operating life, utility and reliability of battery operated equipment such as wrist watches, pocket calculators and cellular phones, hearing aids, implantable cardiac pacemakers and a myriad of portable military equipments used by individual foot soldiers [Keonjian64] [Meindl69]. Perhaps no segment of the electronics industry has a growth potential as explosive as that of the personal digital assistant (PDA) which has been characterized as a combined pocket cellular phone, pager, e-mail terminal, fax, computer, calendar, address directory, notebook, etc. [Degrauwe94] [Kohyama94] [Singh94] [Molhi94] [Chandrakasan92]. To satisfy the needs of the PDA for low power electronics, comprehensive approaches are proposed that include use of the lowest possible supply voltage coupled with architectural, logic style, circuit and CMOS technology optimizations. The antecedents of these concepts are strikingly evident in publications from the 1960's [Keonjian64] [Meindl69], in which several critical principles of low power design were formulated and codified [Meindl69]. The first of these was simply to use the lowest possible supply voltage, preferably a single cell battery. The second guideline was to use analog techniques wherever possible particularly in order to avoid the large standby power drain of then available bipolar digital circuits, although the micropower potential of CMOS was clearly articulated by G. Moore *et al.* in 1964 [Moore64].

A third key principle of micropower design that was convincingly demonstrated quite early is the advantage of selecting the smallest geometry, highest frequency transistors available to implement a required circuit function, e.g. a wideband amplifier, and then scaling down the quiescent current until the transistor gain-bandwidth product f_T just satisfies the relevant system performance requirements. (The manifestation of this concept in current CMOS technology is to seek the available technology with the smallest minimum feature size in order

to reduce the capacitance that must be charged/discharged in a switching transition.) Bipolar transistor gain bandwidth product is given by [Meindl69]

$$f_T = g_m/2\pi(C_{be}+C_{jc}) \qquad (1)$$

where $g_m = qI_c/kT$ is the transconductance, I_c is the quiescent collector current,
$C_{be} = C_{de} + C_{je}$ is the base-emitter capacitance including both junction capacitance
C_{je} and minority carrier diffusion capacitance C_{de}, and C_{jc} is the collector junction capacitance. As illustrated in Figure 2.2, suppose that required circuit performance demands a transistor gain bandwidth product $f_T = 120$ MHz which can be
satisfied by device A at a collector current $I_{CA} = 0.20$mA or by device B at a collector current $I_{CB} = 6.0$mA. The choice of device A for low power design is clear.
Moreover, for low current operation of both devices,

$$f_T \cong (1/2\,\pi)\,(qI_c\,/\,kT)\,1/(C_{je}+C_{jc}) \qquad (2)$$

is directly proportional to I_c thus indicating the clear advantage of maximizing
gain-bandwidth product per unit of quiescent current drain in all transistors used
in analog information processing functions. For example, this concept applies in
the design of RF receiver circuits for pocket telephones. It clearly suggests a
receiver architecture that minimizes use of high frequency front end analog electronics.

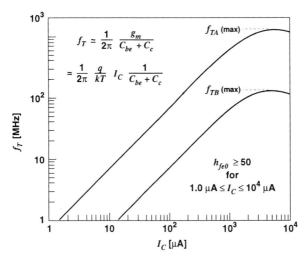

Figure 2.2: Transistor gain-bandwidth product versus quiescent collector
current. $V_{CE} = 3.0$V for transistors A and B [Meindl69].

A fourth generic principle of low power design that was clearly articulated in antiquity is the advantage of using "extra" electronics to reduce total power drain [Meindl69]. This trade-off of silicon hardware for battery hardware was demonstrated e.g. for a multi-stage wideband amplifier in which total current drain was reduced by more than an order of magnitude by doubling the number of stages from two to four while maintaining a constant overall gain-bandwidth product [Meindl69]. This concept is rather analogous to the approach of scaling down the supply voltage of a CMOS subsystem to reduce its power drain and speed and then adding duplicate parallel processing hardware to restore the throughput capability of the subsystem at an overall savings in power drain [Molhi94] [Chandrakasan92].

A final overarching principle of low power design that was rigorously illustrated for a wide variety of circuit functions including DC, audio, video, tuned and low noise amplifiers, non-linear mixers and detectors and harmonic oscillators as well as bipolar and field effect transistor digital circuits is that micropower design begins with a judicious specification of the required system performance and proceeds to the optimal implementation that fulfills the required performance at minimum power drain [Meindl69].

The advent of CMOS digital technology removed quiescent power drain as an unacceptable penalty for broadscale utilization of digital techniques in portable battery operated equipment. Since the average energy dissipation per switching cycle of a CMOS circuit is given by $E=CV^2$ where C is the load capacitance and V is the voltage swing, the obvious path to minimum power dissipation is to reduce C by scaling down minimum feature size and especially to reduce V. The minimum allowable value of supply voltage V for a static CMOS inverter circuit was derived by R. Swanson and the author in 1972 [Swanson72] as

$$V_{smin} \geq 2\text{-}4\, kT / q \tag{3}$$

Early experimental evidence[Swanson72] supporting this rigorous derivation is illustrated in Figure 2.3, which is a graph of the static transfer characteristic of a CMOS inverter for supply voltages as small as $V_s=0.10$ volts and matched MOSFET's whose threshold voltages of $V_t \cong \pm 0.16V$ were controlled by ion implantation. Further discussion of this topic is presented in Section 2.3.

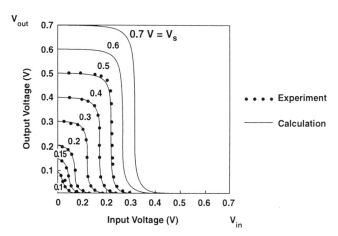

Figure 2.3: Static CMOS inverter transfer characteristic [Swanson72].

During the 1970's a variety of new micropower techniques were introduced [Vittoz94][Meindl86] and by far the most widely used product exploiting these techniques was and is the electronic wristwatch [Vittoz94]. A striking early application of power management occurred in implantable telemetry systems for biomedical research. It was the use of a $15\mu W$ 500 kHz monolithic micropower command receiver as an RF controlled switch to connect/disconnect a single 1.35v mercury cell power source to/from an implantable biomedical telemetry system. The fabrication processes used to produce the receiver chip were optimized to yield high value diffused silicon resistors [Hudson72]. Entire implantable units including active and passive sensors for biopotential, dimension, blood pressure and flow, chemical ion concentrations, temperature and strain were designed and implemented with power conservation as the primary criterion for optimization [Meindl86]. In many respects, the overall system operation of an implantable telemetry unit and its desk-side external electronics subsystem for data processing, display and storage as illustrated in Figure 2.4 is similar to but much smaller in scale than the operation of a modern PDA and its backbone network [Chandrakasan94].

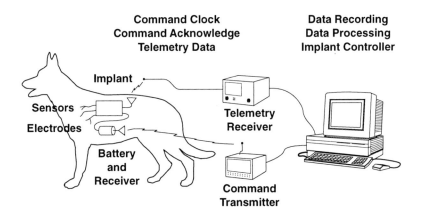

Figure 2.4: Implantable telemetry system[Meindl86].

In the 1980's, the increasing level of power dissipation in mainstream microprocessor, memory and a host of application specific integrated circuit chips prompted an industry wide shift from NMOS and NPN bipolar technologies to CMOS in order to alleviate heat removal problems. The greatly reduced average power drain of CMOS chips provided a relatively effortless interim solution to the problems of low power design. However, the relentless march of microelectronics to higher packing densities and larger clock frequencies has, during the early 1990's, brought low power design to the forefront as a primary requirement for mainstream microelectronics which is addressed in the remainder of this chapter.

2.3 Theoretical Limits

2.3.1 Fundamental Limits

The three most important fundamental limits on low power GSI are derived from the basic physical principles of thermodynamics, quantum mechanics and electromagnetics[Keyes75]. Consider first the limit from thermodynamics. Suppose that the node N illustrated in Figure 2.5 is imbedded in a complex microprocessor chip and that between N and ground G, there is an equivalent resistance of value R. Immediately from statistical thermodynamics, it can be shown that the mean square open circuit noise voltage across R is given by [Sears53]

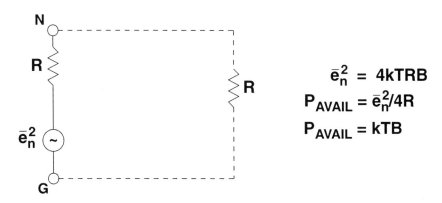

Figure 2.5: Model for derivation of fundamental limit from thermodynamics.

$$\overline{e}_n{}^2 = 4kTRB \tag{4}$$

and consequently the available noise power is

$$P_{avail} = kTB \tag{5}$$

where k is Boltzmann's constant, T is absolute temperature and B is the bandwidth of the node. Now, it is reasonable to assert that if the information represented at the node is to be changed from a zero to a one, or vice versa, then the average signal power P_s transferred during the switching transition should be greater than (or at least equal to) the available noise power P_{avail} by a factor $\gamma \geq 1$ or

$$P_s \geq \gamma P_{avail} \tag{6}$$

One can then derive an expression for the switching energy E_s transferred in the transition,

$$E_s \geq \gamma kT \tag{7}$$

Clearly, Boltzmann's constant k and absolute temperature T are independent of any materials, devices or circuits associated with the node. Consequently, E_s represents a fundamental limit on binary switching energy. For reasons to be cited in the discussion of circuit limits, $\gamma = 4$ will be assumed at this point so that at $T=300°K$, $E_s \geq 1.66\text{x}10^{-20}$ joules $= 0.104eV$. One can interpret this limit as the energy required to move a single electron through a potential difference of 0.104V, which is a Lilliputian energy expenditure compared with current practice which involves energies greater by a factor of about 10^7. One advantage of larger

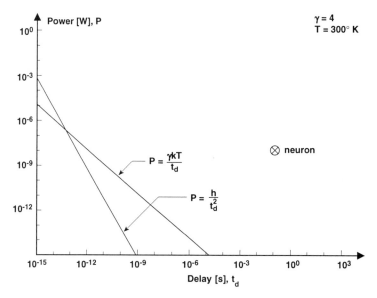

Figure 2.6: Average power transfer P during a switching transition versus transition interval t_d for fundamental limits

switching energies is that the probability of error due to internal thermal noise, $Pr(E>E_s)$, described by a Boltzmann probability distribution function,

$$Pr(E>E_s) = \exp(-E_s/kT) \tag{8}$$

decreases exponentially as E_s/kT increases.

The second fundamental limit on low power GSI is derived from quantum mechanics and more specifically from the Heisenberg uncertainty principle [Haken84], which can be interpreted as follows. A measurable energy charge ΔE associated with a switching transition must satisfy the condition

$$\Delta E \geq h \,/\, \Delta t \tag{9}$$

where h is Planck's constant and Δt is the transition time. Consequently, one can show that

$$P \geq h/(\Delta t)^2 \tag{10}$$

is the required average power transfer during a switching transition of a single electron wave packet. Figure 2.6 illustrates the fundamental limits from thermodynamics and quantum mechanics in the power-delay plane. Switching transi-

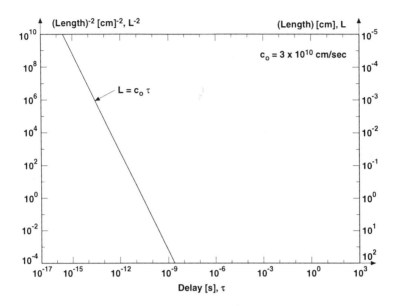

Figure 2.7: Square of reciprocal length $(1/L)^2$ of an interconnect versus interconnect circuit response time τ for the fundamental limit from electromagnetics.

tions to the left of their loci are forbidden, regardless of the materials, devices or circuits used for implementation. The zone of opportunity for low power GSI lies to the right of these limits. The point of intersection of the two limiting loci is of interest because it defines the boundary $t_d = h/\gamma \, kT$ between the region where an electron can be treated as a classical particle and the region where it must be treated as a wave packet.

The fundamental limit based on electromagnetics simply dictates that the velocity of propagation v of a high speed pulse on a global interconnect must be less than the speed of light in free space c_0 or

$$v = L/\tau \leq c_0 \qquad (11)$$

As illustrated in Figure 2.7, the speed of light limit prohibits operation to the left of the $L = c_0\tau$ locus for any interconnect regardless of the materials or structure used for its implementation.

2.3.2 Material Limits

At the second level of the hierarchy, material limits are independent of the macroscopic geometrical configuration and dimensions of particular device structures. The principal material of interest is Si, which is compared here with GaAs. The primary properties of a semiconductor which determine its key material limits are i)carrier mobility μ, ii) carrier saturation velocity v_s , iii) self-ionizing electric field strength ξ_c, and iv) thermal conductivity K. The material limit on switching energy ($E=Pt_d$) can be calculated as the amount of energy stored in a cube of Si of dimension $\Delta x=V_o/\,\xi_c$ with a voltage difference V_o across two of its parallel faces, created by an electric field nearly equal to the self-ionizing value ξ_c. Thus,

$$Pt_d = E = \varepsilon_{Si}V_o^{3}/2\xi_c \qquad (12)$$

The minimum switching time for this stored energy is taken as the transit time of a carrier through the cube or

$$t_d \geq V_o/v_s\xi_c \qquad (13)$$

For V_o=1.0V and v_s=10^7cm/sec, t_d=0.33ps for Si and 0.25 ps for GaAs. Thus, Si bears only a 33% larger electron transit time per unit of potential drop than GaAs

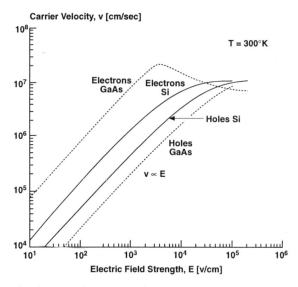

Figure 2.8: Carrier velocity versus electric field strength E for electrons and holes in Si and GaAs.

for large values of electric field strength typical for GSI. This small disadvantage is a consequence of two factors: The nearly equal saturation velocities of electrons in Si and GaAs as shown in Figure 2.8 as well as the 33% larger breakdown field strength ξ_c of GaAs. Figure 2.8 also illustrates the nearly six-fold advantage in electron velocity and therefore mobility that GaAs enjoys at small values of electric field strength (e.g. <500v/cm). In previous generations of technology operating at small values of ξ, it was carrier mobility rather than saturation velocity at large values of ξ (e.g. >50,000v/cm), that was the principal determinate of high speed capability, which is no longer the case.

The switching energy limit for Si is illustrated in Figure 2.9 for a potential swing $V_o=1.0V$, which is presumed to be a minimum acceptable value. For larger values of V_o, constant electric field reverse scaling is engaged to solve Equations 12 and 13 simultaneously for the locus of minimum switching times

$$P = \varepsilon_{Si} \xi_c^{2} \, v_s^{3} \, t_d^{2} / 2 \qquad (14)$$

designated by Equation 13 in Figure 2.9.

Figure 2.10 illustrates an isolated generic device which is hemispherical in shape with a radius r_i and located in a chip that is mounted on an ideal heat sink at

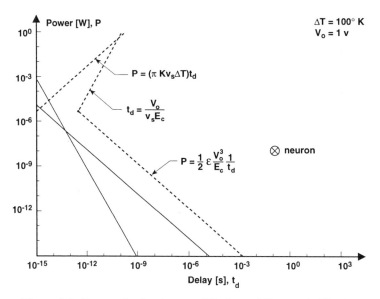

Figure 2.9: P vs. t_d for fundamental limits and Si material limits.

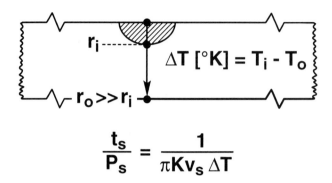

$$\frac{t_s}{P_s} = \frac{1}{\pi K v_s \Delta T}$$

Figure 2.10: Model for derivation of material limit based on heat conduction.

a temperature T_0. The heat conducted away from the device to the heat sink is given by

$$P = \pi K v_s \Delta T t_d \qquad (15)$$

where K is the semiconductor thermal conductivity and ΔT is the temperature difference between the device and the heat sink. Substituting representative values indicates that $t_d/P=0.212$ns/W for Si and 0.692 ns/W for GaAs for $\Delta T=100°$K. This sample calculation indicates that GaAs suffers a switching time per unit of heat removal that is over 300% greater than the corresponding value for Si, when switching time is limited by substrate thermal conductivity, which is about three times larger for Si than GaAs.

If the generic device illustrated in Figure 2.10 is surrounded by a hemispherical shell of SiO_2 of radius r_s representing an SOI structure, the equivalent thermal conductivity K_{EQ} of the composite structure is given by [Bhavnagarwala]

$$K_{EQ} = (K_{ox}K_{Si}r_s/r_i)\{K_{si}[(r_s/r_i)-1]+K_{ox}\}^{-1} \qquad (16)$$

Note that as $r_i \to r_s$, $K_{EQ} \to K_{Si}$ and as $K_{ox} \to K_{Si}$, $K_{EQ} \to K_{Si}$. For $K_{ox} \cong 0.01 K_{Si}$ and r_s=1.5 , 2, 4r_i, $K_{EQ} \cong 0.029$, 0.02, 0.013K_{si} which indicates a severe reduction in equivalent thermal conductivity of the SOI structure relative to bulk Si.

Figure 2.9 illustrates a second forbidden zone of operation imposed by the characteristics of Si as a material. Operation to the left of the loci of the three material limits defined by Equations 12, 13 and 15 is proscribed for any Si device.

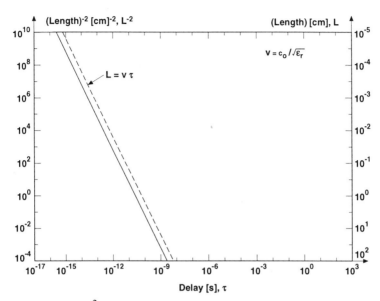

Figure 2.11: $(1/L)^2$ versus τ for fundamental limit and material limit with polymer dielectric replacing free space.

The primary interconnect material limit is defined by substituting a polymer, with a relative dielectric constant $\varepsilon_r \cong 2$, as the insulator replacing free space in the fundamental speed of light limit, as illustrated in Figure 2.11 for the $L=v\tau$ locus. In essence, both the fundamental and material limits assume a uniform lossless transmission line in a homogeneous dielectric.

The three Si material limits illustrated in Figure 2.9 assume that the distance over which bulk carrier transport occurs is greater than several mean free path lengths. For shorter distances, the possibility of quasi-ballistic or velocity overshoot effects is best treated in a particular device context [Fischetti91] [Assaderaghi93].

2.3.3 Device Limits

Device limits are independent of the particular circuit configuration in which a transistor or an interconnect is applied. The most important device in modern microelectronics is the MOSFET [Baccarani84] and the most critical limit on it is its allowable minimum effective channel length L_{min}[Ratnakumer82]. Consider a family of MOSFETs in which all parameters are held constant except effective channel length L, which is allowed to take on a wide range of values, e.g. 3.0μ ≥ L ≥ 0.03 μ. As L is reduced within this range, eventually so-called short channel effects, are manifest [Yau74]. The source and drain depletion regions begin to capture ion charge in the central region of the channel that is strictly under gate control for longer channels. The salient result of such short channel effects which are aggravated as drain voltage increases [Troutman79], is that the threshold voltage V_t is reduced, subthreshold leakage current increases and the MOSFET no longer operates effectively as a switch.

In order to achieve L_{min}, both gate oxide thickness (T_{ox}) and source/drain junction depth (X_j) should be as small as possible [Agrawal93][Fiegna93]. Gate leakage currents due to tunneling limit T_{ox} [Hu94] and parasitic source/drain resistance limits X_j [Ono93]. In addition, low impurity channels with abrupt retrograde doping profiles are highly desirable for control of short channel effects and high transconductance in bulk MOSFETs[Antoniades91]. The use of dual gates on opposite sides of a channel ostensibly provides the ultimate structure to contain short channel effects [Hisamoto90] [Frank92] [Tanaka94]. Figure 2.12 illustrates six different MOSFET structures that have been analyzed consistently to determine their short channel behavior [Agrawal93] for a very aggressive set of parameters including T_{ox}=3.0nm for all devices; X_j=5.0nm for shallow junctions; X_j=50 nm, 100 nm for deep junctions; silicon layer thickness, d=5.5 nm for the

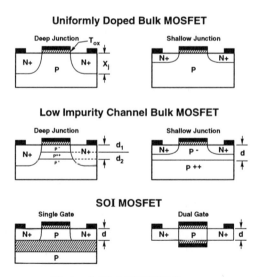

Figure 2.12: MOSFET Structures.

SOI single gate device; and silicon channel thickness, d=10.9nm for the dual gate device. Based on both analytical solutions and numerical solutions, using PISCES and DAVINCI, of the two and three dimensional Poisson equation, the threshold voltage roll-off due to scaling of effective channel length is illustrated in Figure 2.13 for each of the MOSFET structures sketched in Figure 2.12. Families of curves such as these support the prospect of achieving shallow junction retrograde channel profile bulk MOSFETs with channel lengths as short as 60 nm [Agrawal93] [Fiegna93] [Ono93] and dual gate or DELTA MOSFETs with channel lengths as short as 30 nm [Agrawal93] [Fiegna93] [Frank92]. An interesting feature of the analytical formulation of threshold voltage change

$$\Delta V_T \sim \exp\{-(1/\pi)(\varepsilon_{ox}/\varepsilon_{Si})(L/T_{ox})\} \tag{17}$$

specifically for the case of deep junctions and uniform doping, but also suggestive of other cases[Agrawal93] [Fiegna93] indicates the importance of thin gate insulators with high permittivity in the reduction of short channel effects. In addition to threshold voltage shift, other typically more manageable factors such as bulk punchthrough, gate induced drain leakage and impact ionization which contribute

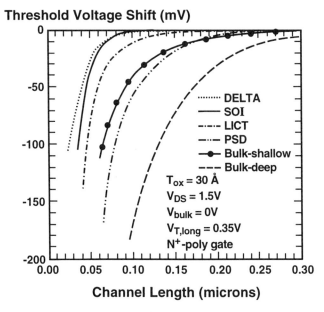

Figure 2.13: Short channel threshold voltage vs. channel length for Si devices shown in Figure 2.12.

to leakage current and the totality of effects that impact reliability must be observed as potential MOSFET limits[Hu94].

Given an allowable minimum effective channel length L_{min}, the switching energy limit for a MOSFET is simply

$$E = Pt_d = (1/2)C_o L_{min}^2 V_o^2 \qquad (18)$$

where C_o is the gate oxide capacitance per unit area corresponding to L_{min}. The smallest possible value of the transition time is

$$t_{dmin} = L_{min}/v_{sc} \qquad (19)$$

where v_{sc} is the saturation velocity of carriers in the channel taken as 8 x 10^6cm/sec for electrons[Toh88]. Assuming minimum feature size $F \cong L_{min}$ and engaging constant electric field reverse scaling, Equations 18 and 19 are solved simultaneously for the locus of minimum transition times

$$P = (1/2) \, (C_o \, L_{min})(V_o/ \, L_{min})^2 \, v_{sc}^3 t_d^2 \qquad (20)$$

which is designated $t_d = L_{min}/v_{sc}$ in Figure 2.14. The MOSFET switching energy limit (Equation 18) and locus of transition time limits are illustrated in Figure 2.14, which includes a third forbidden zone of operation to the left of these loci, for all MOSFETs with channel lengths larger than the conservative value $L_{min} = 0.1\mu$. The proximity of the material (Equations 13 and 14) and device (Equations 19 and 20) loci for minimum transition times in Figure 2.14 reflects the condition that the electric field strength ξ and carrier velocity v_{sc} assumed for the MOSFETs are pressing the material limits of Si. A channel saturation velocity of 8 x 10^6cm/sec at a tangential field strength of 200,000v/cm in a 60nm channel is quite likely to be somewhat underestimated and more refined values that consider velocity overshoot are needed[Assaderaghi93][Frank92].

The key device limit on interconnects is represented by the response time of a canonical distributed resistance-capacitance network driven by an ideal voltage source. The response time of such a network to a unit step function is given implicitly in the complex frequency(s) domain as [Bakoglu90]

$$v_o(s) = 1/s \cosh[sRC]^{1/2} \qquad (21)$$

where R and C are the total resistance and capacitance respectively and the 0 to 90% response time is $\tau = 1.0RC$. For a simple RC lumped element model of the

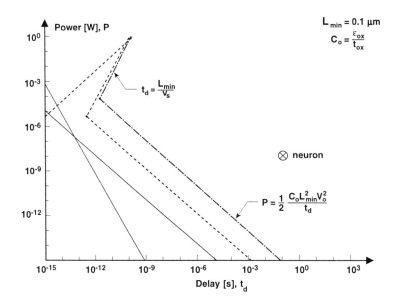

Figure 2.14: P versus t_d for fundamental limits, Si material limits and MOSFET device limits.

distributed network, the 0 to 90% response time is $t=2.3RC$. Neglecting fringing effects,

$$\tau \cong (\rho/H_\rho)(\varepsilon / H_\varepsilon)L^2 \qquad (22)$$

where ρ/H_ρ is the conductor sheet resistance in ohms/square and $(\varepsilon /H_\varepsilon)$ is the sheet capacitance in farads/cm^2. Figure 2.15 illustrates Equation 22 for equal metal and insulator thicknesses $H_\rho=H_\varepsilon=H=0.3$ and 1.0 μ. A third forbidden zone is evident. No polymer-copper interconnect with a thickness H smaller than 0.3, 1.0 μ can operate to the left of the 0.3, 1.0 μ locus which represents a contour of constant distributed resistance-capacitance product [Davis].

Further exploration is needed of MOSFET limits to take account of velocity overshoot and random dopant ion placement as well as other effects and of interconnect limits including, for example, inductance and electromigration.

2.3.4 Circuit Limits

The proliferation of limits as one ascends the hierarchy necessitates an increasing degree of selectivity in choosing those to be investigated. At the

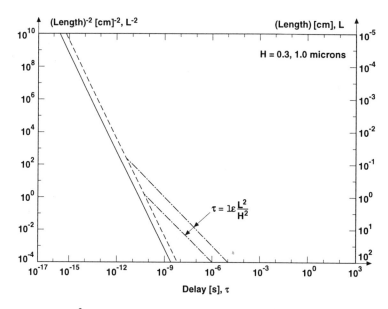

Figure 2.15: $(1/L)^2$ versus τ for fundamental limit, material limit and device limits on interconnects for a polymer-copper technology.

fourth level, circuit limits are independent of the architecture of a particular system. The initial issue to consider is which logic circuit configurations are the most promising for low power microelectronics. The candidates include e.g. GaAs direct coupled field effect transistor logic (DCFL), Si bipolar transistor emitter coupled logic (ECL), mainstream CMOS and BiCMOS. Of all the logic families now in use, it appears that common static CMOS has the most promise for low power GSI because a) it has the lowest standby power drain of any logic family, b) it has the largest operating margins, c) it is the most scaleable and d) it is the most flexible in terms of the circuit functions it can implement. For these reasons, this discussion hereafter focuses exclusively on CMOS logic.

The first and foremost circuit requirement that must be met by a logic gate is commonly taken for granted. In pursuing limits this practice cannot be followed. It is important to recognize that signal quantization or the capability to distinguish zeros from ones virtually without error throughout a large digital system is the quintessential requirement of a logic gate. For static CMOS logic this quantization requirement translates into the necessity for an incremental voltage amplification (a_v) which is greater than unity in absolute value at the transition point P_T of the static transfer characteristic of the gate where input and output

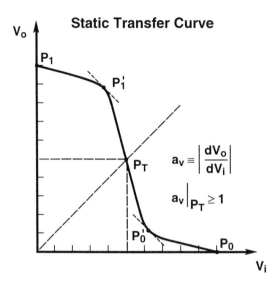

Figure 2.16: Static transfer characteristic of a non-ideal CMOS inverter.

signals are equal, as illustrated in Figure 2.16. This is a heuristic requirement that can be "seen" by considering the need for $|a_v| > 1$ in order to restore zero and one levels in an iterative chain of inverters with an arbitrary input level for the initial stage.

An interesting derivative of the requirement for $|a_v| > 1$ is that the minimum supply voltage V_{ddmin} for which a CMOS inverter can fulfill the requirement is[Swanson72]

$$V_{dd} \geq (2kT/q)[1+C_{fs}/(C_o+C_d)] \ln(2+C_o/C_d) \qquad (23)$$

$$\geq 2\text{-}4kT/q \approx 0.1\text{V} \ @T = 300°\text{K}$$

where C_{fs} is channel fast surface state capacitance, C_d is channel depletion region capacitance and C_o is gate oxide capacitance per unit area in each case. This result which was derived prior to 1972 provides a rationale for selecting a switching energy limit $E_s = \gamma kT$ at the fundamental level of the hierarchy (Equation 7) by postulating that a signal, carried by a single electron charge q through a potential difference ΔV, requires an energy $q\Delta V = \gamma kT$ and therefore a minimum potential swing $\Delta V = \gamma kT/q$ defined by Equation 23. Given that the presumed minimum acceptable signal swing for defining both the material (Equation 12) and device (Equation 18) switching energy limits is $V_o = 1.0\text{V}$, the pregnant question is, "why not set $V_o = V_{ddmin} = 0.1V$?" The simple answer to this question is that to do so would require a threshold voltage V_t so small that drain leakage current in the off-state of the MOSFET would be entirely too large for most applications. In considering logic and memory circuit behavior at low supply voltages [Nagata92] [Nakogame91] [Ishibaski92] [Sakata94] [Nakagome93] [Tauer93] [Mu94] [Burr94], a value of supply voltage $V_{dd} = V_o = 1.0V$ appears to be a broadly acceptable compromise between small dynamic power and small static power dissipation, although confirmation of Equation 23 in a low power system with $V_{dd} = 200mV$ is a prominent recent development [Burr94].

Assuming negligible static power drain due to MOSFET leakage currents, a generic circuit limit on CMOS technology is the familiar energy dissipation per switching transition

$$E = Pt_d = (1/2) C_c V_o^2 \qquad (24)$$

where C_c is taken as the total load capacitance of a ring oscillator stage, including output diffusion capacitance, wiring capacitance and input gate capacitance for an inverter which occupies a substrate area of $100F^2$ and is implemented with conservative 0.1μ technology. Assuming carrier velocity saturation in the MOSFETs, an approximate value of the drain saturation current is

$$I_{ds} \approx Z C_o v_{sc} (V_g - V_t) \qquad (25)$$

where Z is the channel width, V_t is threshold voltage and the gate voltage, $V_g = V_o$. The intrinsic gate delay can be calculated as

$$t_d = (1/2) [C_c / Z v_{sc} C_o][V_o/(V_o - V_t)] \qquad (26)$$

assuming that the product $Zv_{sc}C_o$ is equal for the N and P channel MOSFETs and that their threshold voltages are matched. Engaging constant electric field reverse scaling and solving Equations 24 and 26 simultaneously for the locus of intrinsic gate delay times,

$$P = 4 (V_o/C_c)^2 (ZC_o v_{sc})^3 t_d^2 \qquad (27)$$

which is designated $t_d \cong (1/2) C_c/ZC_o v_{sc}$ in Figure 2.17. The CMOS circuit switching energy limit (Equation 24) and the locus of intrinsic gate delay times (Equation 27) are illustrated in Figure 2.17 which includes a fourth forbidden zone, to the left of these loci, for all CMOS circuits using bulk technology with feature sizes larger than $F = 0.1\mu$.

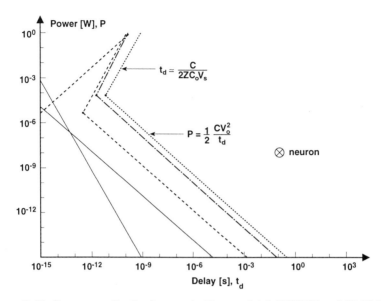

Figure 2.17: P versus t_d for fundamental , Si material, MOSFET and CMOS circuit limits.

The fourth generic circuit limit applies to a transistor driving a global interconnect presented as a distributed resistance-capacitance network extending e.g. between opposite corners of a chip. The response time of this global interconnect circuit is [Bakoglu85]

$$\tau \cong (2.3 \, R_{tr} + R_{int}) C_{int} \qquad (28)$$

where R_{tr} is the output resistance of the transistor driver and R_{int} and C_{int} are the total resistance and capacitance, respectively of the global interconnect. To prevent excessive delay due to wiring resistance, the circuit should be designed so that $R_{int} < 2.3 R_{tr}$ giving

$$\tau \cong 2.3 \, R_{tr} C_{int} \cong 2.3 \, R_{tr} c_{int} L \qquad (29)$$

where c_{int} is the capacitance per unit length of the interconnect. The distributed capacitance of a nearly lossless or TEM-mode transmission line can be expressed as $c_{int} = 1/v \, Z_o$ where $v = c_o [\varepsilon_r]^{-1/2}$ is the wave propagation velocity of the line, ε_r is the relative permittivity of its dielectric, $Z_o \cong [\mu_o / \varepsilon_o \varepsilon_r]^{1/2}$ is its characteristic impedance and $c_o = 1/[\mu_o \varepsilon_o]^{1/2}$. As $2.3 \, R_{tr} \to Z_o$ then $\tau \to L/v$, the minimum possible response time of the interconnect as defined by the material limit illustrated in

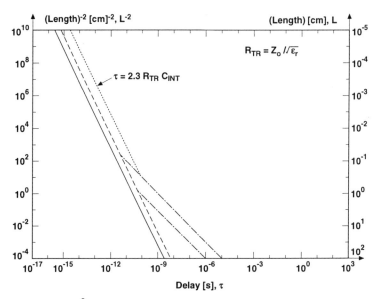

Figure 2.18: $(1/L)^2$ versus τ for fundamental, material, device and circuit limits on interconnect.

Figures 2.11 and 2.18. The price that is paid for 2.3 $R_{tr} \rightarrow Z_o$ is an increase in driver MOSFET channel width Z, substrate area and dynamic power drain compared to drivers with larger values of R_{tr}. The point of this simplified discussion is twofold: 1) the driver-interconnect combination should be designed so that R_{int}<2.3R_{tr} which may entail conductors with cross-sectional dimensions much larger than a minimum feature size F, and 2) the selected value of R_{tr} should then be the maximum possible while still meeting the necessary performance requirement for τ. This global interconnect response time limit is illustrated in Figure 2.18. The region to the left of the

$$\tau \cong 2.3 \, R_{tr} \, C_{int} = 2.3 \, (R_{tr}/Z_o)(L/v) \qquad (30)$$

locus is a forbidden zone for driver resistances larger than the designated value for the locus. Although the interconnect models engaged in the preceding discussion are rather elementary, they provide a clear picture of circuit limits which govern global interconnect performance.

Further exploration is needed of limits imposed by other important circuit configurations and by new device structures.

2.3.5 System Limits

System limits are the most numerous and nebulous ones in the hierarchy. They depend on all other limits and above all they are the most restrictive ones in the hierarchy. Consequently, it is imperative that these predominant limits be carefully considered. Among the innumerable constraints arising from the fact that each different chip design has its own unique set of limits, there are five inescapable generic system limits that are elucidated in this discussion. These limits are set by: 1) the architecture of the chip, 2) the power-delay product of the CMOS technology used to implement the chip, 3) the heat removal or cooling capacity of the chip package, 4) the cycle time requirements imposed on the chip, and 5) its physical size. To illustrate these generic limits it is necessary to select a particular example for a case study, which is intended to be broadly applicable. In keeping with the intent to explore opportunities for low power GSI, salient boundary conditions that are assumed for the case study are: 1) a generic architecture equivalent in complexity to one billion logic gates, i.e. $N_{gT}=10^9$, 2) CMOS technology with a conservative minimum feature size, $F=0.1$ μ, 3) a package cooling coefficient, $Q=50w/cm^2$ 4) a clock frequency $f_c=1.0GHz$ and 5) a single chip implementation.

A block diagram of the system architecture is illustrated in Figure 2.19. It is conceived as a two-dimensional systolic array[Kung82][Fortes87][Stone91] of 1024 identical macrocells, each consisting of a number of gates $N_g=976,562$. Communication between macrocells is assumed to occur only at the physical boundaries of adjacent macrocells. Each macrocell is assumed to receive an unskewed clock signal distributed by a balanced five-level H-tree net-

32x32 Cellular Array of 10^9 Gates

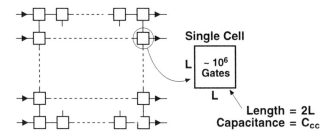

Figure 2.19: Systolic array system block diagram.

work[Bakoglu90] [Chin92], to the geometric center of the macrocell. Both logic and timing signals are communicated over arbitrary paths within a square macrocell of dimension L so the maximum path length for clock skew is L and the maximum logic signal path length is $2L$. The macrocell is represented as a random logic network described by Rent's rule[Landrum71].

$$N_p = K_p N_g^p \qquad (31)$$

where N_p = the number of signal lines entering or exiting the macrocell, Rent's coefficient K_p=0.82 and Rent's exponent p=0.45 which are empirically determined values for microprocessors[Bakoglu90]. Using Rent's rule as the basis of a stochastic analysis, the average length of an interconnect in gate pitches $\overline{R}$ can be calculated as [Donath79]

$$\overline{R}_{rl} = (2/9) \{7 (N_g^{p-0.5}-1)(4^{p-0.5}-1)^{-1} - (1-N_g^{p-1.5})(1-4^{p-1.5})^{-1}\}$$
$$(1-4^{p-1.5}) (1-N_g^{p-1})^{-1} \qquad (32)$$

for $p \neq 0.5$. For the microprocessor-like macrocell $\overline{R}_{rl} \cong 6$. Thus, the total wire length loading a gate in the random logic network is

$$l_{rl} = \overline{R}_{rl} FO [A_{rl}]^{1/2} \qquad (33)$$

where FO=3 is the fan-out and gate area A_{rl}=200F^2 is assumed to be limited by transistor packing density. This places a stringent demand on local wiring area which requires a logic gate dimension[Bakoglu90]

$$[A_{rl}]^{1/2} = \overline{R}_{rl} FO \, p_w/e_w n_w p_w \qquad (34)$$

where n_w is the number of wiring levels, p_w is the wiring pitch and e_w is the wiring efficiency factor. For n_w=4, p_w=0.2μ, and e_w=0.75, transistor packing density limits logic gate area.

As illustrated in Figure 2.20 the system switching energy limit is defined by a composite gate which characterizes the critical path of a macrocell. For a logic signal this path is assumed to consist of 1) a chain of n_{cp} random logic gates and 2) one macrocell corner-to-corner global interconnect of length $2L$ [Masaki93][Sai-Halasz94]. Therefore, the prorata switching energy of the composite gate is given by

$$E = Pt_d = (1/2) C_{rl} [1+C_{cc}/n_{cp}C_{rl}]V_o^2 \qquad (35)$$

System Switching Energy Limit

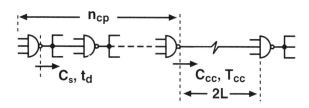

$$P_s t_{ds} = a C_s \left[1 + \frac{C_{cc}}{n_{cp} C_s} \right] V_o^2$$

Figure 2.20: Critical path used to define the system switching energy limit.

where C_{rl} is the total capacitance loading a random logic gate including MOSFET diffusion capacitance, wiring capacitance for a total interconnect length l_{rl} (Equation 33), and MOSFET gate capacitance, and C_{cc} is the total capacitance of the corner-to-corner interconnect circuit. The effective propagation delay time of the composite gate is defined as

$$t_d = t_{drl} (1 + T_{cc}/n_{cp} t_{drl}) \tag{36}$$

where t_{drl} is the delay time of a random logic gate and T_{cc} is the response time of the corner-to-corner interconnect circuit. In Equation 35, P is the average power dissipation of a composite gate during the propagation delay time t_d.

As illustrated in Figure 2.21, the system heat removal limit is defined by the requirement that the average power dissipation of a composite gate $\overline{P}$ must be less than the cooling capacity of the packaging or

$$\overline{P} = aE / T_c \le QA \tag{37}$$

where E as in Equation 35 is the switching energy of a composite gate, $a \le 1$ is the probability that the gate switches during a clock cycle, $T_c = 1/f_c$ is the clock period, $Q[\text{watts/cm}^2]$ is the package cooling coefficient and A is the substrate area occupied by the critical path composite gate. It is assumed that the composite gate area A consists of the area occupied by a random logic gate A_{rl} plus a prorata share of the area of the corner-to-corner driver circuit A_{cc} and that $A_{rl}/A_{cc} = C_{rl}/C_{cc}$ which gives

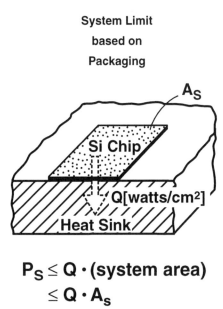

$$P_S \leq Q \cdot \text{(system area)}$$
$$\leq Q \cdot A_s$$

Figure 2.21: System heat removal limit based on packaging.

$$A = A_{rl}(1+C_{cc}/n_{cp}C_{rl}) \tag{38}$$

The cycle time can be expressed as

$$T_c = s_{cp}n_{cp}t_d \tag{39}$$

where it is to be shown that $s_{cp} \geq 1$ accounts for a small clock skew[Eble]. Combining Equations 35, 37, 38, and 39,

$$P \leq (s_{cp}n_{cp}/a)\, Q\, A_{rl}(1+C_{cc}/n_{cp}C_{rl}) \tag{40}$$

gives the maximum allowable value of P, that is permitted by the cooling capability of the package and therefore the minimum composite gate delay t_d as defined by Equation 35. Assuming constant electric field reverse scaling and solving Equations 35 and 40 simultaneously gives the locus of minimum composite gate delays

$$P = (1+C_{cc}/n_{cp}C_{rl})(1/2\, C_{rl}V_o^2)^{-2}\, [(s_{cp}n_{cp}/a)\, QA_{rl}]^3 t_d^2 \tag{41}$$

which is designated as $\overline{P} \leq QA$ in Figure 2.22.

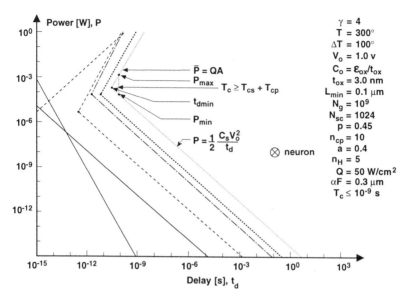

Figure 2.22: P versus t_d for fundamental, material, device, circuit and system limits.

The system cycle time limit is given by

$$T_c \geq T_{cs}+T_{cp} \tag{42}$$

where T_{cs} is the maximum clock skew within a macrocell, $T_{cp}=n_{cp}t_d$ is the critical path delay and t_d is given by Equation 36. From Equation 42,

$$t_{dmax} \leq (T_c-T_{cs})/n_{cp} \tag{43}$$

is the maximum allowable value of critical path composite gate delay that enables the required cycle time. If the allowed clock skew is $T_{cs}= \eta\, T_c$ (e.g. $\eta = 0.1$) then referring to Equation 39, $s_{cp} = 1/(1-\eta) \approx 1.11$. To calculate t_d as given by Equation 36 at the system level, appropriate values are used in Equation 32 to compute the random logic gate delay t_{drl} and in Equation 28 for both logic and clock global interconnects assuming $2.3R_{tr}=R_{int}$ and, referring to Equation 22, $H_\rho=H_\varepsilon=0.3\mu$. In calculating both T_{cs} and T_{cc} the macrocell size is taken as $L^2 \cong N_g A_{rl}$.

The system critical path composite gate switching energy, heat removal and timing limits are illustrated in Figure 2.22. Operation to the left of the switching energy locus designated $P_s t_d =1/2C_s V^2{}_o$ and the heat removal locus designated $\overline{P} \leq QA$ is forbidden and operation to the <u>right</u> of the timing locus $T_c \geq T_{cs}+T_{cp}$ is also forbidden. The allowable design space for this particular macrocell is the small triangle with vertices 1) t_{dmin} corresponding to minimum achievable propagation delay for the composite gate and therefore to the maximum performance design, 2) P_{min} corresponding to the lowest composite gate switching power that provides the required clock frequency f_c=1.0GHz and 3) P_{max} corresponding to the most mature technology or largest minimum feature size and chip size that provides f_c=1.0GHz. The three sides of the triangle correspond to contours of 1) constant switching energy E reflecting the performance level of the MOSFET and interconnect technologies, 2) constant heat removal capacity Q reflecting the performance level of the packaging technology and 3) constant clock period T_c reflecting the performance level required by the design.

At the system level, $(1/L)^2\tau$ plane limits focus on the longest interconnects since typically they impose the most stringent demands on performance. As illustrated in Figure 2.23, the response time of the longest global interconnect, i.e. a logic signal path, of length 2L is designated by $T_c \geq T_{cs}+T_{cp}$ and the actual length of its path is designated by $L^2_{min}=N_gA_g$. The longest global interconnect cannot have a slower response time nor a smaller length than designated by these two limits. The forbidden zone of operation for the longest interconnects lies external

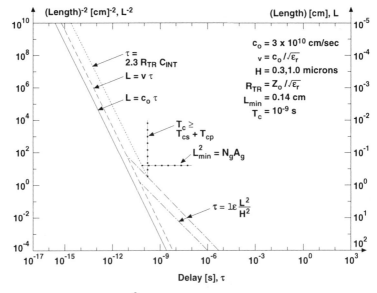

Figure 2.23: $(1/L)^2$ versus τ for all levels of the hierarchy.

to the small triangle, two of whose sides are defined by the preceding limits. The size of this triangle appears to be almost vanishingly small, particularly as a result of the stringent demands of a 1.0 GHz clock frequency. For smaller values of f_c, the size of the triangle can be enlarged at the expense of reduced performance.

The distinctive feature of the preceding treatment of system limits is that it seeks to describe the unbounded range of options of system architecture (not to mention algorithms) in terms of the absolute minimum number of parameters that enable a concise definition of the generic physical limits on system performance and hence a revealing juxtaposition of these system limits with the full hierarchy of limits which governs opportunities for GSI. A salient feature of the system representation is the definition of a critical path and from that the derivation of a composite logic gate which performs canonical computational operations. Only the first rudimentary results of this approach to low power system simulation, as illustrated in Figures 2.22 and 2.23 are available at this juncture.

2.4 Quasi-Adiabatic Microelectronics

The fundamental opportunity of quasi-adiabatic microelectronics is based on the second law of thermodynamics, which can be stated as follows: In any thermodynamic process that proceeds from one equilibrium state to another, the entropy of a system plus its environment either remains unchanged or increases[Halliday93]. Entropy change can be expressed as

$$dQ/T = dS \geq 0 \qquad (44)$$

where dQ is the heat added to the system, T is its absolute temperature and dS is the change in its entropy or atomic disorder. In a computational process, it is only those steps that discard information and therefore increase entropy $(dS>0)$ which have a lower limit on energy dissipation or heat generation $(dQ>0)$ imposed by the second law of thermodynamics[Landauer61][Landauer88]. Consequently, the intriguing prospect of inventing quasi-adiabatic computational technology offers the possibility of reducing power dissipation to levels below those imposed by limits on the non-adiabatic processes discussed in Section 2.3.

To elucidate this principle, consider the circuit operation illustrated in Figure 2.24 in which the capacitor C is charged through the resistor R from a voltage source v_{in}. If v_{in} changes as a step function, the energy dissipated in R while charging C to a voltage V_o is given by

Quasi-Adiabatic Switching

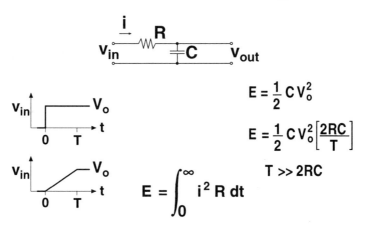

Figure 2.24: Quasi-Adiabatic Switching.

$$E_{d1} = 1/2\ C\ V_o^2 \qquad (45)$$

However, if v_{in} changes as a very slowly varying ramp function of rise time $T_d \gg 2RC$, then the energy dissipated in R is given by

$$E_{d2} \cong 1/2\ CV_o^2[2t_d/T_d] \qquad (46)$$

where $t_d = RC$. Since $2t_d/T_d \ll 1$, $E_{d2} \ll E_{d1}$. In fact, one might say that E_{d2} describes an asymptotically vanishing amount of energy as $T_d \to \infty$ [Younis94]. The reduction in energy dissipation is a consequence of maintaining at all times a quasi-equilibrium condition for which $v_{in} \to v_{out}$ and $i \to 0.0$ so that

$$E_d = \int_0^\infty i^2 R dt \to 0 \qquad (47)$$

Moreover, the discharge of C through R must be achieved through a very slowly varying ramp function whose fall time is $T_d \gg 2RC$. And, the source voltage generator providing v_{in} must include highly efficient resonant circuits to enable recycling a major fraction of the transferred energy.

The two key requirements for quasi-adiabatic or asymptotically zero dissipation digital microelectronics are summarized by Equations 44 and 47: informa-

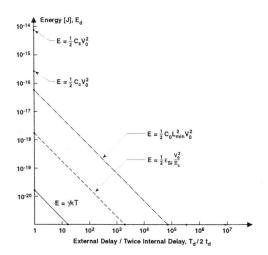

Figure 2.25: Energy dissipation versus the ratio of twice internal transition time $2t_d$ to external transition time T_d.

tion cannot be destroyed and quasi-equilibrium operation must prevail [Halliday93] [Younis94]. These requirements can be reflected in a hierarchy of limits on quasi-adiabatic microelectronics as illustrated in Figure 2.25, which graphs the energy dissipation during a switching transition E_d versus the ratio of external transition time T_d to twice internal transition time $2t_d=2RC$ or E_d vs $T_d/2t_d$. For logarithmic scales, diagonal lines in the $E_d T_d/2t_d$ plane represent loci of constant switching energy transfer E, which is precisely the case for the Pt_d plane. Consequently, it becomes clear that Pt_d plane limits serve as helpful benchmarks in assessing performance of quasi-adiabatic microelectronics. At the fundamental level of the hierarchy $E_d=\gamma kT$ at $T_d/2t_d=1$ corresponds to the funda-mental limit from thermodynamics (Equation 7) as previously shown in the Pt_d plane by Figure 2.6. At the material level, $E_d=1/2\ \varepsilon_{Si} V_0^3/\xi_c$ at $T_d/2t_d=1$ as given by Equation 12 and Figure 2.9 and for the device level $E_d=1/2\ C_0 L_{min}^2 V_0^2$ at $T_d/2t_d=1$ as given by Equation 18 and Figure 2.14. The capability to illustrate fundamental, material and device limits in the E_d vs $T_d/2t_d$ plane is predicated on the assumption that the associated switching behavior can be enabled by means that are unspecified. While this assumption serves to add insight at the first three levels of the hierarchy, it should not be casually engaged at the fourth level because unlike the unchanging materials and device structures of the second and third levels, the circuit configurations used for non-adiabatic operation, such as static CMOS, must change very significantly for quasi-adiabatic operation[Younis94][WLPD94][De].   Consequently, without identifying specific quasi-adiabatic circuit topologies and system architectures, the circuit and system limits that would be displayed in Figure 2.25 must be held in abeyance. Invention is expected to abbreviate the delay in completing the E_d vs $T_d/2t_d$ plane hierarchy in which quasi-adiabatic operation will improve on the conventional non-adia-batic circuit and system level benchmarks defined by Equations 26 and 34 respec-tively and illustrated as single points on the $T_d/2t_d=1$ axis in Figure 2.25 [De].

The current surge of interest in quasi-adiabatic circuit and system tech-niques[WLPD94] underscores the importance of low power microelectronics and thus the establishment of wholistic approaches to minimization of energy expen-diture as exemplified by the hierarchy of limits defined in this discussion.

2.5 Practical Limits

In dealing with practical limits, the key question is "How many transistors can we expect to fabricate in a single silicon chip that will prove to be useful at some specific time in the future?" To gain insight into this issue, the number of transistors per chip N can be elegantly expressed in terms of three macrovari-ables: $N=F^{-2} D^2 PE$ [Meindl93][Meindl83].

The evolution of average minimum feature size F for state-of-the-art microchips is described in simplified form in Figure 2.26, which is a graph of F

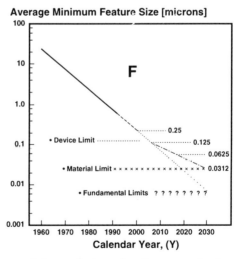

Figure 2.26: Average minimum feature size F versus calendar year Y.

versus calendar year Y. In 1960, F was about 25μ. By 1980, it had scaled down to 2.5 μ. If the historical rate of evolution continues throughout the 1990's, F will be about 0.25μ in the year 2000. Following that, Figure 2.13 illustrates three possible scenarios: 1) the 0.25 micron or pessimistic scenario, 2) the 0.125μ or realistic scenario, and 3) the 0.0625μ or optimistic scenario. The pessimistic scenario simply projects no further reduction of F beyond 0.25μ based on the adverse expectation that the cost per function or the cost per logic circuit of a microchip will reach a minimum for the design, manufacturing, testing and packaging technologies required by the 0.25μ generation. This scenario seems unlikely at this time except to pessimists. The realistic scenario projects a further reduction of F at the historic rate until the later years of the first decade of the next century. Then saturation occurs at 0.125μ again based on the economic expectation that the cost per function of a microchip will reach a minimum for the 0.125μ generation especially because it could be the last generation for which deep ultraviolet microlithography will suffice. The optimistic scenario projects further reduction of F at a slower rate resulting in about 0.0625-0.0500 μ average minimum feature size during the second decade of the millennium, and then saturation. The slower rate of reduction and saturation of F at 0.0625 microns could be caused by a combination of factors, including astronomical capital costs particularly due to introduction of a radically different microlithography technology (e.g. using X-rays) and a soft collision with the physical limits on dimensions of MOSFETs finally imposing a minimum cost per function on the 0.0625 μ generation. The author's estimate is that CMOS microchips with minimum feature sizes in the 0.0625 μ range will be widely used.

Historically, the advantages of larger chip area have been reduced cost per function, improved performance, enhanced reliability and smaller size and weight at the module, board or box level for microelectronic equipment. The evolution of the square root D of microchip area or chip size is illustrated in simplified from in Figure 2.27. In 1960, D was about 1.2 mm; in 1980, about 6.5-7.0 mm; and, if the recent historic rate of increase continues throughout the 1990's, D will reach a range around 25 mm in the year 2000. Thereafter, three possible scenarios are again illustrated by segments F, G and H. Scenario F pessimistically projects saturation of D at about 25 mm based on a maximum silicon wafer diameter of 200 mm. A realistic scenario for the 2000-2010 period is that 300 mm wafers will be commonly used and that chip sizes up to 40 mm will be economic. Beyond this a long range optimistic scenario projects 400 mm wafers and over 50 mm chips.

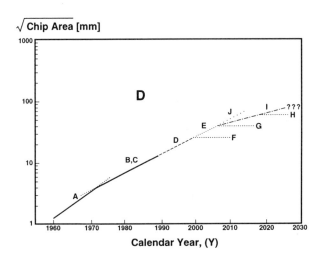

Figure 2.27: Square root of die area D versus calendar year Y.

The third macrovariable that contributes to the growing number of transistors per chip is their packing efficiency PE, the number of transistors per minimum feature area. The most prominent feature of the evolution of PE, presented in simplified form in Figure 2.28, is the abrupt change in the slope of the locus which occurred in the early 1970s. Its cause was the unavailability of silicon real estate on the chip. Prior to about 1972, PE was increased simply by moving transistors and metal interconnects closer together. Since 1972, improvements in PE have been achieved by extending into the third dimension through increasing the number of mask levels in a chip manufacturing sequence. This trend toward clever use of the third dimension is not expected to change. It is interesting that about 2010, PE approaches unity; that is the areal packing efficiency is projected as one transistor per minimum feature area, which is truly a three-dimensional microchip.

A simplified composite curve illustrating the number of transistors per chip N versus calendar year is shown in Figure 2.29. This graph more than any other chronicles the progress of the microchip from its inception in 1959, until 1995 and beyond. The pessimistic scenario denoted by segment F projects a one-billion transistor chip or GSI by the year 2000, a forecast first proposed by the author in 1983[Meindl83]. The realistic scenario projects over 100 billion transistors per chip before the year 2020.

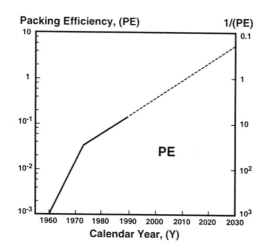

Figure 2.28: Packing efficiency *PE* versus calendar year *Y*. Note that packing efficiency is defined as the number of transistors per minimum feature area.

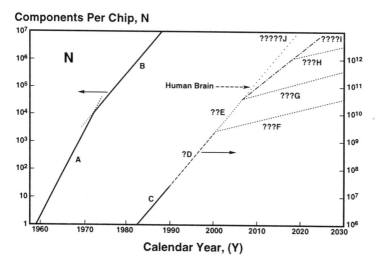

Figure 2.29: Number of transistors per chip *N* versus calendar year *Y*.

One can also graph switching energy or power-delay product (Pt_d) versus calendar year as illustrated for CMOS technology in Figure 2.30, again for three possible future scenarios[De]. Finally, the chip performance index *CPI* can be calculated as the quotient, of N and Pt_d or $CPI=N/Pt_d$. As illustrated in Figure

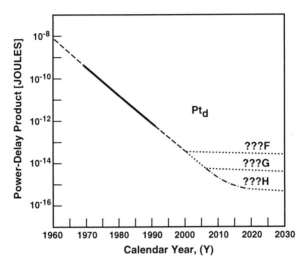

Figure 2.30: System level power-delay product Pt_d versus calendar year Y.

2.31, the *CPI* has grown by about twelve decades since 1960 and is realistically projected to grow by about another six decades before 2020. This enormous rate of both productivity and performance enhancements is unprecedented in technological history.

2.6 Conclusion

Historically there can be no doubt that the predominant pair of forces influencing the explosive growth in the number of transistors per chip has been the technological push of a continuous reduction in the cost per transistor or electronic function performed by a microchip coupled with the pull of ever expanding markets and revenues. The paramount issue confronting these positive trends has been, is, and will be the concomitant exponential growth in the capital cost of a new high volume manufacturing line needed for each successive generation of microchips[Barrett93]. While this economic issue is well beyond the scope of the current discussion, one relevant hypothesis is explored.

The hierarchy of theoretical limits on microelectronics established over the past decade and more and summarized in this discussion does not indicate that the pessimistic or the realistic or even the optimistic projections of minimum feature size F, die area D^2, packing efficiency PE, number of transistors per chip N and

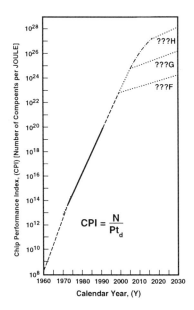

Figure 2.31: Chip performance index $CPI = N/Pt_d$ versus calendar year Y.

chip performance index N/Pt_d cannot be achieved. In other words, physical limits per se do not appear to be "show stoppers" over the next two decades. Moreover, assuming that the cost per electronic function performed by a microchip continues to decline, it does appear that market demand will continue to escalate over the next two decades simply because the capacity of the microchip to provide cost effective solutions to the myriad problems of the information revolution is virtually unlimited within this timeframe. Consequently, the paramount issue is unchanged: Will there continue to be sufficient economic incentives to risk the ever growing capital investments required for further reduction of the cost per function of microchips? It is feasible that the response will also be unchanged, especially if the manufacturing cost goals of Sematech are fulfilled [Spencer94]. Within the time interval addressed in this discussion, fundamental, material, device, circuit and system physical limitations may well permit and virtually unbounded market opportunities may well stimulate development of the highly expensive manufacturing technology that will enable continuous reduction, although perhaps at a smaller than historic rate, in the cost per function of microchips. Consequently, it is imperative that we continue to pursue as deep an understanding as possible of the hierarchy of physical limits that govern future opportunities for GSI. The National Technology Roadmap for Semiconductors prepared under the leadership of Sematech, the Semiconductor Research Corporation and the Semiconductor Industries Association is a laudable contribution toward this effort.

REFERENCES:

[Agrawal93] B. Agrawal, V.K. De, and J.D. Meindl, "Opportunities for Scaling MOSFET's for GSI," *ESSDERC*, pp. 919-926, 1993.

[Antoniades91] D. A. Antoniades and J.E. Chung, "Physics and Technology of Ultra Short Channel MOSFET's," *IEEE IEDM*, pp. 21-24, 1991.

[Assaderaghi93] F. Assaderaghi, "Observation of Velocity Overshoot in Silicon Inversion Layers," *IEEE EDL*, Vol. 14, No. 10, pp. 484-486, Oct. 1993.

[Baccarani84] G. Baccarani, *et al.*, "Generalized Scaling Theory and Its Application to a 0.25 Micron MOSFET," *IEEE TED*, Vol. ED-31, April 1984.

[Bakoglu85] H.B. Bakoglu and J.D. Meindl, "Optimal Interconnection Circuits for VLSI," *IEEE TED*, Ed-37-(5), pp. 903-909, May 1985.

[Bakoglu90] B. Bakoglu, Circuits, *Interconnections and Packaging for VLSI*, Addison Wesley, Inc., pp. 198-200, 1990.

[Barrett93] C.R. Barrett, "Silicon Valley, What Next," MRS Bulletin, pp. 3-10, July 1993.

[Bhavnagarwala] Azeez Bhavnagarwala, Private Communication.

[Burr94] J. Burr and J. Shott, "A 200 mV Self-Testing Encoder/Decoder Using Standard Ultra-Low Power CMOS," *IEEE ISSCC*, pp. 84-85, Feb. 1994.

[Chandrakasan92] A.P. Chandrakasan, S. Sheng, and R.W. Brodersen, "Low-Power CMOS Digital Design," *IEEE JSSC*, Vol. 27, No. 4, pp. 473-484, April 1992.

[Chandrakasan94] A.P. Chandrakasan, A. Burstein, and R.W. Brodersen, "A Low Power Chipset for Portable Multimedia Applications," *IEEE ISSCC*, pp. 82-83,1994.

[Chin92] K. Chin *et al.*, "IBM Enterprise System/9000 Clock System: A Technology and System Perspective," *IBM J. Research Development*, Vol. 37, No. 5, pp. 867-874, Sept. 1992.

[Davis] Jeffrey Davis, Private Communication.

[Degrauwe94] M. Degrauwe *et al.*, "Low Power/Low Voltage: Future Needs and Envisioned Solutions," *IEEE ISSCC*, pp. 98-99, 1994.

[Donath79] W.E. Donath, "Placement and Average Interconnection Lengths of Computer Logic," *IEEE Transaction on Circuits and Systems*, CAS-26, pp. 272-277, April 1979.

[Eble] J.W. Eble, Private Communication.

[Fiegna93] C. Fiegna *et al.*, "A New Scaling Method for 0.1-0.25 Micron MOSFET," *Symposium on VLSI*, pp. 33-34, 1993.

[Fischetti91] M.V. Fischetti and S.E. Laux, "Monte Carlo Simulation of Transport in Technologically Significant Semiconductors-Part II: Submicrometer MOSFET's," *IEEE TED*, Vol. 38, No. 3, pp. 650-660, Mar. 1991.

[Fortes87] J.A.B. Fortes and B.W. Wah, "Systolic Arrays - From Concept to Implementation," *IEEE Computer*, pp. 12-17, July 1987.

[Frank92] D.J. Frank *et al.*, "Monte Carlo Simulation of a 30nm Dual Gate Mosfet: How Short Can Si Go?" *IEEE IEDM*, pp. 553-556, 1992.

[Haken84] H. Haken and H.C. Wulf, Springer-Verlog, *Atomic and Quantum Physics*, pp. 83-85, 1984.

[Halliday93] D. Halliday, D. Resnick, and J. Walker, *Fundamentals of Physics*, Fourth Edition, J. Wiley and Sons, New York, 1993.

[Hisamoto90] D. Hisamoto *et al.*, "A Fully Depleted Lean Channel Transistor (DELTA)-A Novel Vertical Ultrathin SOI MOSFET," *IEEE EDL*, Vol. 11, No. 1, pp. 36-38, Jan. 1990.

[Hu94] C. Hu, "MOSFET Scaling in the Next Decade and Beyond," *Semicon International*, pp. 105-114, June 1994.

[Hudson72] P.H. Hudson and J.D. Meindl, "A Monolithic Micropower Command Receiver," *IEEE JSSC*, Vol. SC-7, No. 2, pp. 125-134, April 1972.

[Ishibaski92] K. Ishibaski *et al.*, "A 1-V TFT-Load SRAM Using a Two-Step Work-Voltage Method," *IEEE JSSC*, Vol. 27, No. 11, Nov. 1992.

[Keonjian64] E. Keonjian, Ed., *Micropower Electronics*, Pergamon Press, London, New York, 1964.

[Keyes75] R.W. Keyes, "Physical Limits in Digital Electronics," *Proceedings of the IEEE*, Vol. 63, No. 5, pp. 740-766, May 1975.

[Kohyama94] S. Kohyama, "Semiconductor Technology Crises and Challenges Toward the Year 2000," *1994 Symposium on VLSI*, pp. 5-8.

[Kung82] H.T. Kung, "Why Systolic Architectures," *IEEE Computer*, pp. 37-46, Jan. 1982.

[Landauer61] R. Landauer, "Irreversibility and Heat Generation in the Computing Process," *IBM Journal Research and Development*, Vol. 5, No. 3, pp. 183-191, July 1961.

[Landauer88] R. Landauer, "Dissipation and Noise Immunity in Computation and Communication," *Nature*, Vol. 335, pp. 779-784, Oct. 27, 1988.

[Landrum71] B.S. Landrum and R.L. Russo, "On a Pin vs. Block Relationship for Partitioning of Logic Graphs," *IEEE Transaction on Computers*, Vol. C-20, pp. 1469-1479, Dec. 1971.

[Masaki93] A Masaki, "Possibilities of Deep-Submicrometer CMOS for Very High Speed Computer Logic," *Proceedings IEEE*, Vol. 81, No. 9, pp. 1311-1324, Sept. 1993.

[Meindl69] J.D. Meindl, *Micropower Circuits*, J. Wiley and Sons, New York, 1969.

[Meindl83] J.D. Meindl, "Theoretical, Practical and Analogical Limits in ULSI," *IEEE IEDM* Technical Digest, pp. 8-13, 1983.

[Meindl86] J.D. Meindl *et al.*, "Implantable Telemetry," in *Methods of Animal Experimentation*, Vol. VII, Edited by W.I. Gay and J.E. Heavner, Academic Press, pp. 37-112, 1986.

[Meindl93] J.D. Meindl, "The Evolution of Solid State Circuits: 1958-1992-20??," *1993 IEEE ISSCC Commemorative Supplement*, pp. 23-26, Feb. 1993.

[Molhi94] S. Molhi and P. Chatterjee, "I-V Microsystems-Scaling on Schedule for Personal Communications," *IEEE Circuits and Devices*, pp. 13-17, March 1994.

[Moore64] G. Moore *et al.*, "Metal-Oxide-Semiconductor Field Effect Devices for Micropower Logic Circuitry," in *Micropower Electronics*, E. Keonjian, Ed., Pergamon Press, London, New York, 1964.

[Moore75] G.E. Moore, "Progress in Digital Integrated Electronics," *IEEE IEDM*, pp. 11-13, 1975.

[Mu94] Y. Mu *et al.*, "An Ultra Low-Power 0.1µm CMOS," *Symposium on VLSI*, pp. 9-10, June 1994.

[Nagata92] M. Nagata, "Limitations, Innovations and Challenges of Circuits and Devices into a Half Micrometer and Beyond," *IEEE JSSC*, Vol. 27, No. 4, pp. 465-472, April 1992.

[Nakagome93] Y. Nakagome, "Sub-1-V Swing Internal Bus Architecture for Future Low-Power ULSI's," *IEEE JSSC*, Vol. 28, No. 4, pp. 414-419, April 1993.

[Nakogame91] Y. Nakogame *et al.*, "An Experimental 1.5v 64mb DRAM," *IEEE JSSC*, Vol. 26, No. 4, pp. 465-472, Apr. 1991.

[Ono93] M. Ono *et al.*, "Sub-59nm Gate Length N-MOSFET's with 10nm Phosphorous S/D Junctions," *IEEE IEDM*, pp. 119-121, 1993.

[Ratnakumer82] K.N. Ratnakumer and J. Meindl, "Short Channel MOSFET Threshold Voltage Model," *IEEE JSSC*, SC-17, pp. 937-947, Oct. 1982.

[Sai-Halasz94] G.A. Sai-Halasz, "High End Processor Trends and Limits," *Interconnect Conference on Advanced Microelectronic Device*, Sendai, Japan, March 3-5, pp. 753-760, 1994.

[Sakata94] T. Sakata *et al.*, "Subthreshold-Current Reduction Circuits for Multi-Gigabit DRAM's," *IEEE JSSC*, Vol. 29, No. 7, pp. 761-769, July 1994.

[Sears53] F.W. Sears, *Thermodynamics*, Addison Wesley, Reading, Mass. 1953.

[Singh94] D. Singh, "Prospects for Low Power Microprocessor Design," *1994 International Workshop on Low Power Design*, Napa, Cal., p. 1, 1994.

[Spencer94] W.J. Spencer, "National Interests in a Global Semiconductor Industry," Distinguished Lecturer Series, Georgia Institute of Technology, Atlanta, GA, Oct. 3, 1994.

[Stone91] H.S. Stone and J. Cocke, "Computer Architecture in the 1990's," *IEEE Computer*, pp. 30-38, Sept. 1991.

[Swanson72] R.M. Swanson and J.D. Meindl, "Ion-Implanted Complementary MOS Transistors in Low-Voltage Circuits," *IEEE JSSC*, Vol. SC-7, No. 2, pp. 146-152, April 1972.

[Tanaka94] T. Tanaka *et al.*, "Ultrafast Low Power Operation of P+N+ Double-Gate SOI Mosfets," *Symposium on VLSI*, pp. 11-12, 1994.

[Tauer93] Y. Tauer *et al.*, "High Performance 0.1mm, CMOS Devices with 1.5V power Supply," *IEEE JEDM*, pp. 127-130, Dec. 1993.

[Toh88] K.Y. Toh *et al.*, "An Engineering Model for Short-Channel MOS Device," *IEEE JSSC*, pp. 950-958, Aug. 1988.

[Troutman79] R.R. Troutman, "VLSI Limitations from Drain Induced Barrier Lowering," *IEEE TED*, Vol. ED-26, pp. 461-469, 1979.

[Vittoz94] E.A. Vittoz, "Low-Power Design: Ways to Approach the Limits," *IEEE ISSCC*, pp. 14-18, 1994.

[De] Dr. Vivek De., Private Communication.

[WLPD94] Proceedings of the *1994 International Workshop on Low Power Design*, Napa, Cal., April, 1994.

[Yau74] L.D. Yau, "A Simple Theory to Predict the Threshold Voltage of Short Channel IGFET's," *Solid State Electronics*, pp. 1059-1063, 1974.

[Younis94] S.G. Younis, Asymptotically Zero Energy Computing Using Split-Level Charge Recovery Logic, Ph.D. Thesis, Dept. of EECS, MIT, June 1, 1994.

3

Sources of Power Consumption

The design of portable devices certainly requires consideration of the peak power consumption for reliability and proper circuit operation, but more critical is the time averaged power consumption which is directly proportional to the battery weight and volume required to operate circuits for a given amount of time. In fact, the approaches which will be presented to minimize the average power consumption will also reduce the peak power consumption and improve reliability.

There are four sources of power dissipation in digital CMOS circuits which are summarized in the following equation:

$$P_{avg} = P_{switching} + P_{short\text{-}circuit} + P_{leakage} + P_{static} =$$
$$\alpha_{0\text{->}1} C_L \bullet V \bullet V_{dd} \bullet f_{clk} + I_{sc} \bullet V_{dd} + I_{leakage} \bullet V_{dd} + I_{static} \bullet V_{dd} \quad (48)$$

$P_{switching}$ represents the switching component of power, where C_L is the load capacitance, f_{clk} is the clock frequency and $\alpha_{0\text{->}1}$ is the node transition activity factor (the average number of times the node makes a power consuming transition in one clock period). In most cases, the voltage swing, V, is the same as the supply voltage, V_{dd}; however, some logic circuits, such as in single-transistor pass-transistor implementations, the voltage swing on some internal nodes may be less. $P_{short\text{-}circuit}$ is due to the direct-path short circuit current, I_{sc}, which arises when both the NMOS and PMOS transistors are simultaneously active, conducting current directly from supply to ground. $P_{leakage}$ is due to the leakage current, $I_{leakage}$, which can arise from reverse bias diode currents and sub-threshold effects, is primarily determined by fabrication technology considerations. Finally, static currents, I_{static}, arise from circuits that have a constant source of current between the power supplies (such as bias circuitry, pseudo-NMOS logic families, etc.). These four components of power consumption are described in detail below.

3.1 Switching Component of Power

The switching component of power arises when energy is drawn from the power supply to charge parasitic capacitors (made up of gate, diffusion, and interconnect). The components of switching activity are described in this section and opportunities for power reduction are explored.

3.1.1 Switching Energy Per Transition

The energy drawn from the supply for each power consuming transition is derived for various types of circuits implemented in CMOS technology.

3.1.1.1 Conventional CMOS Circuits with Rail-to-Rail Swing

The switching or dynamic component of power consumption arises when the output of a CMOS circuit is charged through the power supply or is discharged to ground. The switching component of energy drawn from the power supply for a $0 \to V_{dd}$ transition at the output of a CMOS gate is given by $C_L V_{dd}^2$, where C_L is the physical load capacitance at the output node and V_{dd} is the supply voltage. Figure 3.1 shows a circuit model for computing the switching component of power. The switching component of energy is relatively independent of the function being performed (i.e the interconnection network of the NMOS and

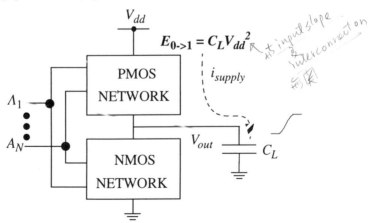

Figure 3.1: Circuit model for computing the dynamic power of a CMOS gate.

PMOS transistors) and the slope or rise times for the input signals of the CMOS gate (i.e., the short-circuit currents are ignored) and is only dependent on the output load capacitance and the power supply voltage.

To compute the average energy consumed per switching event for the generic CMOS gate shown in Figure 3.1, consider any input transition which causes a $0 \rightarrow V_{dd}$ transition at the output of the CMOS gate. For this particular transition (ignoring short-circuit currents), the instantaneous power and energy drawn can be computed as follows:

$$P(t) = \frac{dE}{dt} = i_{supply} \cdot V_{dd} \quad (49)$$

where i_{supply} is the instantaneous current being drawn from the supply voltage V_{dd} and is equal to,

$$i_{supply} = C_L \frac{dV_{out}}{dt} \quad (50)$$

Therefore, the energy drawn from the power supply for a $0 \rightarrow V_{dd}$ transition at the output node is given by:

$$E_{0 \rightarrow 1} = \int_0^T P(t)\, dt = V_{dd} \int_0^T i_{supply}(t)\, dt = V_{dd} \int_0^{Vdd} C_L dV_{out} = C_L \cdot V_{dd}^2 \quad (51)$$

From Equation 51, it is clear that the energy drawn from the power supply for a $0 \rightarrow V_{dd}$ transition at the output is $C_L V_{dd}^2$ regardless of the waveform at the output of the CMOS gate. For this transition, the energy stored in the output load capacitor, E_{cap}, is given by:

$$E_{cap} = \int_0^T P_{cap}(t)\, dt = \int_0^T V_{out} i_{cap}(t)\, dt = \int_0^{Vdd} C_L V_{out} dV_{out} = \frac{1}{2} C_L \cdot V_{dd}^2 \quad (52)$$

Therefore, half of the energy drawn from the power supply is stored in the output capacitor and half of the energy is dissipated in the PMOS network. For the V_{dd} -> 0 (1->0) transition at the output, no charge is drawn from the supply and the energy stored in the capacitor ($1/2\ C_LV_{dd}{}^2$) is dissipated in the pull-down NMOS network (i.e $E_{1->0} = 0$).

3.1.1.2 Circuits With Reduced Voltage Swings

The energy drawn from the power supply per switching event for a standard CMOS gate which has rail to rail output swing was calculated in Equation 51. This equation requires slight modification to account for logic that does not have rail to rail swing, such as those that use single transistor NMOS pass gates as seen in Figure 3.2.

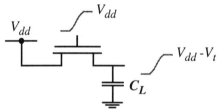

Figure 3.2: Switching power for reduced swing logic.

For this case, the output rises only to V_{dd} -V_t, rather than all the way to the rail voltage and the energy drawn from the supply for a 0 ->1 transition at the output is given by:

$$E_{0 \rightarrow 1} = C_L \bullet V_{dd} \bullet \left(V_{dd} - V_t \right) \tag{53}$$

3.1.1.3 Charge Sharing

Dynamic logic circuits suffer from charge sharing problems which result in extra parasitic switching power. For N-tree dynamic logic (with PMOS precharge transistor), the precharge operation charges the output to V_{dd}. During the evaluation operation, even if the logic function would not result in a transition the voltage will drop on the output load capacitance, C_L, by an amount ΔV, due to the parasitic capacitances, C_{int}, that are connected to the output through NMOS switches.

$$\Delta V = V_{dd} - \frac{C_L}{C_L + C_{int}} \cdot V_{dd} \tag{54}$$

Figure 3.3 shows a simple NAND gate example. Here input A goes high during the evaluate period while node B remains LOW. Therefore, the charge on C_L is redistributed between C_L and C_{int} during the evaluate period.

Therefore, during the next precharge period, the output has to be pre-charged back to V_{dd}. Therefore, the power consumed for this operation is obtained similar to the analysis of the previous section and is given by:

$$P = C_L \cdot V_{dd} \cdot \Delta V = C_L \cdot V_{dd} \cdot \left(V_{dd} - \left(\frac{C_L}{C_L + C_{int}} \cdot V_{dd} \right) \right)$$

$$= C_L \cdot V_{dd}^2 \cdot \frac{C_{int}}{C_L + C_{int}} \tag{55}$$

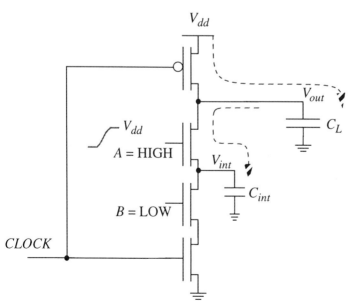

Figure 3.3: Charge sharing in dynamic circuits result in "extra" switching power.

3.1.2 Components of Node Capacitance

Figure 3.4 shows the various components of node capacitance for a CMOS inverter driving another identical stage. For the sake of analysis, the various parasitic components are lumped into a single output load capacitance. The various components of the physical capacitance are summarized in this section from [Rabaey95].

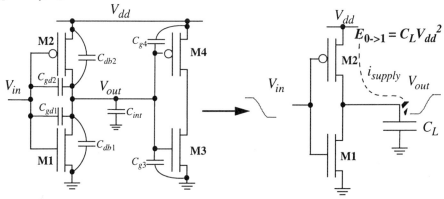

Figure 3.4: Model for computing the parasitic load capacitance of a CMOS gate.

The total physical node capacitance can be broken into three components, which often have roughly the same contribution to the total, the gate capacitance C_{gate}, the diffusion capacitance C_{diff}, and the interconnect capacitance C_{int}:

$$C_L = C_{gate} + C_{diff} + C_{int} \qquad (56)$$

Gate Capacitances (C_{gate}), C_{g3} and C_{g4}: The gate-to-channel capacitance of a MOS device consists of three components: the capacitance between the gate and the source (C_{gs}), the gate and the drain (C_{gd}), and the gate and the bulk region (C_{gb}). All three components are non-linear and their value depends upon the operation region. In the cut-off mode, no channel exists and the total capacitance $C_{ox}\text{-}WL_{eff}$ appears between gate and bulk (where $C_{ox} = \varepsilon_{ox}/t_{ox}$ is the gate capacitance per unit area and L_{eff} is the effective channel length). In the triode region, an inversion layer is formed, which acts as a conductor between source and drain. Consequently, $C_{gb} = 0$ as the bulk electrode is shielded from the gate by the channel. For this region, by symmetry $C_{gs} \approx C_{gd} \approx C_{ox}W\,L_{eff}/2$. Finally, in the saturation mode, the channel is pinched off. The capacitance between gate and drain, hence, is approximately zero, and so is the gate to bulk capacitance. A careful analysis of the channel charge taking into account the potential variations over the channel indicates that C_{gs} averages 2/3 $C_{ox}WL_{eff}$. Although these expressions are approximations, they are adequate for initial design estimates. The capacitance values are summarized in Table 3.1.

Table 3.1 Average channel capacitances of a MOS transistor for different operation regions.

Operation Region	C_{gb}	C_{gs}	C_{gd}
Cutoff	$C_{ox}WL_{eff}$	0	0
Triode	0	$C_{ox}WL_{eff}/2$	$C_{ox}WL_{eff}/2$
Saturation	0	$(2/3)C_{ox}WL_{eff}$	0

The total gate capacitance of the loading transistors (M3 and M4) is lumped into a capacitor connected between V_{out} and GND. For power estimation of digital circuits, the gate capacitances are approximated as $C_g = C_{ox} W L_{eff}$.

Overlap Capacitance, C_{gd1} and C_{gd2}: Ideally, the source and drain implants should end at the edge of the gate oxide. However, in reality, the source and drain implants tend to extend below the oxide by an amount x_d (as seen in Figure 3.5), the lateral diffusion. As a result, the effective channel of the transistor L_{eff} becomes shorter than the drawn length by an amount of $2x_d$. It also gives rise to a parasitic capacitance between gate and source (drain), which is called the *overlap capacitance*. This overlap capacitance is given by:

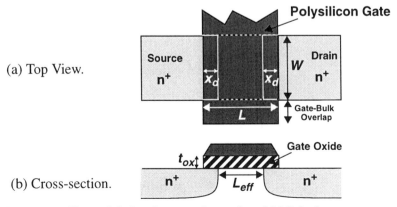

(a) Top View.

(b) Cross-section.

Figure 3.5: Overlap capacitance for a MOS device.

$$C_{gs} = C_{gd} = C_{ox} x_d W = C_O W \tag{57}$$

Since x_d is a technology determined parameter, it is typically combined with the oxide capacitance, to yield the overlap capacitance per unit width C_O.

In the lumped capacitor model, the gate to drain capacitors for M1 and M2 are replaced by a capacitor to ground whose contribution is counted twice, due to the *Miller Effect:* the effective voltage change over the gate-drain capacitor during a low-high or high-low transition is actually twice the output voltage swing, as the terminals of the capacitor are moving in opposite directions. Since x_d is a technology parameter the gate to drain capacitance is given by $C_{gd} = 2\ C_O\ W$ (where C_O is the overlap capacitance per unit width).

Drain/Source Diffusion Capacitances (C_{diff}), C_{db1} and C_{db2}: There are two components to the drain/source junction capacitance - bottom plate junction capacitance, $C_{bottom-plate}$, and the side-wall junction capacitance, C_{sw}.

$$C_{diff} = C_{bottom-plate} + C_{sw} \tag{58}$$

Figure 3.6 shows the cross section of a MOS device which illustrates the two components. The bottom plate capacitance is the reverse biased pn-junction formed between the drain (or source) and the lightly doped substrate. This capacitance value depends heavily on the applied voltage. In the lumped model, the non-linear capacitor is replaced by a linear one, that has the same change in charge as the non-linear one for the voltage range of interest. A multiplicative factor K_{eq} is introduced to relate the linearized capacitor to the value of the junction capacitance under zero bias conditions, $C_{area} = K_{eq}\ C_{j0}$, where C_{j0} is the junction capacitance per unit area under zero bias condition and K_{eq} is:

$$K_{eq} = \frac{-\phi_0^m}{\left(V_{high} - V_{low}\right) \bullet (1-m)} \left[\left(\phi_0 - V_{high}\right)^{1-m} - \left(\phi_0 - V_{low}\right)^{1-m}\right] \tag{59}$$

where m is called the grading coefficient (equal to 1/2 for abrupt junction and 1/3 for the linear or graded junction), ϕ_0 is the built-in potential (typically around 0.6V for a standard CMOS technology), and V_{low} and V_{high} are the low and high voltage values corresponding to the swing on the non-linear capacitor. The bottom-plate capacitance is given by $C_{bottom-plate} = C_{area} W L_s$ where C_{area} is the equivalent capacitance per unit area.

The sidewall capacitance is formed between the source (or drain) region and the heavily doped stopper implant with level N_A^+(for P-well). The sidewall capacitance per unit area will be larger due to the higher doping levels and the side-wall capacitance is given by, $C_{sw} = C_{jsw}(W + 2L_s)$, where C_{jsw} is the equivalent side wall capacitance per unit length (x_j is a fixed value for a given technology).

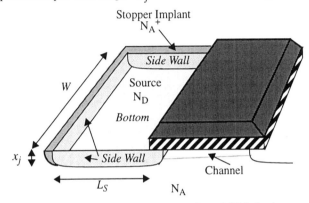

Figure 3.6: Junction capacitance for a MOS device.

Interconnect Capacitance, C_{int}: C_{int} is the parasitic capacitance of wiring used to interconnect logic circuits. The interconnect component in general is a function of the placement and routing, which for a chip level design is affected by the number of modules and busses, locality of data communication, etc. There are three main components to interconnect capacitance: area capacitance (parallel plate capacitance component), fringing field component, and wire-wire capacitance [Bakoglu90].

The parallel plate capacitance component can be modeled as shown in Figure 3.7. The parallel plate interconnect capacitance is given by:

$$C_{p-p} = \frac{\varepsilon_{ins}}{t_{ins}} \bullet W \bullet L \qquad (60)$$

If SiO_2 is used as the insulating material then $\varepsilon_{ins} = 3.9\varepsilon_o$. Typically wires are routed over field oxide which is much thicker than gate oxide hence resulting in a smaller capacitance per unit area. Also, wires on higher layers (such as Metal 2 relative to Metal1) will have a lower capacitance per unit area (see Section 3.1.2.1).

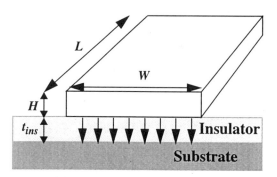

Figure 3.7: Parallel plate capacitance model of interconnect capacitance.

Clearly, from Equation 60, by increasing the dielectric thickness, the parallel-plate capacitance can be reduced; however, this is not effective if the insulator thickness becomes comparable to the interconnect width and thickness because of the effects of fringing fields [Schaper83]. As shown in Figure 3.8, the capacitance of a single wire can be modeled as a parallel plate capacitor with width equal to *W-H/2* (the *H/2* term is to incorporate second order effects [Yuan82]) and a cylindrical wire with a diameter equal to *H*. An empirical formula for interconnect capacitance has been derived per unit length as:

$$C_{int} = \varepsilon_{ox} \left\{ \frac{W}{t_{ox}} - \frac{H}{2t_{ox}} + \frac{2\pi}{\ln\left(1 + \frac{2t_{ox}}{H}\left(1 + \sqrt{1 + \frac{H}{t_{ox}}}\right)\right)} \right\} \tag{61}$$

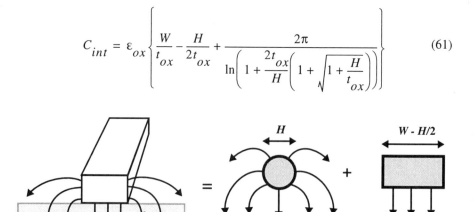

(a) Fringing fields   (b) Model of fringing field capacitance

Figure 3.8: Fringe field contribution to the total interconnect capacitance
[Glasser85].

Figure 3.9 shows a plot of the interconnect capacitance which includes fringe effects as a function of W/t_{ox} for two H/t_{ox} ratios [Schaper83]. As seen from this plot, the interconnect capacitance stops decreasing when $W < t_{ox}$ and approaches an asymptotic value of 1pF/cm and approaches the parallel plate value, C_{p-p}, for $W >> t_{ox}$.

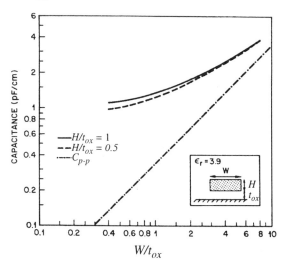

Figure 3.9: Interconnect capacitance including fringing effect as a function of W/t_{ox} [Schaper83]. (© 1983 IEEE)

The third component of capacitance that is significant at small line widths is the capacitance between neighboring lines. In order to keep the interconnect RC constant of the wires small, the thickness of the wires have not scaled as fast as other dimensions. However, to increase packing density, the distance between wires have scaled. This increases the capacitance between the wires. Figure 3.10 shows the total capacitance, C_{TOTAL} and its two sub-components - C_{GROUND}, the capacitance to ground which is the parallel-plate (C_{p-p}) and fringe combined, and C_X which is the wire-to-wire capacitance [Schaper83]. The parallel-plate capacitance is also shown for reference. When the wire width (W) is much larger than the insulator thickness (t_{ox}), wire capacitance to ground is larger than the wire-to-wire capacitances. When W is smaller than H, the capacitance between the wires dominate. The minimum is obtained when $W/H=1.75$.

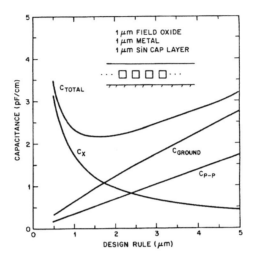

Figure 3.10: Interconnect capacitance including wire-to-wire capacitance
[Schaper83]. (© 1983 IEEE)

3.1.2.1 Interconnect Capacitance Parameters for a 1.2μm CMOS Technology

Table 3.2 shows the area and perimeter capacitances for different interconnect components of a Mosis technology with the feature size $\lambda = 0.6$ (which corresponds to a 1.2μm technology).

Table 3.2 Area and perimeter capacitances for a 1.2μm CMOS technology.

Interconnect Capacitance Component	Area Capacitance AF/λ^2	Fringing Capacitance AF/λ
Metal1 to Substrate	13	35
Metal2 to Substrate	7	34
Poly to Substrate	23	30
Metal2 to Metal1	24	54
Metal1 to Poly	25	47
Metal2 to Poly	10	41

Consider a Metal1 wire whose length is 100λ with a minimum width of 3λ (the area is 300λ^{2}and perimeter is 2 * (100 + 3) = 206λ), where λ is a technology scaling factor. Table 3.3 shows the capacitance for a 1.2μm technology (ignoring any wire to wire capacitance). Also shown in this table is the percentage contribution of perimeter capacitance to the total capacitance. As seen from this table, the fringe component is a significant portion of the total capacitance.

Table 3.3 Area and perimeter capacitances for Metal1 wire of length 100λ by 3λ.

Technology, λ	Area Cap	Perimeter Capacitance	Total Capacitance	$C_P / (C_P + C_A)$
0.6	3.9fF	7.2fF	11.1fF	65%

As described earlier, the wire to wire capacitance can be quite significant and cannot be ignored. For a 1.2μm CMOS process, the wire to wire capacitance has been determined experimentally from test structures to be 30AF/λ for Metal1 and 38AF/λ for Metal2 [Kingsbury94].

3.1.2.2 Capacitance Breakdown for Example Circuits

Table 3.4 shows the breakdown between the various components of switched capacitance for several cells from a low-power cell-library. Table 3.5 shows the breakdown of node capacitance switched for datapaths implemented using two different implementation approaches (tiled datapath and standard cells). This data was obtained from modifying a switch-level simulator (IRSIM) to isolate parts of the capacitance. As will be shown in the next chapter, the data about the breakdown between the various physical capacitance components is critical to optimizing transistor sizes for low power.

Table 3.4 Capacitance breakdown at the module level.

MODULE	GATE	DIFFUSION	INTERCONNECT
Adder (Ripple Carry)	30%	45%	25%
Adder (CSA)	37%	31%	32%
TSPC Counter	32%	32%	36%

Table 3.4 Capacitance breakdown at the module level.

MODULE	GATE	DIFFUSION	INTERCONNECT
LOG Shifter (8 bit shift by 4)	15%	42%	43%
Comparator	33%	38%	29%

Table 3.5 Capacitance breakdown at the datapath level.

MODULE	GATE	DIFFUSION	INTERCONNECT
Adder Chain (7 adders)	38%	38%	24%
Wave Digital Filter	31%	29%	40%
Address Generator (Standard Cell Block)	56%	24%	20%
Video NTSC Sync Generator (Standard Cell Block)	45%	25%	30%

3.1.3 Definition of Node Transition Activity Factor, α

The energy drawn for each $0 \to V_{dd}$ transition at the output of a CMOS gate is $C_L V_{dd}^2$. If a single $0 \to V_{dd}$ transition is made every clock cycle at a rate f_{clk}, then the power is $C_L V_{dd}^2 f_{clk}$. However, this is usually not the case with the node transition rate usually less than f_{clk}, but can be greater as well, as will be shown.

In order to handle the transition rate variation statistically, consider N clock periods and let $n(N)$ be the number of $0 \to V_{dd}$ output transitions in the time interval $[0,N]$ in Figure 3.1. The total energy drawn from the power supply for this interval, E_N, is given by:

$$E_N = C_L \cdot V_{dd}^2 \cdot n(N) \qquad (62)$$

The average power consumption corresponds to the average number of switching transitions for an extended period of time and is given by:

$$P_{avg} = \lim_{N \to \infty} \frac{E_N}{N} \cdot f_{clk} = \left(\lim_{N \to \infty} \frac{n(N)}{N} \right) \cdot C_L \cdot V_{dd}^2 \cdot f_{clk} \qquad (63)$$

The limit term in the above equation is the expected (average) value of the number of transitions per clock cycle or the node transition activity factor, $\alpha_{0 \to 1}$.

$$\alpha_{0 \to 1} = \lim_{N \to \infty} \frac{n(N)}{N} \qquad (64)$$

The average power can then be expressed as:

$$P_{avg} = \alpha_{0 \to 1} \, C_L V_{dd}^2 f_{clk} \qquad (65)$$

Since internal nodes of a gate may also make transitions, e.g. V_{int} in Figure 3.11, the transition activity must be calculated for all nodes in a circuit. The total power of a circuit is found by summing over all circuit nodes, i, yielding:

$$P_{total} = \sum_{i=1}^{number\ of\ nodes} \alpha_i C_i V_{dd}^2 f_{clk} \qquad (66)$$

If we include the possibility that individual nodes will swing to a voltage V_i which is less than V_{dd}, then the total power expression is:

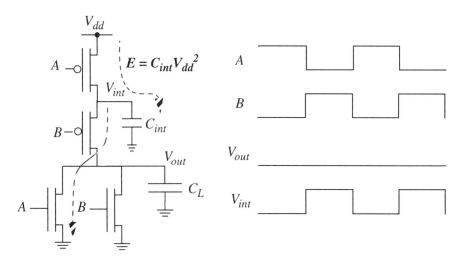

Figure 3.11: Switching energy analysis must include internal node switching.

$$P_{total} = \left(\overset{number\ of\ nodes}{\underset{i=1}{\sum}} \alpha_i C_i V_i \right) V_{dd} f_{clk} \tag{67}$$

3.1.4 Influence of Logic Level Statistics and Circuit Topologies on the Node Transition Activity Factor, α

There are two components to switching activity: a static component (which does not take into account the timing behavior and is strictly a function of the topology and the signal statistics) and a dynamic component (which takes into account the timing behavior of the circuit).

3.1.4.1 Type of logic function

The amount of transition activity is a strong function of the logic function (NOR, XOR, NAND, etc.) being implemented. For a static logic design style, the static transition probability (which is computed strictly based on the boolean function and is not a function of the timing skew) assuming independent inputs is the probability that the output will be in the ZERO state in one cycle multiplied by the probability that the output will be in the ONE state in the next clock cycle:

$$P_{0 \to 1} = P_0 \bullet P_1 = P_0 \bullet \left(1 - P_0 \right) \tag{68}$$

where p_0 is the probability that the output will be in the ZERO state and p_1 is the probability that the output will be in the ONE state. Assuming that the inputs are independent and uniformly distributed, any N-input static gate will have a transition probability that corresponds to:

$$P_{0 \to 1} = \frac{N_0}{2^N} \bullet \frac{N_1}{2^N} = \frac{N_0 \bullet \left(2^N - N_0 \right)}{2^{2N}} \tag{69}$$

where N_0 is the number of ZERO entries in the truth table for the output of the N-input function and N_1 is the number of ONE entries in the truth table for the output of the N-input function.

To illustrate, consider a static 2-input NOR gate whose truth table is shown in Table 3.6. Assume that only one input transition is possible during a clock cycle and also assume that the inputs to the NOR gate have a uniform input distribution of high and low levels.

Table 3.6 Truth table of a 2 input NOR gate.

A	B	Out
0	0	1
0	1	0
1	0	0
1	1	0

This means that the four possible states for inputs A and B (00, 01, 10, 11) are equally likely. For a NOR gate, the probability that the output is in the ZERO state is 3/4 and that it will in the ONE state is 1/4. From Table 3.6 and Equation 69, the ZERO to ONE output transition probability of a 2-input static CMOS NOR gate is given by:

$$p_{0 \to 1} = \frac{N_0 \bullet \left(2^N - N_0\right)}{2^{2N}} = \frac{3 \bullet \left(2^2 - 3\right)}{2^2 \bullet 2} = \frac{3}{16} \tag{70}$$

The state transition diagram annotated with transition probabilities is shown in Figure 3.12. Note that the output probabilities are no longer uniform.

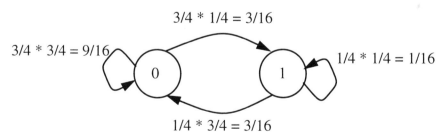

3/4 * 1/4 = 3/16

3/4 * 3/4 = 9/16

0   1

1/4 * 1/4 = 1/16

1/4 * 3/4 = 3/16

Figure 3.12: State transition diagram for a 2 input NOR gate with transition probabilities.

Table 3.7 shows the truth table for another example gate, a 2-input static XOR gate.

Table 3.7 Truth table of a 2 input XOR gate.

A	B	Out
0	0	0
0	1	1
1	0	1
1	1	0

The ZERO to ONE output transition probability of a 2-input static CMOS XOR gate is given by:

$$p_{0 \to 1} = \frac{N_0 \bullet \left(2^N - N_0\right)}{2^{2N}} = \frac{2 \bullet \left(2^2 - 2\right)}{2^{2 \bullet 2}} = \frac{1}{4} \tag{71}$$

Figure 3.13 shows the transition probability for NAND, NOR, and XOR gates as a function of the number of inputs. For an XOR, the truth table output will have 50% 1's and 50% 0's and therefore assuming a uniform input distribution, the output will always have a transition probability of 1/4, independent of the number of inputs. For a NAND or NOR, the truth table will have only one 0 or 1 respectively for the output regardless of the number of inputs. Evaluating Equation 69,

$$p_{0 \to 1} = \frac{\left(2^N - 1\right) \bullet \left(2^N - \left(2^N - 1\right)\right)}{2^{2N}} = \frac{\left(2^N - 1\right)}{2^{2N}} \tag{72}$$

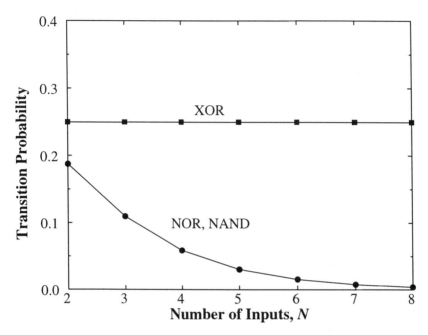

Figure 3.13: Transition probability for different gates as function of the
number of inputs.

3.1.4.2 Type of Logic Style

For a given function, the type of logic style (static CMOS, dynamic logic,
dual rail self-timed implementation) has a strong impact on the node transition
probability. In the previous section the transition probability for a static gate was
shown to be $p_0 \, p_1 = p_0 \, (1 - p_0)$. For dynamic logic, the output transition probability does not depend on the state (history) of the inputs but rather on just the signal
probabilities. For an N-tree structure (in which the P-network is replaced by a single PMOS precharge transistor) dynamic gate, the output will make a 0 to 1 transition during the precharge phase only if the output was discharged by the N-tree
logic during the evaluate phase (ignoring any charge sharing). The ZERO to ONE
transition probability for an N-tree structure is therefore

$$p_{0 \to 1} = p_0 \tag{73}$$

where p_0 is the probability that the output is in the ZERO state. For uniformly
distributed inputs, this means that the transition probability is:

$$P_{0 \to 1} = \frac{N_0}{2^N} \qquad (74)$$

where N_0 is once again the number of ZERO entries in the truth table.

To illustrate the activity for a dynamic gate, once again consider a 2 input NOR gate. An N-tree dynamic NOR gate is shown in Figure 3.14 along with the static counterpart. For the dynamic implementation, power is consumed during the precharge operation for the times when the output capacitor was discharged the previous cycle. For equi-probable input, there is then a 75% probability that the output node will discharge immediately after the precharge phase, implying that the activity for such a gate is 0.75 (i.e $P_{NOR}= 0.75 \, C_L V_{dd}^2 f_{clk}$). From the truth table shown in Table 3.6 and using Equation 74, the transition probability is given by,

$$P_{0 \to 1} = \frac{N_0}{2^N} = \frac{3}{4} \qquad (75)$$

The corresponding activity is a lot smaller, 3/16, for a static implementation (as computed in the previous sub-section). Note that for the dynamic case, the activity depends only on the signal probability, while for the static case the

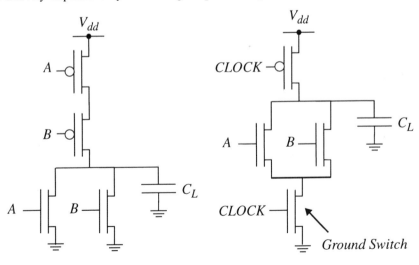

Figure 3.14: Static NOR vs. N-tree based dynamic NOR.

transition probability depends on previous state. If the inputs to a static CMOS gate do not change from the previous sample period, then the gate does not switch. This is not true in the case of dynamic logic in which gates can switch. For a dynamic NAND gate, the transition probability is 1/4 (since there is a 25% probability the output will be discharged) while it is 3/16 for a static implementation.

Yet another logic style (often used in self-timed circuits) is the dual rail coding logic style Differential Cascade Voltage Switch Logic (DCVSL). A generalized DCVSL gate is shown Figure 3.15 [Heller84].

It is a pre-charged logic family that is very similar to domino logic. The gate also has an NMOS tree which implements the required function but has two complementary outputs. Similar to domino logic, there is a pre-charge phase (I is LOW) when both OUT and $OUTB$ are precharged to LOW. During evaluation, since one output is guaranteed to go high, there is a guaranteed transition for each input transition; that is, the transition probability is 1 regardless of the input statistics and thus potentially power inefficient.

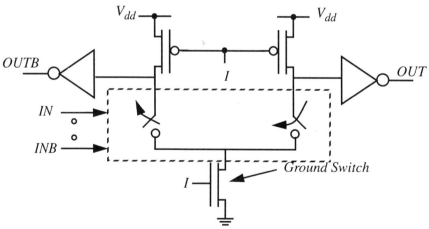

Figure 3.15: A generalized DCVSL gate.

Table 3.8 shows the summary of activity factor for a 2-input NOR gate implemented using different logic styles for independent and uniformly distributed inputs.

Table 3.8 Output transition probability comparison of different logic styles.

Logic Style	$p_{0->1}$
Static CMOS	3/16
Dynamic CMOS	3/4
DCVSL	1

3.1.4.3 Including Signal Statistics

Signal statistics have a very strong influence on power consumption. In the previous sections, the NOR gate was analyzed assuming random inputs. However, signals in a circuit are typically not random and this attribute can be exploited to reduce the power consumption. For a CMOS logic gate, it was shown earlier in Equation 68 that the probability of transition is $p_0\ p_1$. The transition probabilities were derived assuming that the inputs were bit-wise uncorrelated and that they were equi-probable. In the analysis to follow, the inputs will still be assumed to be uncorrelated, but the equi-probable assumption will not be assumed. To illustrate the influence of signal statistics on power consumption, consider once again a 2 input static NOR gate, and let p_a and p_b be the probabilities that the inputs A and B are ONE. In this case the probability that the output node is a ONE is given by:

$$p_1 = (1-p_a)\ (1-p_b) \tag{76}$$

Therefore, the probability of a transition from 0 to 1 is:

$$p_{0->1} = p_0\ p_1 = (1-(1-p_a)\ (1-p_b))\ (1-p_a)\ (1-p_b) \tag{77}$$

Figure 3.16 shows the transition probability as a function of p_a and P_b. From this plot, it is clear that understanding the signal statistics and their impact on switching events can be used to significantly impact the power dissipation.

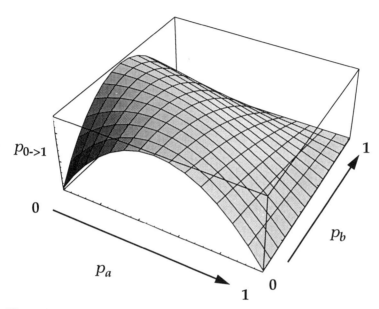

Figure 3.16: Transition probability for a 2 input NOR as a function of input statistics.

Table 3.9 shows the transition probabilities for other logic gates as a function of the input signal probabilities.

Table 3.9 Output transition probabilities for various static logic gates.

Function	$p_{0 \rightarrow 1}$
AND	$(1 - p_a p_b) \, p_a \, p_b$
OR	$(1 - p_a)(1 - p_b)(1 - (1 - p_a)(1 - p_b))$
XOR	$(1 - (p_a + p_b - 2 p_a p_b))(p_a + p_b - 2 p_a p_b)$

Figure 3.17 shows the transition probability for a 2-input XOR gate as a function on the input probabilities.

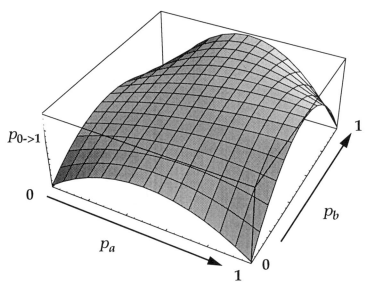

Figure 3.17: Transition probability for a 2 input XOR gate.

3.1.4.4 Inter-signal Correlations

In the previous sections, the transition probability of a gate was computed assuming that the inputs are independent. The evaluation of the switching activity becomes more involved when complex logic networks have to be analyzed because when signals propagate through a number of logic layers, they often become correlated or 'colored'. The logic shown in Figure 3.18 will be used to illustrate the effect of signal correlations on switching activity. First consider the circuit shown in Figure 3.18a and assume that the primary inputs, A and B, are uncorrelated and are uniformly distributed. Node C will have a signal probability of $1/2$ and a $0->1$ transition probability of $1/4$. The probability that the node Z undergoes a power consuming transition is determined from the expression shown in Table 3.9.

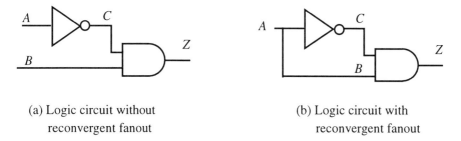

(a) Logic circuit without   (b) Logic circuit with
    reconvergent fanout   reconvergent fanout

Figure 3.18: Example illustrating the effect of signal correlations.

$$p_{0->1} = (1 - p_a \, p_b) \, p_a \, p_b = (1 - 1/2 \bullet 1/2) \, 1/2 \bullet 1/2 = 3/16 \tag{78}$$

The computation of the probabilities used for the circuit of Figure 3.18a is straightforward: signal and transition probabilities are evaluated in an ordered fashion, progressing from the input to the output node. This approach however, has two major limitations. First, it does not deal with circuits with feedback as found in sequential circuits. Second, it assumes that the signal probabilities at the input of a gate are independent. This is rarely the case in actual circuits, where reconvergent fanout often causes inter-signal dependencies.

Consider, for example, the logic network of Figure 3.18b. The inputs to the AND gate (C and B) are interdependent, as B is also a function of A (they are identical). It is easy to see that the approach to compute probabilities presented previously fails under these circumstances. Traversing from inputs to outputs yields a transition probability of 3/16 for node Z (similar to the previous analysis). This value for transition probability is clearly false, as a simple logic minimization shows that the network can be reduced to $Z = C \bullet B = A \bullet \overline{A} =$ logical zero.

To get the precise results in the progressive analysis approach, its is essential to take signal inter-dependencies into account. This can be accomplished with the aid of conditional probabilities. This will be explained with the aid of this simple example. For an AND gate, Z equals 1 if and only if B and C are equal to 1.

$$p_Z = p(Z=1) = p(B=1, C=1) \tag{79}$$

where $p(B=1, C=1)$ represents the probability that B and C are equal to 1 simultaneously. If B and C are independent, as with the circuit shown in Figure 3.18a, $p(B=1, C=1)$ can be decomposed into $p(B=1) \bullet p(C=1)$, and this yields the expression for the AND-gate, derived earlier: $p_Z = p(B=1) \bullet p(C=1) = p_B \, p_C$. If a dependency between the two exists (as is the case in Figure 3.18b), a conditional probability has to be employed, such as

$$p_Z = p(C=1|B=1) \bullet p(B=1) \tag{80}$$

The first factor in Equation 80 represents the probability that $C=1$ given that $B=1$. The extra condition is necessary as C is dependent upon B. Inspection of the network shows that this probability is equal to 0, since C and B are logical inversions of each other, resulting in the signal probability for Z, $p_Z = 0$.

Deriving those expressions in a structured way for large networks with reconvergent fanout is complex, especially when the networks are also sequential and contain feedback loops, and therefore computer support is essential. The goal is to estimate or analyze the transition probabilities at the nodes of a network. To be meaningful, the analysis program has to take in a typical sequence of input signals, as the power dissipation will be a strong function of statistics of those signals.

3.1.4.5 Dynamic Transitions (Glitching or Hazards)

The manner in which logic gates are interconnected can have a strong influence on the overall switching activity. So far, the emphasis has been on the static component of activity which does not take into account the timing behavior and is strictly a function of the topology and the signal statistics. Another important component of switching power is the dynamic component which takes into account the timing behavior of the circuit. The dynamic component can cause the node transition activity, $\alpha_{0->1}$, to be greater than one (note that $\alpha_{0->1}$ is the same as $p_{0->1}$ used in the previous sections if there are no dynamic or glitching transitions).

A node in a circuit implemented using static logic can have multiple transitions inside a clock cycle before settling to the correct logical value. The number of extra transitions or glitching transitions is a function of the logic depth, the signal skew due to different arrival times of the input, and the signal patterns. In the worst case, the number of "extra" transitions (also called glitching transitions) can grows as $O(N^2)$, where N is the logic depth.

Figure 3.19 shows an adder circuit with a depth of $N+1$, for which the worst case transition activity will be computed. Assume that the primary inputs A_0-A_N, B_0-B_N, and C_0 all arrive at the same time (this is a reasonable assumption if they are all output from registers and the wiring skew is not significant). Assume that each level of logic takes a fixed unit time to evaluate. Let arrival time be defined as the time at which a signal arrives at a logic gate. The first level of logic (level 0) will make at most one transition based on C_0, A_0, and B_0 and therefore C_1 will have an arrival time at the level 1 logic block at time $T=1$. The second level of logic (level 1) will first evaluate based on C_1 at time $T=0$ and the value of the new inputs A_1 and B_1 (whose arrival time is at $T=0$). This will cause C_2 to have a transition at $T=1$. When C_1 arrives at time $T=1$, the level 1 logic will

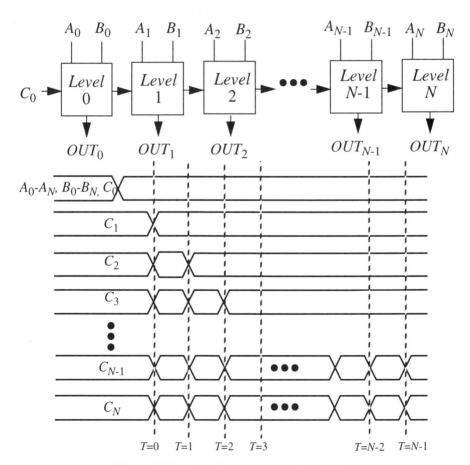

Figure 3.19: Dynamic component of switching power.

re-evaluate with the same A and B inputs (which are stable from time $T=0$ onwards) and will cause another transition at C_2. That is, node C_2 will have 2 new arrival times in a clock period for a worst case input pattern. This can cause three transitions at node C_3 since the logic block of level 3 will evaluate with inputs A_3 and B_3 at time $T=0$, and with three values of C_3 at times $T=0$, $T=1$, and $T=2$.

It is easy to see that for the Nth logic stage, the output will transition N times in the worst case. The number of extra transitions for all N stages is $N(N+1)/2$, and the number of extra glitching transitions in the worst case grows as $O(N^2)$. In reality, the transition activity due to glitching will be less since the worst input pattern will occur infrequently.

The actual waveforms illustrating the glitching behavior in static CMOS circuits is shown in Figure 3.20, which is the SPICE simulation of a static 16-bit adder, with all bits of input A and the CIN of the LSB going from "zero" to "one", and with all the bits of input B set to "zero". For all bits, the resultant sum should be zero; however, the propagation of the carry signal causes a "one" to appear briefly at most of the outputs. These spurious transitions dissipate extra power over that strictly required to perform the computation. Note that some of the bits only have partial glitching. This is due to the intertial delay of the logic gate - the timing skew between the inputs must be greater than a certain value for the glitch to propagate through the gate. The number of these extra transitions is a function of input patterns, internal state assignment in the logic design, delay skew, and logic depth. Though it is possible with careful logic design to eliminate these transitions (for example using balanced paths as described earlier), a major advantage of dynamic logic is that it intrinsically does not have this problem, since any node can undergo at most one power-consuming transition per clock cycle.

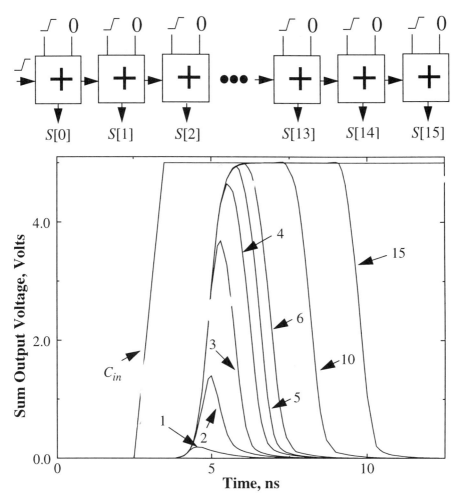

Figure 3.20: Waveforms for a 16-bit adder demonstrating glitching behavior.

In making power trade-offs between logic structures, it is important to consider both the static and dynamic transition activities simultaneously. A circuit can have a lower static transition probability while having a larger dynamic transition probability relative to another circuit. To illustrate this point consider two alternate implementations of $F = A \bullet B \bullet C \bullet D$ as shown in Figure 3.21.

First consider the static behavior assuming that all primary inputs (A,B,C,D) are uncorrelated and random (i.e., $p_{1\ (a,b,c,d)}= 0.5$). For an AND gate, the probability that the output is in the 1 state is given by:

$$p_1 = p_a\, p_b \qquad\qquad (81)$$

Therefore the probability that the output will make a 0 to 1 transition is given by:

$$p_{0\to1} = p_0\, p_1 = p_0\, (1\text{-}p_0) = (1\text{-}p_a\, p_b)\, p_a\, p_b \qquad\qquad (82)$$

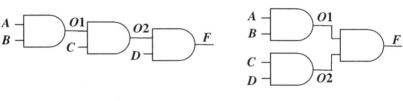

Chain structure   Tree structure

Figure 3.21: Simple example to demonstrate the influence of circuit topology on activity.

Given this, the static signal and transition probabilities can be computed for the two topologies shown in Figure 3.21 which is summarized in Table 3.10.

Table 3.10 Probabilities for tree and chain topologies.

	$O1$	$O2$	F
p_1 (chain)	1/4	1/8	1/16
$p_0 = 1\text{-}p_1$ (chain)	3/4	7/8	15/16
$p_{0\to1}$ (chain)	3/16	**7/64**	15/256
p_1 (tree)	1/4	1/4	1/16
$p_0 = 1\text{-}p_1$ (tree)	3/4	3/4	15/16
$p_{0\to1}$ (tree)	3/16	**3/16**	15/256

The results indicate that the chain implementation will have an overall lower switching activity than the tree implementation for random inputs. However, as mentioned before, it is also important to consider the timing behavior to accurately make power trade-offs. Table 3.11 shows both the relative transition probabilities (from Table 3.10) and activities for the two topologies. Assuming equal capacitance on all nodes, both circuit topologies consume approximately the same power when considering both the static and the dynamic components.

Table 3.11 Node transition activities for tree and chain topologies.

	$O1$	$O2$	F
$p_{0->1}$ (chain) $/p_{0->1}$ (tree)	1	0.58	1
$\alpha_{0->1}$ (chain) $/ \alpha_{0->1}$ (tree)	1	0.83	1.47

3.1.5 Word Level Signal Statistics Influencing Activity, α

A very important attribute of signal processing applications which can be used in minimizing the switched capacitance, is the correlation which can exist between values of a temporal sequence of data, since switching should decrease if the data is slowly changing (highly correlated). This is in contrast to general pur-

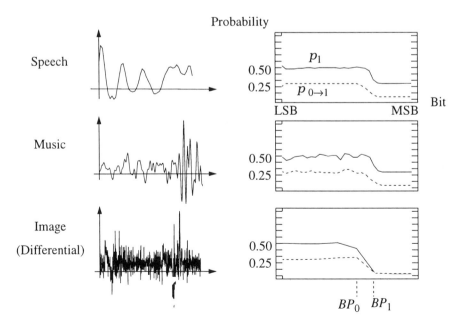

Figure 3.22: Data in signal processing applications is often correlated.

pose computation, where the data being processed tends to be random. To illustrate the correlation in signal processing applications, consider the word level transition characteristics of three common DSP signals: speech, music, and image.

It has been shown that there exists a direct relationship between bit level probabilities and word level statistics [Landman93]. This correlation is illustrated in Figure 3.22 for the three input signals. For each input, the right hand portion of the figure shows both signal and transition probabilities for each bit of the input data word (two's-complement assumed). Clearly, these signals follow a similar pattern which can be exploited to yield a simple piecewise linear model with breakpoints BP_0 and BP_1. The important features of this model - the values of the breakpoints and the signal and transition probabilities - can all be extracted from three statistical parameters: the mean, μ; the variance, σ^2; and the lag one correlation coefficient, ρ_1. The first order correlation coefficient for a signal X with variance σ^2 is given by,

$$\rho_1 = cov(X_t, X_{t+1})/\sigma^2 = E[(X_t\text{-}E[X])(X_{t+1}\text{ -}E[X])] /\sigma^2 \qquad (83)$$

where $E[X]$ is the expected value of the signal X and t is time.

In the lower-order region from the *LSB* to BP_0, as might be expected, the bits are uncorrelated in space and time and are essentially independent of the data distribution, having signal probabilities of 1/2 and therefore the 0->1 transition probabilities of 1/4:

$$p_{lsb's}(1) = 1/2, \; p_{lsb's}(0{\rightarrow}1) = 1/4 \qquad (84)$$

where $p_{lsb's}(1)$ is the probability that the lsb's are 1. The value of the breakpoint denoting the end of this low-order region is related to the spread, or variance, of the signal distribution and has been found by curve fitting to be:

$$BP_0 = \log_2(3\sigma/32) \qquad (85)$$

Strictly speaking, this formula is exact only for the signal probability curve. The breakpoint for the transition probability curve, although typically nearby, experiences an offset towards the *LSB* for highly correlated input signals:

$$\Delta BP_0 = \log_2(1\text{-}\rho_1^2)^{0.5} \qquad (86)$$

In two's-complement representation, the purpose of the high-order bits is that of sign extension. As a result, the bits in this region exhibit complete dependence, having signal and transition probabilities that are functions of the mean, variance, and first-order correlation coefficient of the data word:

$$p_{msb's}(1) = p(-) = F_1(\mu/\sigma) \tag{87}$$

$$p_{msb's}(0 \rightarrow 1) = p(+ \rightarrow -) = F_{01}(\mu/\sigma, \rho_1) \tag{88}$$

where $p_{msb's}(1)$ is the probability that the msb's are 1 which is equal to the probability that the signal is negative, $p(-)$ (for two's complement representation). The exact probabilities depend, of course, on the distribution of the signal; however, noting that many typical DSP inputs are closely approximated by Gaussian processes, the univariate and bivariate normal distribution functions can be substituted for F_1 and F_{01}, respectively. The breakpoint for the sign extension region is determined by the maximum extent of the signal distribution into either positive or negative values and is given specifically by:

$$BP_1 = \log_2(|\mu|+3\sigma) \tag{89}$$

In the middle region between BP_0 and BP_1, the correlation of the bits falls between the extremes represented by the lower and higher-order regions and as a result a linear approximation for the probabilities in this transition region models the situation well.

These signal correlations can be exploited to minimize the bit transition activity through data coding on busses, optimized representation for arithmetic, optimized time-multiplexing, ordering of operations, and activity driven placement (these techniques will be described in Chapter 7).

3.1.6 Influence of Voltage Scaling

Since power consumption in CMOS circuits is proportional to the square of the supply voltage (assuming that the dynamic component of power consumption dominates), it is clear voltage reduction will have a significant impact on power; indeed, reducing the supply voltage is the key to low-power operation, even after taking into account the modifications to the system architecture which is required to maintain the computational throughput. A review of circuit behavior (delay and energy characteristics) as a function of scaling supply voltage and feature sizes will be presented. By comparison with experimental data, it is found that simple first order theory yields an amazingly accurate representation of the various dependencies over a wide variety of circuit styles and architectures.

3.1.6.1 Impact on Delay and Power-Delay Product

As noted in Equation 51, the energy per transition is proportional to V^2. This is seen from Figure 3.23, which is a plot of two experimental circuits which exhibit the expected V^2 dependence. Therefore, it is only necessary to reduce the supply voltage for a *quadratic* improvement in the power-delay product of a logic family.

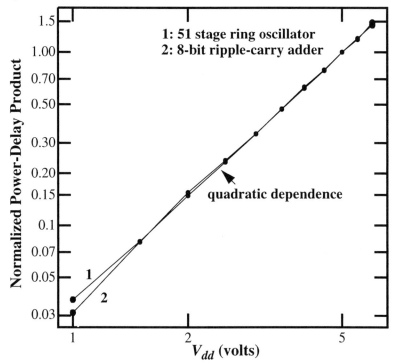

Figure 3.23: Power-delay product exhibiting square law dependence
for two different circuits.

Unfortunately, this simple solution to low-power design comes at a cost. As shown in Figure 3.24, the effect of reducing V_{dd} on circuit delay is shown for a variety of different logic circuits, that range in size from 56 to 44,000 transistors spanning a wide variety of functions, all exhibiting essentially the same dependence (see Table 3.12). Clearly, we pay a speed penalty for a V_{dd} reduction, with the delays drastically increasing as V_{dd} approaches the sum of the threshold voltages of the devices. Even though the exact analysis of delay is quite complex if the non-linear characteristic of a CMOS gate are taken into account, it is found

that a simple first-order derivation adequately predicts the experimentally determined dependence and is given by:

$$T_d = \frac{C_L \times V_{dd}}{I} = \frac{C_L \times V_{dd}}{\dfrac{\mu C_{ox}(W/L)\left(V_{dd}-V_t\right)^2}{2}} \tag{90}$$

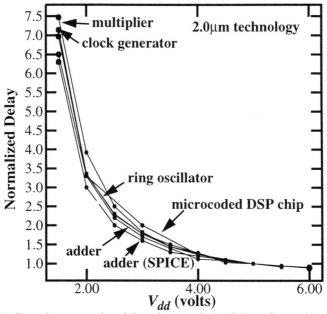

Figure 3.24: Data demonstrating delay characteristics follow first order theory.

Table 3.12 Details of components used for the study in Figure 3.24.

Component (all in 2μm)	# of transistors	Area	Comments
Microcoded DSP Chip	44802	94mm^2	20-bit datapath
Multiplier	20432	12.2mm^2	24x24 bits
Adder	256	0.083mm^2	Conventional Static

Table 3.12 Details of components used for the study in Figure 3.24.

Component (all in 2μm)	# of transistors	Area	Comments
Ring Oscillator	102	$0.055mm^2$	51-stages
Clock Generator	56	$0.04mm^2$	Cross-coupled NOR

Through experimental measurements and SPICE simulations, the energy and delay performance for several different logic styles and topologies were evaluated using an 8-bit adder as a reference; the results are shown on a log-log plot in Figure 3.25. Table 3.13 describes the circuits used for analysis in Figure 3.25. It is clear that the power-delay product (or the energy per transition) improves as delays increase (through reduction of the supply voltage), and therefore a CMOS circuit is most effective when it operates at the *slowest* possible speed. Since the objective is to reduce power consumption while maintaining the overall system throughput, compensation for these increased delays at low voltages is required. Of particular interest in this figure is the range of energies required for a transition at a given amount of delay. The best logic family analyzed (over 10 times better than the worst that was investigated) was the pass gate family, CPL if a reduced value for the threshold is assumed [Yano90].

Figures 3.23, 3.24, and 3.25 suggest that the delay and energy behavior as a function of V_{dd} scaling for a given technology is "well-behaved" and relatively independent of logic style and circuit complexity. We will use this result during our optimization of architecture for low-power by treating V_{dd} as a free variable and by allowing the architectures to vary to retain constant throughput. By exploiting the monotonic dependencies of delay and energy versus supply voltage that hold over wide circuit variations, it is possible to make relatively strong predictions about the types of architectures that are best for low-power design. Of course, as mentioned previously, there are some logic styles such as NMOS pass-transistor logic without reduced thresholds whose delay and energy characteristics would deviate from the ones presented above, but even for these cases, though the quantitative results will be different, the basic conclusions will still

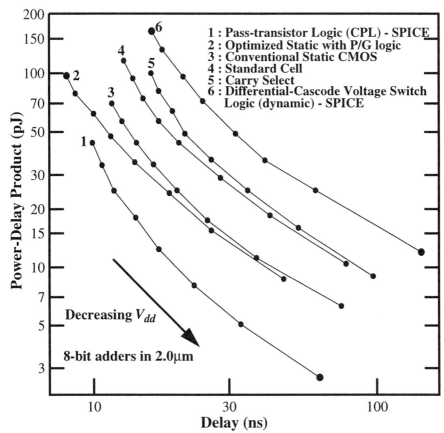

Figure 3.25: Data showing improvement in power-delay product at the cost of speed for various circuit approaches.

hold. The next chapter will describe the architecture driven voltage scaling strategy that allows voltage reduction while keeping throughput constant.

Table 3.13 Description of circuits used for study in Figure 3.25.

Adder Number in Figure 3.25	Short Description of Each Adder
1	Differential pass transistor logic made of single-transistor NMOS pass-gates (as opposed to NMOS and PMOS connected in parallel). The circuit is described in [Yano90]. A reduced threshold voltage is assumed for the pass transistors.

Table 3.13 Description of circuits used for study in Figure 3.25.

Adder Number in Figure 3.25	Short Description of Each Adder
2	Optimized static logic with propagate and generate circuitry. The sum uses complementary pass-transistor circuits while the carry logic is implemented using conventional CMOS.
3	A conventional CMOS implementation of an adder without propagate/generate logic [Weste88].
4	A standard cell implementation of a conventional adder. The devices are significantly bigger and the interconnect capacitance is larger since the placement is not as structured as a tiled datapath approach [Brodersen88].
5	A carry select implementation realized using conventional CMOS gates [Brodersen88].
6	A DCVSL adder (Differential Cascode Voltage Swing Logic) is a precharged implementation that uses dual-rail coding. This circuit is typically used in self-timed circuits [Jacobs90]. Due to the differential nature of this topology, there is a guaranteed transition on every bit for each access resulting in high switching activity.

3.2 Short-circuit Component of Power

The previous section analyzed the switching component of power consumption which corresponds to the amount of energy required to charge parasitic capacitors. The switching component of power is independent of the rise and fall times at the input of logic gates. Finite rise and fall times of the input waveforms however result in a direct current path between V_{dd} and GND which exist for a short period of time during switching. Specifically, when $V_{Tn} < V_{in} < V_{dd} - |V_{Tp}|$ holds for the input voltage, there will be a conductive path open between V_{dd} and GND because both the NMOS and PMOS devices are ON. Such a path never exists in dynamic circuits, as precharge and evaluate transistors should never be on simultaneously as this would lead to incorrect evaluation.

Short-circuit currents are, therefore, a problem solely encountered in static designs.

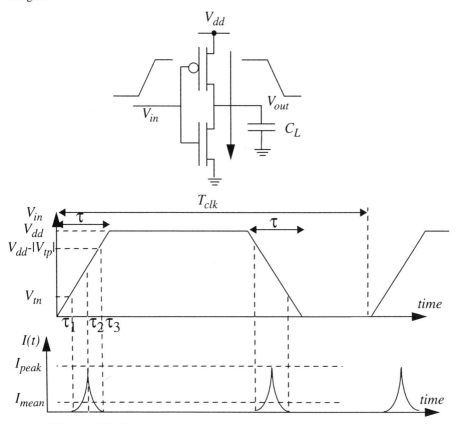

Figure 3.26: Current behavior with no output load [Veendrick84].

Figure 3.26 shows the behavior of an inverter assuming no output load. On a low-to-high transition at the input, the NMOS will start to conduct when V_{in} is equal to V_{tn}, and the PMOS will stop conducting when V_{in} is equal to $V_{dd}-|V_{tp}|$. Under zero capacitive load conditions, all the current that is drawn from the supply goes to short-circuit power.

Let τ be the rise time of the input signal, and let $V_t = V_{tn} = |V_{tp}|$. Also, the effective transistor strengths be equal for the NMOS and PMOS; let $\beta = \beta_n W_n = \beta_p W_p$. From τ_1 to τ_2, the NMOS device is in saturation since the output voltage will be greater than $V_{in}-V_t$. Then the current for the NMOS is given by:

$$I = \frac{\beta}{2}\left(V_{in} - V_t\right)^2 \qquad \text{for } 0 < I < I_{max} \tag{91}$$

The current will reach a maximum, I_{max} at $V_{in} = V_{dd}/2$ due to the symmetric assumption. The mean current can be computed [Veendrick84] to determine the average short-circuit current that is drawn from the supply.

$$I_{mean} = 2 \bullet \frac{2}{T_{clk}} \bullet \int_{\tau_1}^{\tau_2} I(t) \, dt = \frac{4}{T_{clk}} \bullet \int_{\tau_1}^{\tau_2} \frac{\beta}{2}\left(V_{in}(t) - V_t\right)^2 dt \tag{92}$$

Here, $V_{in}(t)$ is given by:

$$V_{in}(t) = \frac{V_{dd}}{\tau} \bullet t \tag{93}$$

and therefore, τ_1 to τ_2 are given by:

$$\tau_1 = \frac{V_t}{V_{dd}} \bullet \tau \tag{94}$$

$$\tau_2 = \frac{\tau}{2} \tag{95}$$

Plugging into Equation 92,

$$I_{mean} = \frac{2\beta}{T_{clk}} \int_{\tau/2}^{V_t \bullet \tau / V_{dd}} \left(\frac{V_{dd} \bullet t}{\tau} - V_t\right)^2 d\left(\frac{V_{dd} \bullet t}{\tau} - V_t\right) \tag{96}$$

which results in:

$$I_{mean} = \frac{\beta}{12 \bullet V_{dd}}\left(V_{dd} - 2V_t\right)^3 \frac{\tau}{T_{clk}} \tag{97}$$

The short-circuit power is then given by:

$$P_{short-circuit} = V_{dd} \bullet I_{mean} = \frac{\beta}{12}\left(V_{dd}-2V_t\right)^3 \frac{\tau}{T_{clk}} \qquad (98)$$

In the case with no capacitive load, the short-circuit current is directly proportional to the risetime and the effective transistor strength, β.

Short circuit currents are significant when the rise/fall time at the input of a gate is much longer than the output rise/fall time. This is because the short-circuit path will be active for a longer period of time. Consider a static CMOS inverter with a low to high input transition. Assume first that the load capacitance is very large, such that the output fall time is significantly larger than the input rise times. Here, the input will go through the transient region before the output starts to change. Since the source-drain voltage of the PMOS device is essentially 0 during that period, the device shuts off without ever delivering any current. The short-circuit current is therefore essentially zero. If the output capacitance is very small and the output fall time is substantially smaller than the input rise time, the drain-source voltage of the PMOS device is close to V_{dd} during most of the transition period, resulting in large short-circuit current (equal to the saturation current of the PMOS) during most of the transient period. Figure 3.27 shows a graph of the short-circuit current as a function of the output capacitance for a fixed input rise time [Veendrick84].

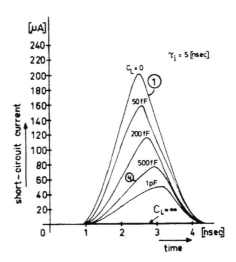

Figure 3.27: Short-circuit current as a function of load capacitance [Veendrick84].
(© 1984 IEEE)

This above analysis implies that short-circuit dissipation is minimized by making the output rise/fall times larger than the input rise/fall times.

On the other hand, making the output rise/fall times too large slows down the circuit and might cause short-circuit current in the fanout-gates. Therefore a good compromise to minimize the total average short-circuit current, it is desirable to have equal input and output edge times [Veendrick84].

Figure 3.28 plots the fraction of energy consumed by short-circuit current versus the ratio of the input rise time, t_{Rin}, to the output rise time, t_{Rout}. The ΔE increases with increasing input edge time. In this case, the power consumed by the short-circuit currents is typically less than 10% of the total dynamic power.

An important point to note is that if the supply is lowered to be below the sum of the thresholds of the transistors, $V_{dd} < V_{Tn} + |V_{Tp}|$, the short-circuit currents can be virtually (since sub-threshold currents will still flow) *eliminated* because both devices cannot conduct simultaneously for any value of the input voltage.

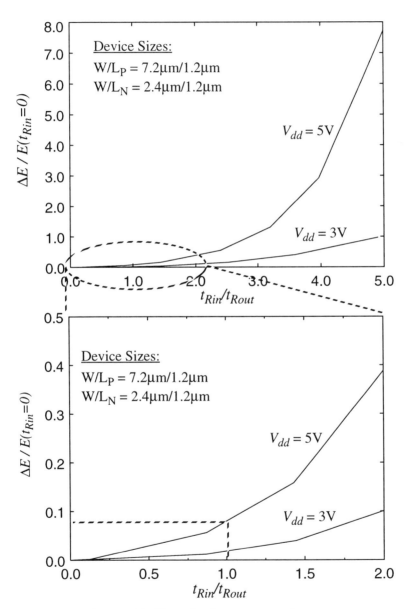

Figure 3.28: Short-circuit component of power.

3.3 Leakage Component of Power

There are two types of leakage currents: reverse-bias diode leakage at the transistor drains, and sub-threshold leakage through the channel of an "off" device. The magnitude of these currents is set predominantly by the processing technology; however, there are some things that a designer can do to minimize their contribution.

3.3.1 Diode Leakage

Diode leakage occurs when a transistor is turned off and another active transistor charges up/down the drain with respect to the former's bulk potential. For example, consider an inverter with a high input voltage, in which the NMOS transistor is turned on and the output voltage is driven low. The PMOS transistor will be turned off, but its drain-to-bulk voltage will be equal to $-V_{dd}$ since the output voltage is at 0V and the bulk for the PMOS is at V_{dd} (as shown in Figure 3.29).

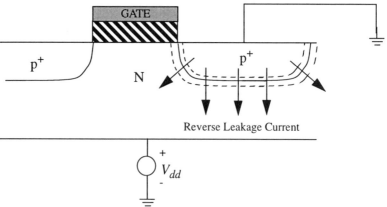

Figure 3.29: Reverse bias diode leakage currents.

The leakage current for the diode is given by

$$I_{leakage} = I_S \left(e^{\frac{V}{V_T}} - 1 \right) \tag{99}$$

where I_s is the reverse saturation current, V is the diode voltage, $V_T = KT/q$ is the thermal voltage. Due to the exponential dependence, even with a small reverse bias voltage across the diode, the leakage current will be relatively independent of the voltage and will equal the reverse saturation current.

The reverse saturation current per unit area is defined as the current density J_s, and the resulting current will be approximately $I_L = A_D J_S$, where A_D is the area of the drain diffusion. For a typical CMOS process, J_S is approximately 1-5 pA/μm^2 (25° C), and the minimum A_D is 7.2 μm^2 for a 1.2 μm minimum feature size. J_s doubles with every 9 degree increase in temperature. For a 1 million transistor chip, assuming an average drain area of 10 μm^2, the total leakage current is on the order of 25 μA. While this is typically a small fraction of the total power consumption in most chips, it could be significant for a system application which spends much of its time in standby operation, since this power is always being dissipated even when no switching is occurring.

3.3.2 Sub-threshold Leakage

The other component is the sub-threshold leakage which occurs due to carrier diffusion between the source and the drain when the gate-source voltage, V_{gs}, has exceeded the weak inversion point, but is still below the threshold voltage V_t, where carrier drift is dominant. In this regime, the MOSFET behaves similarly to a bipolar transistor, and the subthreshold current is exponentially dependent on the gate-source voltage V_{gs} (as seen from Figure 3.30). The current in the subthreshold region is given by [Sze81]:

$$I_{ds} = K e^{\left(V_{gs} - V_t\right)/\left(nV_T\right)}\left(1 - e^{-\frac{V_{ds}}{V_T}}\right) \tag{100}$$

where K is a function of the technology, V_T is the thermal voltage (KT/q) and V_t is the threshold voltage and $n = 1 + \Omega\, t_{ox}/D$, where t_{ox} is the gate oxide thickness, D is the channel depletion width, and $\Omega = \varepsilon_{si}/\varepsilon_{ox}$. For $V_{ds} \gg V_T$, $(1-e^{-Vds/VT}) \approx 1$; that is, the drain to source leakage current is independent of the drain-source voltage V_{ds}, for V_{ds} approximately larger than 0.1V.

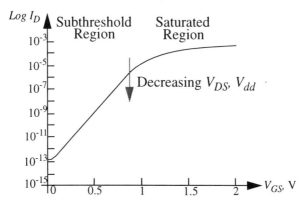

Figure 3.30: Sub-threshold leakage component of power.

Associated with this is the subthreshold slope S_{th}, which is the amount of voltage required to drop the subthreshold current by one decade. The subthreshold slope can be determined by taking the ratio of two points in this region (from Equation 100):

$$\frac{I_1}{I_2} = e^{\left(V_1 - V_2\right)/\left(nV_T\right)} \tag{101}$$

which results in $S_{th} = nV_Tln$ (10). At room temperature, typical values for S_{th} lie between 60 to 90 mV/(decade current), with 60 mV/dec being the lower limit. Clearly, the lower S_{th} is, the better, since it is desirable to have the device "turn-off" as close to V_t as possible.

As a reference, for an $L=1.5\mu$, $W=70\mu$ NMOS device, at the point where V_{gs} equals V_t, with V_t defined as where the surface inversion charge density is equal to the bulk doping, approximately 1μA of leakage current is exhibited, or .014μA/micron of gate width [Sze81]. The issue is whether this extra current is negligible in comparison to the time-average current during switching. For a CMOS inverter (PMOS: $W=4\mu$, NMOS: $W=8\mu$), the current was measured to be 64μA over 3.7nsec at a supply voltage of 2V. This implies that there would be a 100% power penalty for subthreshold leakage if the device were operating at a clock speed of 25 MHz with an activity factor = 1/6th, i.e. the devices were left idle and leaking current 83% of the time. It is not advisable, therefore, to use a true zero threshold device, but instead to use thresholds of at least 0.2V, which provides for at least two orders of magnitude of reduction of subthreshold current. The topic of optimal threshold voltage selection is presented in the next chapter.

3.4 Static Power

Migrating from NMOS technology to CMOS, has resulted in logic circuits that primarily consume power only during switching. There are two situations, however, when circuits dissipate power in steady state operation (i.e. non transient operation): degenerated voltage levels feeding into complementary static gates and pseudo NMOS circuit styles.

3.4.1 Reduced Voltage Levels Feeding CMOS Gates

Figure 3.31 shows an NMOS pass transistor driving a CMOS inverter. With V_{dd} applied to both the gate and the drain of the NMOS pass transistor, node B will rise to V_{dd}-V_{tn}. In steady state, this input will cause the inverter output to be forced low. The PMOS device will however be weakly ON (since $|V_{gs}$-$V_t|$ is approximately equal to 0), conducting static current from the supply to ground. This might consume significant power if the circuit is idle most of the time, and in this case, a weak level restoring PMOS transistor must be introduced from the output node to node B to pull it all the way to the rail.

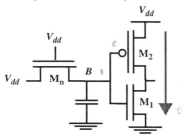

Figure 3.31: Degenerated voltage level feeding a CMOS gate results in DC power.

3.4.2 Pseudo-NMOS Logic Style

Pseudo-NMOS circuits also dissipate static power. In this logic implementation style, the PMOS pullup network of the CMOS implementation is replaced by a single PMOS device whose gate is grounded and is therefore always ON. When the output is driven low by a resistive path from the output node to ground though the NMOS network, there is a conducting path from the supply to ground. This implementation style can reduce power only for implementations that require complex logic switching at high frequencies. Figure 3.32 shows an example of such a gate -- a wide AND-OR-Invert gate. To implement this in full static CMOS would require several times the area to implement the stacked PMOS transistors. The extra PMOS transistors would also increase the capacitance on the input nodes, which would load down the previous gates. There is usually a large area savings as well. This is primarily applicable for higher-frequency circuits, where the savings of dynamic power is proportionally higher. If the circuit is idle most of the time (when the circuit is clocked at low frequencies), then the static power will tend to increase the total power consumption.

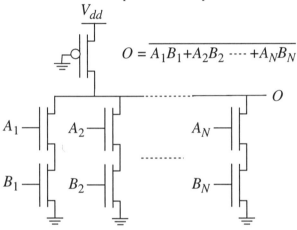

Figure 3.32: Implementing complex logic using pseudo-NMOS.

3.5 Summary

There are four components to power consumption in CMOS technology arising from switching currents, short-circuit currents, leakage currents and finally static currents. For "properly" designed circuits the switching component will dominate and will contribute to more than 90% of the total power consumption, making it the primary target for power reduction.

The switching component of power for a CMOS gate with load capacitor C_L is given by $\alpha C_L V_{dd}^2 f$ where α is node transition activity, C_L is physical load capacitance, V_{dd} is supply voltage and f is operating frequency. There are two components to node transition activity: transitions due to the static behavior of the circuit (in which the transitions are determined strictly from the boolean logic ignoring timing skew) and transitions that occur due to the dynamic nature of the circuit (that is due to timing skew). The node transition activity factor is a function of the logic function being implemented, the logic style, the circuit topologies, signal statistics, signal correlations, and the sequencing of operations. A system level approach which involves optimizing algorithms, architectures, logic design, circuit design and physical design can be used to minimize the switched capacitance and is described in detail in Chapter 7. The components of physical capacitance include the transistor parasitics including the gate and diffusion capacitances and the interconnect capacitance. The physical capacitance can be minimized through choice of substrate (like SOI), layout optimization, device sizing (as described in the next chapter) and choice of logic style. The choice of supply voltage has the greatest impact on the power-delay product, which is the amount of energy required to perform a given function. It is only necessary to reduce the supply voltage to quadratically improve the power-delay product. Unfortunately, a reduction in circuit speed is associated with supply voltage scaling. However, if the goal is to reduce the power consumption for realizing a fixed functionality or if the goal is to increase the MIPS/Watt in general purpose computing for a fixed MIPS level, then various architectural schemes can be used for voltage reduction.

The second component of power is the short-circuit power (also called direct path power) which exists when there is a direct conducting path from supply to ground. Through proper choice of transistor sizes, the short-circuit power can be kept to less than 10%. Alternatively, operating the circuits at a supply voltage less than the sum of the NMOS and PMOS threshold voltages will essentially eliminate any short-circuit currents. The architecture-driven voltage scaling strategy described in the next chapter results in supply voltages which satisfy this condition.

The third component of power is due to leakage currents. There are two types of leakage currents: reverse-bias diode leakage at the transistor drains, and sub-threshold leakage through the channel of an "off" device. The subthreshold leakage occurs due to carrier diffusion between the source and the drain when the gate-source voltage, V_{gs}, has exceeded the weak inversion point, but is still below the threshold voltage V_t, where carrier drift is dominant. In this regime, the MOSFET behaves similarly to a bipolar transistor, and the subthreshold current is exponentially dependent on the gate-source voltage V_{gs}. An important figure of merit for a low-power technology is the subthreshold slope S_{th}, which is the amount of voltage required to drop the subthreshold current by one decade. The lower the subthreshold slope, the better since the devices can be turned-off as close to V_t as possible.

The final component of power is the static currents found in circuits that do not have rail-to-rail swing feeding other circuits, pseudo-NMOS logic styles, and in analog bias circuits. The static currents must be minimized as much as possible. For example, in SRAM sense amplifiers, pulsed circuits should be used to minimize static currents.

REFERENCES:

[Bakoglu90] H.B. Bakoglu, *Circuits, Interconnections, and Packaging for VLSI*, Addison-Wesley, Menlo Park, CA, 1990.

[Brodersen88] R.W. Brodersen *et al.*, LagerIV Cell Library Documentation, Electronics Research Laboratory, University of California, Berkeley, June 23, 1988.

[Glasser85] L.A. Glasser and D.W. Dobberpuhl, *The Design and Analysis of VLSI Circuits*, Addison-Wesley Publishing Co., Reading, Mass., 1985.

[Heller84] L. Heller and W. Griffin, "Cascode voltage switch logic : A differential CMOS logic Family," *ISSCC*, pp. 16-17, Feb. 1984.

[Jacobs90] G. Jacobs and R.W. Brodersen, "A Fully Asynchronous Digital Signal Processor Using Self-Timed Circuits," *IEEE Journal of Solid-State Circuits*, pp. 1526-1537, December 1990.

[Kingsbury94] B. Kingsbury and J. Wawrzynek, Private Communication.

[Landman93] P. E. Landman, J. M. Rabaey, "Power Estimation for High Level Synthesis," *EDAC '93*, Paris, pp. 361-366, Feb. 1993.

[Rabaey95] J. Rabaey, *Digital Integrated Circuits: A Design Perspective*, Prentice Hall, Englewood Cliffs, N.J., to be published in 1995.

[Schaper83] L.W. Schaper and D.I. Amey, "Improved Electrical Performance Required for Future MOS Packaging," *IEEE Transaction on Components, Hybrids, and Manufacturing Technology*, vol. CHMT-6, pp. 282-289, Sept. 1983.

[Sze81] S. Sze, *Physics of Semiconductor Devices*, John Wiley & Sons, 1981.

[Veendrick84] H.J.M. Veendrick, "Short-Circuit Dissipation of Static CMOS Circuitry and Its Impact on the Design of Buffer Circuits," *IEEE Journal of Solid-State Circuits*, Vol. SC-19, pp. 468-473, August 1984.

[Weste88] N. Weste and K. Eshragian, *Principles of CMOS VLSI Design: A Systems Perspective*, Addison-Wesley, MA, 1988.

[Yano90] K. Yano *et al.*, "A 3.8ns CMOS 16x16 Multiplier Using Complementary Pass Transistor Logic," *IEEE Journal of Solid-State Circuits*, pp. 388-395, April 1990.

[Yuan82] C.P. Yuan and T.N. Trick, "A Simple Formula for the Estimation of the Capacitance of Two-dimensional Interconnects in VLSI Circuits," *IEEE Electron Device Letters*, vol. EDL-3, pp. 391-393, Dec. 1982.

4

Voltage Scaling Approaches

Since the dominant component of power consumption for a properly designed CMOS circuit is proportional to the square of the supply voltage, operating circuits at the lowest voltage is the key to minimizing the energy consumed per operation. However, the individual circuit elements run slower at lower supply voltages (Figure 3.25) and this must be compensated for through appropriate architectural design. One important class of applications are those which have no advantage in exceeding a bounded computation rate, as found in real-time signal processing. The strategies presented in this chapter also have some applicability to the maximum throughput situation of general purpose computing though additional system level trade-offs must be made.

A survey of previous approaches to supply-voltage scaling will be presented, which are focused on maintaining reliability, maintaining performance, and optimizing the speed-energy characteristics for a given technology. Then the coupling of voltage scaling with transistor sizing will be given, which yields conclusions about the optimum transistor sizing strategy for low-power. The effect of scaling the threshold voltage on the power supply voltage and power consumption is then presented. This is followed by an architecture-driven approach, from which an "optimal" supply voltage is derived.

4.1 Reliability-Driven Voltage Scaling

One approach to the selection of an optimal power supply voltage for deep-submicron technologies is based on optimizing the trade-off between speed, long-term reliability, and power dissipation [Davari88]. Circuit speed is increased in designs with higher electric fields owing to increased carrier concentration and velocity. Constant-voltage scaling - the most commonly used technique - results in higher electric fields that create hot carriers. As a result of this, the devices degrade with time (including changes in threshold voltages, degradation of transconductance, and increase in subthreshold currents), leading to eventual breakdown.

One approach to reduce the number of hot carriers is to change the device structure to reduce the fields inside or to limit the spatial extent of the high field regions so that carriers do not gain enough energy from the field to cause difficulty [Watts89]. Since high fields occur only near the drain, the drain doping profile can be tailored such that the potential gradient is more gradual. One solution to reducing the number of hot carriers is to change the physical device structure, such as the use of LDD (lightly doped drain), usually at the cost of decreased performance. Figure 4.1 shows the schematic diagram of the drain doping profile of a conventional MOSFET and an LDD device.

A lightly doped implant is introduced between the heavily doped n+ region and the channel. This structure can be obtained by two independent implantations, with one of them offset by an oxide sidewall on the gate structure. This structure reduces the peak electric field on the drain side of the channel improving reliability. The main problem with LDD is however an increase in the series resistance of the drain and source which results in reduced performance.

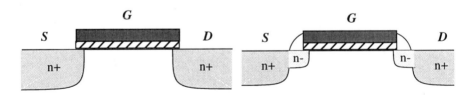

(a) Conventional Structure   (b) Lightly doped drain (LDD)

Figure 4.1: Drain doping profile for a conventional device and an LDD device.

Thus the reliability driven power supply voltage selection is based on a trade-off between performance and reliability. An empirical relationship between channel hot carriers (CHC) limited voltage and device series resistance (added due to the drain modification of LDD structures) has been developed from measurements on a variety of junction technologies [Davari88]. Using this relationship, a model of circuit performance of a technology as a function of power supply voltage and CHC margin can be determined. Figure 4.2 shows a plot of simulated delay vs. power supply voltage. The solid line represents the delay curve for a fixed device structure. The dotted line describes the case where the hot carrier margin is kept fixed by altering the device structure (i.e., varying the series resistance of the source/drain junctions).

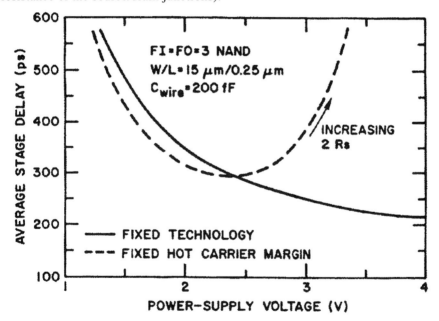

Figure 4.2: Reliability-driven supply voltage selection [Davari88]. (© 1988 IEEE)

The delay starts to increase with increasing supply voltage since the parasitic resistance of the LDD structure must be increased to maintain the CHC margin and this limits the circuit performance. Therefore a reliability driven optimum supply voltage can be found; for this 0.25µm technology used, an optimal voltage of 2.5V was found by choosing the minimum point on the delay vs. V_{dd} curve.

Clearly it does not make sense from a power perspective to operate above the optimum voltage since throughput degrades and the energy consumed is increased. However, operating below the reliability driven optimum voltage is desirable but results in loss of device speed, which can be compensated through architectural design.

4.2 Technology-Driven Voltage Scaling

Another approach for voltage selection has been proposed based on considering the effects of device level properties involving velocity saturation. The simple first order delay analysis presented in Section 3.1.6 is reasonably accurate for long channel devices. However, as feature sizes shrink below 1.0μ, the delay characteristics as a function of lowering the supply voltage deviate from the first order theory presented since it does not consider carrier velocity saturation under high electric fields [Bakoglu90]. As a result of velocity saturation, the current is no longer a quadratic function of the voltage but linear; hence, the current drive is significantly reduced and is approximately given by

$$I = WC_{ox}\left(V_{dd} - V_t\right)v_{max} \qquad (102)$$

where v_{max} is the maximum velocity. Given this and the equation for delay as $C_L V_{dd} / I$, we see that the delay for submicron circuits is relatively independent of supply voltages at high electric fields.

A "technology" based approach proposes choosing the power supply voltage based on maintaining the speed performance for a given submicron technology [Kakumu90]. By exploiting the relative independence of delay on supply voltage at high electric fields, the voltage can be dropped to some extent for a velocity-saturated device with very little penalty in speed performance. This implies that there is little advantage to operating above a certain voltage. This idea has been formalized by Kakumu et al., yielding the concept of a "critical voltage" which provides a lower limit on the supply voltage [Kakumu90].

Figure 4.3 shows the model for computing the delay of circuit under high electric field effects and the "critical voltage" as defined by Kakumu. The model used for the MOS devices under high electric field is the one proposed by Sodini [Sodini84].

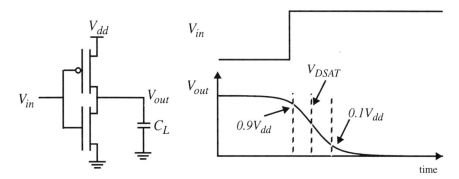

Figure 4.3: Model for computing delay and critical voltage.

$$I_D = \frac{W_{eff}\mu_{eff}C_{ox}\left(V_G' - \frac{V_{DS}}{2}\right)V_{DS}}{L_{eff}\left(1 + \frac{V_{DS}}{E_C L_{eff}}\right)} \qquad (V_{DS} \le V_{DSAT}) \tag{103}$$

$$I_D = \frac{W_{eff}\mu_{eff}C_{ox}\left(V_G' - V_{DSAT}\right)E_C}{2} \qquad (V_{DS} > V_{DSAT}) \tag{104}$$

where

$$V_{DSAT} = \frac{E_C L_{eff} V_G'}{E_C L_{eff} + V_G'} \tag{105}$$

and V_G' is the gate to source voltage minus the threshold voltage and E_c is the critical electric field at which the carrier velocity is saturated and is equal to $2v_{sat}/\mu_{eff}$. The delay of the circuit is the time taken for the output to transition from $0.9V_{dd}$ to $0.1V_{dd}$. For this range, the NMOS transistor goes through two regions: from $0.9V_{dd}$ to V_{DSAT} (saturation) and from V_{DSAT} to $0.1V_{dd}$ (linear). Let these times be defined as τ_1 and τ_2. The delays are then determined as:

$$\tau_1 = \frac{2C_L\left(0.9V_{dd} - V_{DSAT}\right)}{W_{eff}\mu_{eff}C_{ox}\left(V_G' - V_{DSAT}\right)E_C} \tag{106}$$

$$\tau_2 = \frac{C_L L_{eff}}{W_{eff} \mu_{eff} C_{ox}} \left(\frac{1}{V_G'} \ln \frac{V_{DSAT}}{0.1 V_{dd}} A + \frac{2}{E_C L_{eff}} \ln A \right) \quad (107)$$

where

$$A = \frac{2V_G' - 0.1 V_{dd}}{2V_G' - V_{DSAT}} \quad (108)$$

To understand the two delay terms some simplifications can be made. For τ_1 it can be assumed that $(0.9 V_{dd} - V_{DSAT})/(V_G' - V_{DSAT})$ is approximately equal to 1. This means that τ_1 is to first order independent of the power supply voltage. For τ_2, the second term in the bracket of Equation 107 is negligible and therefore τ_2 can be approximated as:

$$\tau_2 \approx \frac{C_L L_{eff}}{W_{eff} \mu_{eff} C_{ox} V_G'} \frac{1}{} \ln \frac{V_{DSAT}}{0.1 V_{dd}} A \quad (109)$$

Since the ln term is nearly constant, it can be seen that τ_2 is inversely proportional to V_{dd}. Therefore, the total delay $= \tau_1 + \tau_2$ contains a term which is independent of supply voltage and one term which is dependent on supply voltage. Figure 4.4 plots the fall time as a function of the power supply voltage on a log-log plot and the voltage dependent and voltage independent terms are identified. When V_{dd} becomes larger than the critical voltage, V_C (the voltage at which, $\tau_1 = \tau_2$), the delay is dominated by τ_1 and does not improve significantly with an increase in voltage; Kakumu and Kinugawa therefore regard V_C as the lower limit of power supply voltage.

If the threshold voltage is taken as $0.2 V_{dd}$, the critical voltage is $1.1 E_c L_{eff}$ where E_c is the critical electric field which results in velocity saturation. For an experimental delay vs. V_{dd} curve, this is the voltage at which the curve approaches a $\sqrt{V_{dd}}$ dependence. For 0.3μ technology, the lower limit on supply voltage (the critical voltage) was found to be 2.43V.

Because of this effect, there is strong movement to a 3.3V industrial voltage standard since at this level of voltage reduction there is not a significant loss of circuit speed [Bell91][Dhale91] and yields a 60% reduction in power when compared to a 5 volt operation [Dhale91].

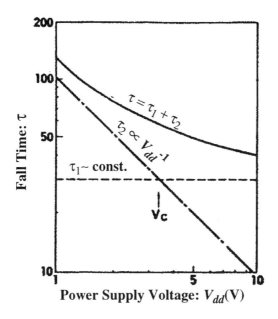

Figure 4.4: Two components of delay and critical voltage [Kakamu90].
(© 1990 IEEE)

4.3 Energy x Delay Minimum Based Voltage Scaling

Another approach has been proposed for voltage reduction which involves minimizing the energy x delay product [Burr91]. For a fixed technology (both feature size and threshold voltage), there is a supply voltage that compromises between the quadratically reduced energy per computation (due to a lower supply voltage) and the increased circuit delays. This "optimum" supply voltage can be computed analytically assuming that the transistors operate mostly in the saturation region and assuming that the threshold voltages for the NMOS and PMOS are equal:

$$Energy \bullet T_d = K \bullet C_L \bullet V_{dd}^2 \bullet V_{dd}/(V_{dd} - V_t)^2 \qquad (110)$$

where T_d is the delay of the circuit. Taking the derivative of the $Energy \bullet T_d$ and solving for the supply voltage results in a minimum (or optimum based on this

strategy) of $3V_t$. This can be seen from the plot of measured normalized *Energy* •
T_d vs. V_{dd} (Figure 4.5) for a ring oscillator. In this case, power consumption is
lowered at the expense of a *reduced* computational throughput. This is not desirable for throughput constrained functions where the hardware has to meet a
required functional throughput.

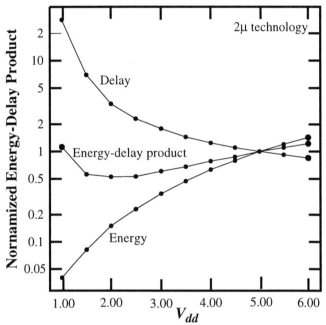

Figure 4.5: Energy-Delay product vs. V_{dd}.

4.4 Voltage Scaling Through Optimal Transistor Sizing

Independent of the choice of logic family or topology, optimized transistor
sizing will play an important role in reducing power consumption. For low power,
as is true for high speed design, it is important to equalize all delay paths so that a
single critical path does not unnecessarily limit the performance of the entire circuit. However, beyond this constraint, there is the issue of to what extent the
(W/L) ratios should be uniformly raised for all the devices, yielding a uniform
decrease in the gate delay and hence allowing for a corresponding reduction in
voltage and power. It is shown in this section, that if voltage is allowed to vary,
the optimal sizing for low power is quite different from that required for high
speed operation.

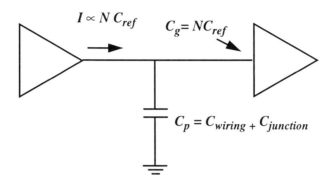

$$I \propto N\,C_{ref} \qquad C_g = N C_{ref}$$

$$C_p = C_{wiring} + C_{junction}$$

Figure 4.6: Circuit model for analyzing the effect of transistor sizing.

In Figure 4.6, a simple two-gate circuit is shown, with the first stage driving the gate capacitance of the second, in addition to the parasitic capacitance C_p due to substrate coupling and interconnect. Assuming that the input gate capacitance of both stages is given by NC_{ref}, where C_{ref} represents the gate capacitance of a MOS device with the smallest allowable (W/L), then the delay through the first gate at a supply voltage V_{ref} is given by:

$$T_N = K\frac{\left(C_p + NC_{ref}\right)}{\left(NC_{ref}\right)}\frac{V_{ref}}{\left(V_{ref} - V_t\right)^2} = K(1 + \beta/N)\frac{V_{ref}}{\left(V_{ref} - V_t\right)^2} \quad (111)$$

where β is defined as the ratio of C_p to C_{ref}, and K represents terms independent of device width and voltage. For a given supply voltage V_{ref}, the speed up of a circuit whose W/L ratios are sized up by a factor of N over a reference circuit using minimum size transistors $(N=1)$ is given by $(1 + \beta/N)/(1 + \beta)$. In order to evaluate the energy performance of the two designs at the same speed, the voltage of the scaled solution is allowed to vary as to keep delay constant. Assuming that the delay scales as $1/V_{dd}$ (ignoring threshold voltage reductions in signal swings) the supply voltage, V_N, where the delay of the scaled design and the reference design are equal is given by:

$$V_N = \frac{(1 + \beta/N)}{(1 + \beta)} V_{ref} \quad (112)$$

Under these conditions, the energy consumed by the first stage as a function of N is given by:

$$Energy(N) = \left(C_p + NC_{ref}\right)V_N^2 = \frac{NC_{ref}(1+\beta/N)^3 V_{ref}^2}{(1+\beta)^2} \quad (113)$$

After normalizing against E_{ref} (the energy for the minimum size case), Figure 4.7 shows a plot of $Energy(N) / Energy(1)$ vs. N for various values of β. When there is no parasitic capacitance contribution (i.e., $\beta = 0$), the energy increases linearly with respect to N, and the solution utilizing devices with the smallest (W/L) ratios results in the lowest power.

At high values of β, when parasitic capacitances begin to dominate over the gate capacitances, the power decreases temporarily with increasing device sizes and then starts to increase, resulting in an optimal value for N. The initial decrease in supply voltage achieved from the reduction in delays more than compensates the increase in capacitance due to increasing N. However, after some point the increase in capacitance dominates the achievable reduction in voltage, since the incremental speed increase with transistor sizing is very small (this can be seen in Equation 111, with the delay becoming independent of β as N goes to infinity). Throughout the analysis we have assumed that the parasitic capacitance is independent of device sizing. However, the drain and source diffusion and perimeter capacitances actually increase with increasing area, favoring smaller size devices and making the above a worst-case analysis.

Also plotted in Figure 4.7 are simulation results using the SPICE simulator from extracted layouts of an 8-bit adder carry chain for three different device (W/L) ratios ($N=1$, $N=2$, and $N=4$). The curve follows the simple first-order model derived very well, and suggests that this example is dominated more by the effect of gate capacitance rather than parasitics. In this case, increasing devices (W/L)'s does not help, and the solution using the smallest possible (W/L) ratios results in the best sizing.

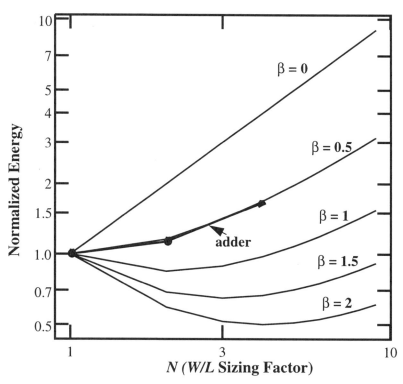

Figure 4.7: Plot of energy vs. transistor sizing factor for various parasitic contributions.

4.5 Voltage Scaling Using Threshold Reduction

Reducing the threshold voltage of the device allows the supply voltage to be scaled down (and therefore lower switching power) without loss in speed. For example, a circuit running at a supply voltage of 1.5V with V_t =1V will have approximately the same performance as the circuit running at a supply voltage of 0.9V and a V_t=0.5V using the simple first order theory of Section 3.1.6. Figure 4.8 shows a plot of normalized delay vs. threshold voltage for various supply voltages.

Since significant power improvements can be gained through the use of low-threshold MOS devices, the question of how low the thresholds can be reduced must be addressed. The limit is set by the requirement to retain adequate noise margins and the increase in subthreshold currents. Noise margins will be

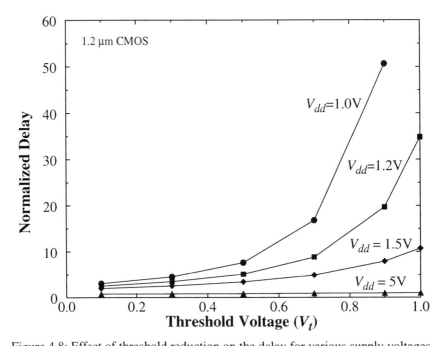

Figure 4.8: Effect of threshold reduction on the delay for various supply voltages.

relaxed in low power designs because of the reduced currents being switched, however, subthreshold currents can result in significant static power dissipation.

Figure 4.9 shows a plot of energy vs. threshold voltage for a fixed through-put for a 16-bit datapath ripple carry adder (which essentially represents the power to perform the operation). Here, the power supply voltage is allowed to vary to keep the throughput fixed. For a fixed throughput (e.g. that obtained at a 20Mhz clock rate), the supply voltage and therefore the switching component of power can be reduced while reducing the threshold voltage. However, at some point, the threshold voltage and supply reduction is offset by an increase in the leakage currents, resulting in an optimal threshold voltage for a given level of logic complexity. That is, the optimum threshold voltage must compromise between improvement of current drive at low supply voltage operation and con-trol of the sub-threshold leakage.

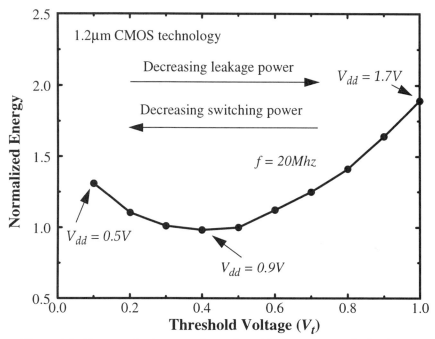

Figure 4.9: Compromise between dynamic and leakage power dissipation through V_t variation.

4.6 Architecture-Driven Voltage Scaling

The "technology" based approaches presented in Sections 4.1 and 4.2 are focusing on reducing the voltage while maintaining device speed, and are not attempting to achieve the minimum possible power. As shown in Figures 3.23 and 3.25, CMOS logic gates achieve lower power-delay products (energy per computation) as the supply voltages are reduced. In fact, once a device is in velocity saturation there is a further degradation in the energy per computation, so in minimizing the energy required for computation, Kakumu's critical voltage for low power therefore provides an *upper* bound on the supply voltage whereas for his analysis it provided a *lower* bound! It now will be the task of the architecture to compensate for the reduced circuit speed, that comes with operating below the critical voltage.

4.6.1 Trading Area for Lower Power Through Hardware Duplication

Figure 4.10 shows a conventional uni-processor implementation of some logic function, of some input IN, $F(IN)$. Let f_{sample} $(= 1/T_{sample})$ be the throughput clock rate for this implementation (or the rate at which the input samples are streamed through) and also let f_{sample} be the fastest clock rate at which the function $F(IN)$ can be clocked (that is, the critical path of $F(IN)$, $T_{critical\text{-}path}$ is equal to T_{sample} at a reference supply voltage of V_{ref}). Let C_{ref} be the capacitance switched every cycle,

$$C_{ref} = C_{in\text{-}reg} + C_F + C_{out\text{-}reg} \qquad (114)$$

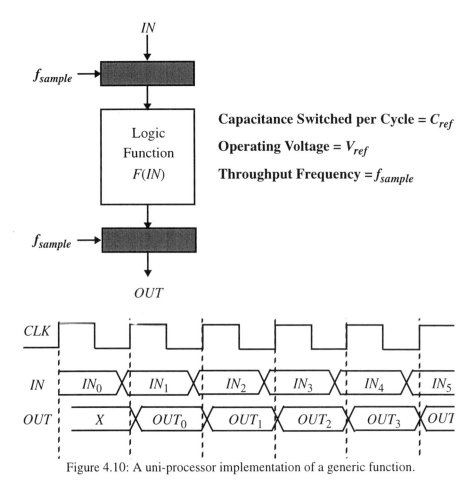

Capacitance Switched per Cycle = C_{ref}

Operating Voltage = V_{ref}

Throughput Frequency = f_{sample}

Figure 4.10: A uni-processor implementation of a generic function.

where $C_{in\text{-}reg}$ is the capacitance switched in the input register, C_F is the capacitance switched to implement the logic function, and $C_{out\text{-}reg}$ is the capacitance of the output register.

The power consumption of this datapath is given by,

$$P_{ref} = C_{ref} V^2_{ref} f_{sample} \qquad (115)$$

Figure 4.10 also shows the timing diagram for the input and output data for this implementation. The output data is valid one clock cycle after the input data has been latched into the input register. The latency is therefore one clock cycle.

Another approach for implementing $F(IN)$ is shown in Figure 4.11, which shows the generalized approach of using parallelism to reduce the clock requirement. Hardware duplication is used in which N processing elements each implementing the function F are computing in parallel. The input is broadcast to all the registers of the N processing elements and gated clocks are used to load each register every N cycles. That is, the clocks to each register are skewed such that IN_0 is loaded in REG0, IN_1 is loaded into REG1, and so on. The output of the N processing elements is multiplexed and sent to an output register which operates at the throughput clock rate (f_{sample}). Since each input register is clocked only at f_{sample}/N, the time available to compute the function F for each input sample, $T_{available} = N/f_{sample}$, is increased by a factor of N over the uni-processor implementation. That is, $T_{available}/T_{critical\text{-}path} = N$. This implies that the supply voltage can be lowered to increase the circuit delays until the critical path is extended to equal to the new slower clock cycle.

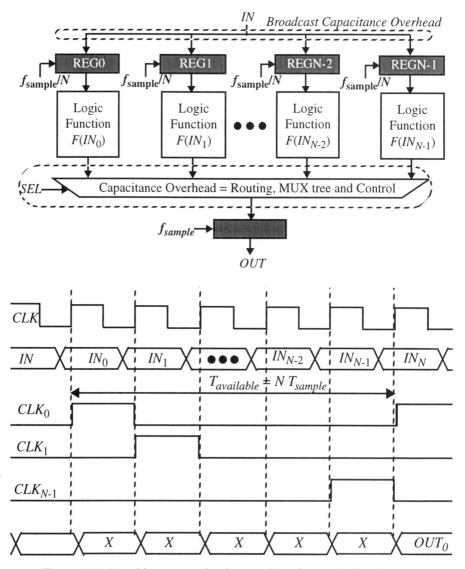

Figure 4.11: A multi-processor implementation of a generic function.

The power supply voltage corresponding to N-level parallelism, V_N, is determined below:

$$T_N = K \frac{V_N}{\left(V_N - V_t\right)^2} = N \bullet T_1 = N \bullet K \frac{V_{ref}}{\left(V_{ref} - V_t\right)^2} \qquad (116)$$

where T_N is the critical path for each computation module of the N processor implementation and T_1 is critical path for the uni-processor. This was derived assuming the long channel delay model discussed in Chapter 3. Therefore the ratio of V_N to V_{ref} is given by:

$$\frac{V_N}{V_{ref}} = N \bullet \frac{\left(V_N - V_t\right)^2}{\left(V_{ref} - V_t\right)^2} = N \bullet \frac{\left(\frac{V_N}{V_{ref}} - \frac{V_t}{V_{ref}}\right)^2}{\left(1 - \frac{V_t}{V_{ref}}\right)^2} = N \bullet \frac{\left(\frac{V_N}{V_{ref}} - \beta\right)^2}{(1-\beta)^2} \qquad (117)$$

where

$$\beta = \frac{V_t}{V_{ref}} \qquad (118)$$

The quadratic relationship of Equation 117 can be solved to obtain V_N as a function of V_{ref} and V_t.

$$\frac{V_N}{V_{ref}} = \frac{2\beta + \gamma + \sqrt{4\beta\gamma + \gamma^2}}{2} \qquad (119)$$

where

$$\gamma = \frac{\left(1 - \beta^2\right)}{N} \qquad (120)$$

If $V_t = 0$, then Equation 119, degenerates to:

$$V_N = \frac{V_{ref}}{N} \qquad (121)$$

Figure 4.12 shows the supply voltage as a function of the number of processors. The best case situation is the bottom curve of Figure 4.12, where $V_t = 0$

and the power supply drops linearly with the supply voltage, as derived in Equation 121. Figure 4.12 also shows the effect of increasing the threshold voltage as modeled in Equation 119. As can be concluded from this figure, a high threshold voltage limits the amount of voltage scaling that can be achieved for given level of parallelism.

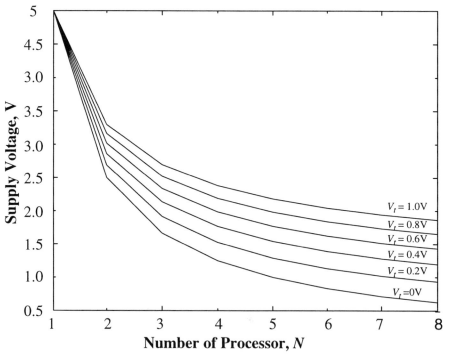

Figure 4.12: Voltage reduction vs. number of processors as a function of V_t.

The parallel architecture shown in Figure 4.11 does have overhead circuitry which requires extra power. Therefore the power consumption of the parallel implementation is:

$$P_N = P_{processing} + P_{overhead} \qquad (122)$$

The processing power, $P_{processing}$, is the power consumed by the input registers, the logic blocks which are clocked at f_{sample}/N, and the output register which is clocked at f_{sample}.

$$P_{processing} = N\left(C_{inreg} + C_F\right)V_N^2 \frac{f_{sample}}{N} + C_{outreg}V_N^2 f_{sample}$$

$$= \left(C_{inreg} + C_F + C_{outreg}\right)V_N^2 f_{sample} = C_{ref}V_N^2 f_{sample} \qquad (123)$$

The overhead power, $P_{overhead}$, consists of three components: routing capacitance due to broadcast of the input which is switched at f_{sample}, the output routing from the processors being clocked at f_{sample}/N, and the multiplexor overhead circuitry operating at the sample rate. The overhead power is given by:

$$P_{overhead} = C_{overhead}(N) V_N^2 f_{sample} \qquad (124)$$

where $C_{overhead}(N)$ is the sum of three capacitance terms: the input routing, the output routing, and the multiplexors. The overhead capacitance, $C_{overhead}$, is a function of the number of processors. For example, the capacitance switched in the input routing is a function of N since the wire length will increase with the number of processors.

If the overhead components of power described above are assumed to be zero, then the power consumption can be arbitrarily reduced by making the computation parallel. The power in this case is given by:

$$P_N = P_{processing} = C_{ref}V_N^2 f_{sample} \qquad (125)$$

The power improvement of the multi-processor implementation relative to a uni-processor implementation under the assumptions of zero circuit overhead is:

$$\frac{P_N}{P_1} = \frac{V_N^2}{V_{ref}^2} \qquad (126)$$

where V_N/V_{ref} can be determined from Equation 119. If $V_t=0$, then the power improvement of the multi-processor implementation relative to a uni-processor implementation is:

$$\frac{P_N}{P_1} = \frac{V_N^2}{V_{ref}^2} = \frac{1}{N^2} \tag{127}$$

This is the limit in power reduction achievable with architecture driven voltage scaling alone. This ignores the effect of leakage currents and the overhead circuitry required to parallelize a circuit.

Figure 4.13 plots the improvement in power as function of N taking into account the limitations to voltage scaling due to finite threshold voltages. As seen from this figure, even assuming ideal conditions for overhead circuitry (where overhead is zero) results in power reductions which saturate as the number of processors is increased for high values of the threshold voltage.

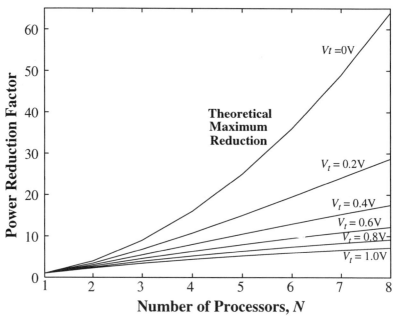

Figure 4.13: Power reduction (P_1/P_N) vs. number of processors as a function of V_t.

There are two key consequences of architecture driven voltage scaling. First, which is quite obvious, is that the area will grow faster than the number of processors, and second, is an increase in latency. The latency is increased over the uni-processor implementation and is equal to N cycles. Fortunately, for most signal processing applications like video compression, throughput is the critical metric and latency can be tolerated.

4.6.2 Optimal Supply Voltage for Architecture Driven Voltage Scaling

In the previous section, we saw that the delay increase due to reduced supply voltages below the critical voltage can be compensated by exploiting parallel architectures. However, as seen in Figure 3.24 and Equation 90, as supply voltages approach the device thresholds, the gate delays increase rapidly. Correspondingly, the amount of parallelism and overhead circuitry increases to a point where the added overhead dominates any gains in power reduction from further voltage reduction, leading to the existence of an "optimal" voltage from an architectural point of view. To determine the value of this voltage, the following model is used for the power for a fixed system throughput as a function of voltage (and hence degree of parallelism) obtained from Equations 122 thru 124.

$$P_N = C_{ref} V_N^2 f_{sample} + C_{overhead}(N) V_N^2 f_{sample} \qquad (128)$$

where N is the number of parallel processors, f_{sample} is the sample frequency, C_{ref} is the capacitance switched by a single processor as defined in Equation 114, $C_{overhead}$ is the overhead capacitance as described in the previous section. The power improvement of a multiprocessor implementation over a uni-processor implementation (i.e., without parallelism) can be expressed as:

$$P_{normalized} = \left(1 + \frac{C_{overhead}(N)}{C_{ref}}\right)\left(\frac{V_N}{V_{ref}}\right)^2 \qquad (129)$$

At very low supply voltages (near the device thresholds), the number of processors (and hence the corresponding overhead in the above equation) typically increases at a faster rate than the V^2 term decreases, resulting in a power increase with further reduction in voltage. The minimum on the Power vs. V_{dd} curve is defined as the "optimum" voltage for architecture driven voltage scaling.

Therefore, if the contribution of overhead circuitry is included in the power consumption, the power vs. number of processors will have an optimum point, parallelizing beyond which will result in an increase of the total power consumption. To illustrate this, let $C_{ref} = 1$ and let $C_{overhead}(N)$ be modelled as a linear function of N.

$$C_{overhead}(N) = mN - m \tag{130}$$

where m is the slope of the overhead function as a function of the number of processors. Figure 4.14 shows the capacitance as a function of the number of processors.

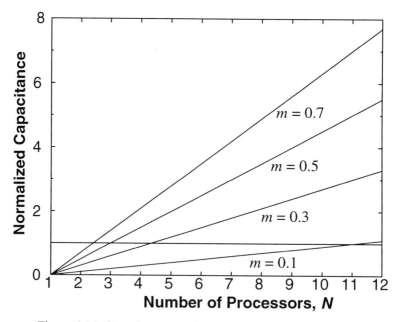

Figure 4.14: Capacitance overhead as a function of N.

The power reduction of an N-processor implementation in this case is given by:

$$\frac{P_N}{P_1} = \left(1 + \frac{C_{overhead}}{C_{ref}}\right)\frac{V_N^2}{V_{ref}^2} = \left(1 + C_{overhead}\right)\frac{V_N^2}{V_{ref}^2} \tag{131}$$

Figure 4.15 shows a plot of power as a function of the number of processors, N, for various values of m. All curves shown are for a $V_t=0.8$V. With $m=0$, there is no overhead for parallelism, and the power reduces as $1/N^2$ as derived earlier in Equation 127. For higher values of m, the power reaches minimum at a certain level of parallelism, resulting in an optimum level of parallelism. The minimum point shifts to the left, as the capacitance overhead factor, m, increases.

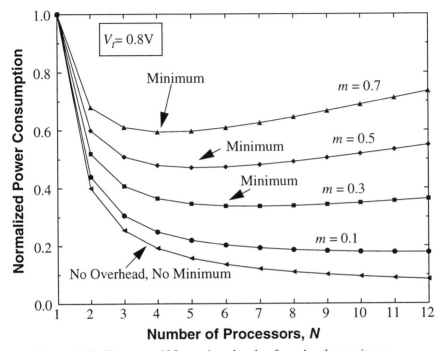

Figure 4.15: Power vs. N for various levels of overhead capacitance.

Figure 4.16 shows the normalized power as a function of V_{dd} for a variety of cases. The lower curve in this figure represents the power dissipation which would be achieved if there were no overhead associated with increased parallelism. For this case, the power is a strictly decreasing function of V_{dd} and the optimum voltage would be set by the minimum value allowed from noise margin constraints. The other curves show the power as a function of the voltage, which is obtained by increasing the level of parallelism. After a certain level of parallelism, the overhead circuitry starts to dominate any gains from voltage reduction and the power starts to increase, and this results in an "optimum" supply voltage for low-power.

It is interesting to note all cases have roughly the same optimum value of supply voltage, approximately $V_{tn} + |V_{tp}|$. Even when the overhead is very high ($m=0.5$), the optimal voltage is only slightly higher. At near the sum of the thresh-

old voltages, the delays and therefore the number of processors start to increase
rapidly with voltage reduction and this results in a steep rise in overhead circuitry.
This rise in overhead dominates and power starts to rise with further reduction of
voltage.

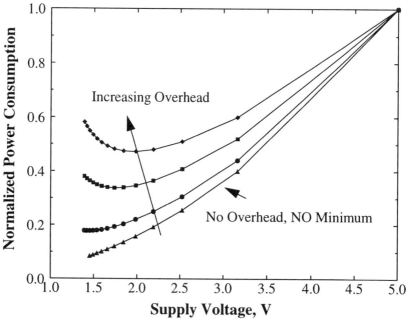

Figure 4.16: Architecture Driven Optimum Supply Voltage.

Figure 4.17 shows the power as function of number of processors, N, for a
fixed value of $m=0.7$, as a function of the threshold voltage. Here, the leakage
currents are ignored for reduced threshold voltages. Reducing V_t shifts the minimum point on the curve to the right.

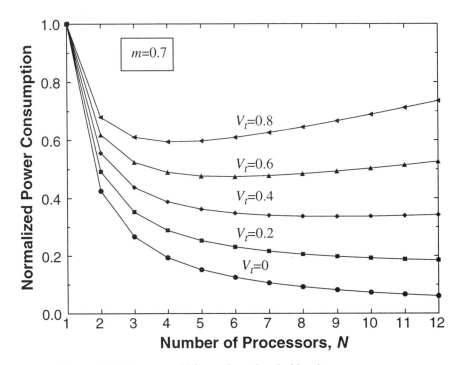

Figure 4.17: Power vs. N for various threshold voltages.

4.6.3 Trading Area for Lower Power Through Hardware Pipelining

An architecture driven voltage scaling strategy was presented earlier in which hardware duplication was used to reduce the clock frequency for the processing module which resulted in scaling of the supply voltage. Here another approach will be presented, which involves using pipelining to reduce the critical path (rather than the clock rate) and the supply voltage.

Figure 4.18 shows an N-stage pipelined approach for implementing the generic function shown on the left side. N-1 registers, all clocked at the sample rate, are introduced in the generic function. This has the effect of reducing the critical path. The critical path for an N-stage pipeline is:

$$T_{critical\ path} = T_{sample}/N \qquad (132)$$

Therefore, the logic between the registers could operate N-times slower while maintaining functional throughput. This implies that the supply voltage can be scaled to slow down the circuit by a factor of N. The voltage at which an N-stage pipelined circuit will have the same throughput as an implementation with no pipelining is derived in Equation 119.

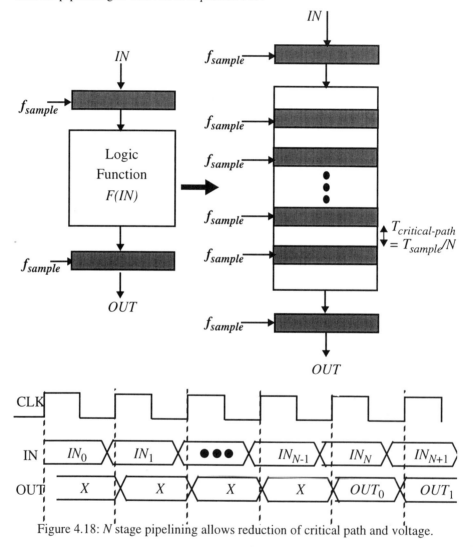

Figure 4.18: N stage pipelining allows reduction of critical path and voltage.

The power consumption of an N-stage pipelined circuit is approximately given by:

$$P_N = \left(C_{ref} + (N-1) C_{reg} \right) V_N^2 f_{sample} \qquad (133)$$

where C_{ref} is the capacitance switched by the original datapath (including the input and output registers) shown on the left of Figure 4.18 and C_{reg} is capacitance switched by each register. Therefore, the power reduction of a N-stage pipelined implementation versus a non-pipelined implementation is:

$$\frac{P_N}{P_1} = \frac{\left(C_{ref} + (N-1) C_{reg} \right) V_N^2 f_{sample}}{C_{ref} V_{ref}^2 f_{sample}} = (1 + \beta (N-1)) \frac{V_N^2}{V_{ref}^2} \qquad (134)$$

where $\beta = C_{reg}/C_{ref}$ and can never be greater than 0.5 since $C_{ref} = 2C_{reg} + C_F$. Figure 4.19 shows the power reduction of an N-stage pipelined architecture for various values of β. For $\beta = 0$, there is no overhead assumed for pipelining.

Notice that, just like parallelism, the latency is increased for the pipelined implementation and is equal to N clock cycles. However, the area overhead for

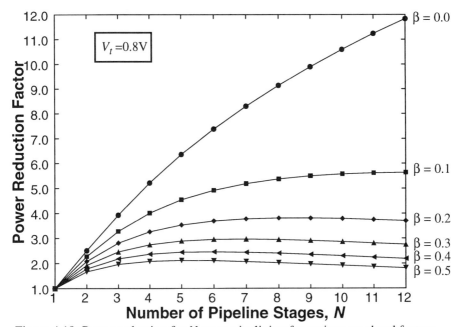

Figure 4.19: Power reduction for N-stage pipelining for various overhead factors.

pipelining is much smaller, since only registers have to be added instead of hardware duplication.

4.6.4 Example: A Simple Adder-Comparator

To illustrate how architectural techniques can be used to compensate for reduced speeds, a simple 8-bit datapath consisting of an adder and a comparator is analyzed assuming a 2.0μm technology [Chandrakasan92]. As shown in Figure 4.20, inputs A and B are added, and the result compared to input C. Assuming the worst-case delay through the adder, comparator and latch is approximately 25ns at a supply voltage of 5V, the system in the best case can be clocked with a clock period of $T = 25$ns. When required to run at this maximum possible throughput, it is clear that the operating voltage cannot be reduced any further since no extra delay can be tolerated, hence yielding no reduction in power. We will use this as the reference datapath for our architectural study and present power improvement numbers with respect to this reference. The power for the reference datapath is given by:

$$P_{ref} = C_{ref} \, V_{ref}^2 \, f_{ref} \qquad (135)$$

where C_{ref} is the total effective capacitance being switched per clock cycle. The effective capacitance was determined by averaging the energy over a sequence of input patterns with a uniform distribution.

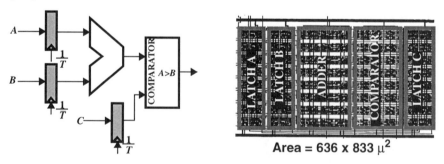

Figure 4.20: A simple datapath with corresponding layout.

One way to maintain throughput while reducing the supply voltage is to utilize a parallel architecture. As shown in Figure 4.21, two identical adder-comparator datapaths are used, allowing each unit to work at half the original rate while maintaining the original throughput. Since the speed requirements for the adder, comparator, and latch have decreased from 25ns to 50ns, the voltage can be dropped from 5V to 2.9V (the voltage at which the delay doubled, from Figure 3.24).

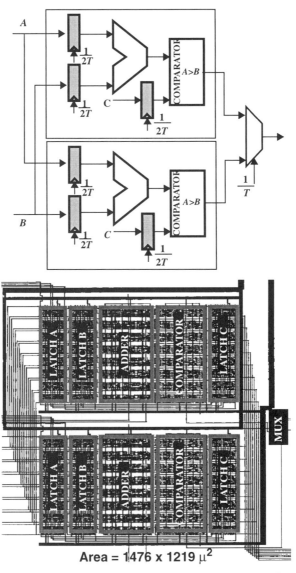

Figure 4.21: Parallel implementation of the simple datapath.

While the datapath capacitance has increased by a factor of 2, the operating frequency has correspondingly decreased by a factor of 2. Unfortunately, there is also a slight increase in the total "effective" capacitance introduced due to the extra routing, resulting in an increased capacitance by a factor of 2.15. Thus the power for the parallel datapath is given by:

$$P_{par} = C_{par} \, V^2_{par} \, f_{par} = \left(2.15 C_{ref}\right)\left(0.58 V_{ref}\right)^2\left(\frac{f_{ref}}{2}\right) \approx 0.36 P_{ref} \quad (136)$$

Another possible approach is to apply pipelining to the architecture, as shown in Figure 4.22. With the additional pipeline latch, the critical path becomes the $\max[T_{adder}, T_{comparator}]$, allowing the adder and the comparator to operate at a slower rate. For this example, the two delays are equal, allowing the supply voltage to again be reduced from 5V used in the reference datapath to 2.9V (the voltage at which the delay doubles) with no loss in throughput. However, there is a much lower area overhead incurred by this technique, as we only need to add pipeline registers. Note that there is again a slight increase in hardware due to the

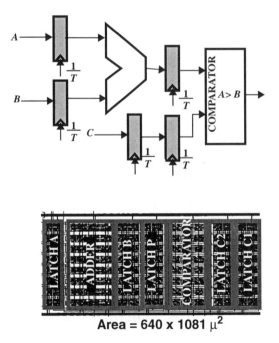

Area = 640 x 1081 μ^2

Figure 4.22: Pipelined implementation of the simple datapath.

extra latches, increasing the "effective" capacitance by approximately a factor of 1.15.

The power consumed by the pipelined datapath is:

$$P_{pipe} = C_{pipe} \, V^2_{pipe} \, f_{pipe} = \left(1.15C_{ref} \right)\left(0.58V_{ref} \right)^2 f_{ref} \approx 0.39 \; P_{ref} \quad (137)$$

With this architecture, power reduces by a factor of approximately 2.5, providing approximately the same power reduction as the parallel case with the advantage of lower area overhead. As an added bonus, increasing the level of pipelining also has the effect of reducing logic depth and hence power contributed due to hazards and critical races.

Furthermore, an obvious extension is to utilize a combination of pipelining and parallelism. Since this architecture reduces the critical path and hence speed requirement by a factor of 4, the voltage can be dropped until the delay increases by a factor of 4. The power consumption in this case is:

$$P_{parpipe} = C_{parpipe} \, V^2_{parpipe} \, f_{parpipe}$$

$$= \left(2.5C_{ref} \right)\left(0.4V_{ref} \right)^2 \left(\frac{f_{ref}}{2} \right) \approx 0.2P_{ref} \quad (138)$$

The parallel-pipeline implementation results in a 5 times reduction in power. Table 4.1 shows a comparative summary of the various architectures described for the simple adder-comparator datapath.

Table 4.1 Results of architecture based voltage scaling

Architecture	Voltage	Area (normalized)	Power (normalized)
Simple	5V	1	1
Parallel	2.9V	3.4	0.36
Pipelined	2.9V	1.3	0.39

Table 4.1 Results of architecture based voltage scaling

Architecture	Voltage	Area (normalized)	Power (normalized)
Pipelined-Parallel	2.0	3.7	0.2

Figure 4.23 shows the power vs. V_{dd} for the parallel and parallel-pipelined datapaths. Curve 1 in this figure represents the power dissipation which would be achieved if there were no overhead associated with increased parallelism. For this case, the power is a strictly decreasing function of V_{dd} and the optimum voltage would be set by the minimum value allowed from noise margin constraints. Curves 2 and 3 are obtained from data from actual layouts and are extensions of the example described earlier in which parallel and parallel-pipeline implementations of the simple datapath were duplicated N times.

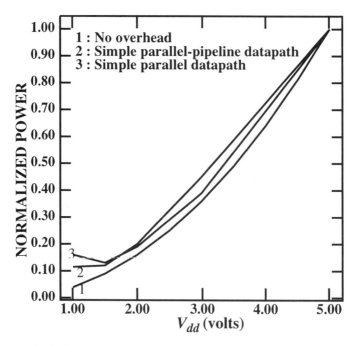

Figure 4.23: Optimum voltage of operation for the add/compare computation.

In Table 4.2 is a summary of the power reduction and normalized areas that were obtained from layouts. The increase in areas gives an indication of the amount of parallelism being exploited. The key point is that the optimal voltage is found to be relatively independent over the cases considered, and occurred around 1.5V for the 2.0μ technology (with V_{tn} =0.7V and V_{tp} = -0.9V); a similar analysis using a 0.5V threshold, 0.8μ process (with an L_{eff} of 0.5μ) resulted in optimal voltages around 1V, with power reductions in excess of a factor of 10. Further scaling of the threshold would allow even lower voltage operation, and hence even greater power savings.

Table 4.2 Normalized Area/Power for various supply voltages for Plots 2,3 in Figure 4.23

Voltage	Parallel Area/Power	Parallel -Pipelined Area/Power
5	1 / 1	1 / 1
2	6 / 0.19	3.7 / 0.2
1.5	11 / 0.13	7 / 0.12
1.4	15 / 0.14	10 / 0.11

4.6.5 Noise Considerations at Reduced Supply Voltages

Reducing the supply voltage raises some issues on noise immunity. There are three important sources of internal noise in integrated circuits: inductive noise due to the simultaneous switching of circuits, resistive noise on the power supply lines, and coupling between neighboring circuits. In this section, the effect of supply voltage scaling on the various components of noise will be analyzed.

Bond wires, package pins, printed circuit board traces, and IC traces all have parasitic inductances. When a CMOS gate switches, a current spike flows through the power bus, which has parasitic inductance from the above mentioned sources. When current changes through an inductor, a voltage is generated across the inductor which is equal to $L * di/dt$. As a result of this voltage drop, there is a fluctuation in the internal power supply, which can cause soft-errors if the voltage drop is large. This effect is commonly referred to as simultaneously switching noise since it most pronounced when many circuits (especially many off-chip

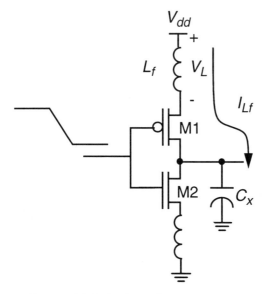

Figure 4.24: Inductive noise on the power supplies.

drivers switch simultaneously). Figure 4.24 shows a model for including bond inductance for computing the internal supply voltage.

Ignoring the threshold effects on current (i.e., assuming that I is proportional to V_{dd}^2 and delay is inversely proportional to V_{dd}), we see that the di/dt scales as V_{dd}^3 (even faster when threshold effects are taken into account), and hence drastically reducing the voltage noise. However, as mentioned earlier, reducing the voltage will require more parallelism (to meet throughput requirements) and will hence increase the number of simultaneously switching elements. Even so, the overall noise reduces in approximately a quadratic fashion with respect to the supply voltage. Also, using advanced packaging can reduce the lead inductances dramatically over conventional packaging (e.g., a MCM package has less than 2nH of lead inductance while a typical wirebond pin grid array package has about 10nH).

Resistive drops occur on the power bus due to its finite resistance. The instantaneous peak resistive drop is equal to $R * I_{peak}$ which is proportional to V_{dd}^2. This component therefore scales quadratically. However, due to increased parallelism at lower voltages, this component only scales linearly.

4.7 Summary

A key design methodology for low-power design is aggressive voltage scaling strategy beyond the conventional 5V to 3.3V scaling. Conventional voltage scaling techniques have used metrics such as maintaining reliability and/or maintaining the circuit speed. Unfortunately, this has been driven by the primary figure-of-merit for computers, that use one of the industry standard microprocessors, which has become the clock rate, and has therefore resulted in architectures which are antagonistic to low power operation - uni-processor implementations which require circuits to operate at the fastest possible rate or at the highest possible voltage. On the other hand, parallel architectures, operating at reduced clock rates, can have the *same* effective performance in terms of instructions/second at substantially reduced power levels. Aggressive voltage scaling can result in more than an order of magnitude reduction in power compared to conventional scaling approaches without loss in performance.

The issue of how far this approach can be taken, is interesting to address. One limit is reached when the noise margin of the devices is degraded (though the reduced clock rates limit ground bounce problems), however, before this limit the overhead in increased parallelism to compensate for the reduced logic speed begins to dominate, increasing the power faster than the voltage reduction decreases it. Through a number of examples, we find that the optimum occurs in the range between 1-1.5volts, resulting in an order of magnitude lower power dissipation than the conventional "low-power" 3.3 volt scaling. A minor modification to the technology which involves threshold voltage reduction is required to scale the power supply voltages into the sub 1-V range, which results in even more power savings.

REFERENCES:

[Bakoglu90] H.B. Bakoglu, *Circuits, Interconnections, and Packaging for VLSI*, Addison-Wesley, Menlo Park, CA, 1990.

[Bell91] T. Bell, "Incredible shrinking computers," *IEEE Spectrum*, pp. 37-43, May 1991.

[Burr91] J. B. Burr and A. M. Peterson, "Energy Considerations in Multichip Module-based Multiprocessors," *ICCD*, pp. 593-600, 1991.

[Chandrakasan92] A. P. Chandrakasan, S. Sheng, and R. W. Brodersen, "Low-power digital CMOS design," *IEEE Journal of Solid State Circuits*, pp. 473-484, April 1992.

[Dahle91] D. Dahle, "Designing High Performance Systems to Run from 3.3V or Lower Sources," *Silicon Valley Personal Computer Conference*, pp. 685-691, 1991.

[Davari88] B. Davari *et al.*, "A High Performance 0.25μm CMOS Technology," *IEEE IEDM*, pp. 56-59, 1988.

[Kakumu90] M. Kakumu and M Kinugawa, "Power-Supply Voltage Impact on Circuit Performance for Half and Lower Submicrometer CMOS LSI," *IEEE Transactions on Electron Devices*, Vol 37, No. 8, pp. 1902-1908, August 1990.

[Sodini84] C. Sodini, P.K. Ko, and J. L. Moll, "The Effect of High Fields on MOS Device and Circuit Performance Devices," *IEEE Transactions on Electron Devices*, vol. ED-31, pp. 1386-1393, Oct 1984.

[Watts89] R.K. Watts (ed.), *Submicron Integrated Circuits*, John Wiley & Sons, NY, 1989.

5

DC Power Supply Design in Portable Systems

Coauthored with Anthony J. Stratakos, Charles R. Sullivan, Seth R. Sanders
University of California at Berkeley

In portable systems, a number of low-voltage, low-power DC voltage supplies are needed. To provide these from a single battery source, some form of voltage conversion is necessary. To facilitate portability and conserve battery capacity, this conversion should be accomplished in minimal space and mass, with the high efficiency more easily realized in larger converters. A monolithic CMOS DC power supply could meet the severe size and efficiency requirements of a hand-held device. This chapter describes a design methodology for such converters.

Section 5.1 illustrates the enhancements to overall system run-time that can be achieved through DC-DC conversion. In Section 5.2, several monolithic CMOS DC-DC converter topologies are introduced, and their relative merits are discussed. Since existing integrated circuit (IC) technology cannot provide inductors or capacitors of suitable value and quality for power conversion, such designs often require several off-chip reactive elements for energy storage and filtering. These components will dominate the overall volume of the converter. Thus, in Section 5.3, design techniques which reduce the physical sizes of these components are shown. To compensate for the increase in dissipation associated with

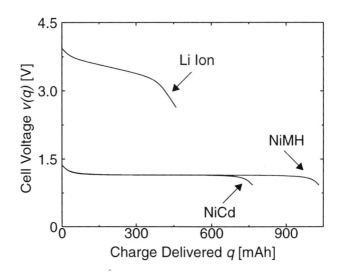

Figure 5.1: Typical low-rate discharge characteristics for AA-type Nickel Cadmium (NiCd), Nickel Metal Hydride (NiMH), and Lithium Ion (Li Ion) cells. Data is approximated from [Caruthers94].

converter miniaturization, several circuit design techniques which can improve the efficiency of the converter are described in Section 5.4.

5.1 Voltage Regulation Enhances System Run-time

Figure 5.1 shows typical low-rate battery discharge curves for three commercially available AA-type secondary battery sources: Nickel Cadmium (NiCd), Nickel Metal Hydride (NiMH), and Lithium Ion (Li Ion). Consider a block of throughput-constrained logic run directly from a NiMH cell and designed to operate down to the end-of-life cell voltage. If the power consumption of the logic is dominated by the dynamic component, and the circuitry is clocked at a frequency $f_{0.9}$ to meet throughput constraints at the minimum cell voltage $v(q) = 0.9$ V, then the circuitry will consume a minimum power at the end of the usable cell life:

$$P_{L-min} = f_{0.9} \cdot C_{eff} \cdot 0.9^2 \qquad (139)$$

Here, C_{eff} is the effective switching capacitance. However, at other points q in the cell discharge characteristic $v(q)$, the power consumption of the circuitry is given by:

$$P_L(q) = f_{0.9} \cdot C_{eff} \cdot v(q)^2 \qquad (140)$$

At initial cell voltage, this is a factor of 2.78 times P_{L-min}, and at nominal cell voltage, a factor of 1.78 times P_{L-min}. Thus, the load is seen to consume greater than minimum power throughout the cell discharge without increased throughput.

If a DC-DC converter, with efficiency:

$$\eta \equiv \frac{P_{out}}{P_{in}} \qquad (141)$$

and zero dropout voltage is inserted between the battery and the load, and the output of the converter is regulated to the end-of-life cell voltage, the logic consumes P_{L-min} independent of the cell voltage, and the power drawn from the cell at any point q in its discharge characteristic is constant and equal to:

$$P(q) = \frac{P_{L-min}}{\eta} \qquad (142)$$

In [Stratakos95] a mathematical model is developed to estimate the impact of DC-DC conversion on system run-time. This analysis considers analog circuitry with supply-independent biasing and throughput-constrained digital CMOS circuitry, and compares system run-time when these loads are run directly from the cell, and from the cell at a minimum voltage through a linear regulator or a switching regulator.

A factor that appears frequently in the comparisons of system run-time is the ratio of the mean battery source voltage (averaged over the delivered charge, q) to the minimum voltage required by the load. For convenience in summarizing the results, the symbol β is used for this ratio:

$$\beta \equiv \frac{\overline{v(q)}}{V_{min}} \qquad (143)$$

In terms of β, Table 5.1 gives the run-time enhancement factor, K, for a linear or digital CMOS load, where K is the run-time relative to the baseline run-time when the load is run directly from the battery source,

$$K \equiv \frac{t_A}{t_{Ao}} \qquad (144)$$

Table 5.1 System Run-Time Enhancement.

Regulator type	Constant-current load	Constant frequency digital CMOS load
Linear	$K = 1$	$K = \beta$
Switching, efficiency η	$K = \eta\,\beta$	$K = \eta\,\beta^2$

The results shown in Table 5.1 can be used to predict the benefits of different regulation schemes for a variety of loads. A linear regulator produces no advantage in system run-time for a constant-current load (e.g. many analog circuits). It should only be used if a stabilized voltage improves the performance of the load circuitry. With a digital load, the linear regulator provides an improvement by the factor β. Regardless of the load type, a switching regulator results in a value of K which is that for a linear regulator, multiplied by an additional factor $\eta\beta$. As long as the efficiency of the regulator is high enough that $\eta\beta > 1$, the switching regulator will give a longer run-time than a linear regulator.

The benefits of a switching regulator are greatest where β is large; that is, where the minimum required load voltage is small compared to the average battery source voltage. This makes intuitive sense, since an unnecessarily high voltage is wasteful of energy. For example, consider the ultra-low-power multimedia chipset introduced in Chapter 9. If this chipset, which can operate at a 1.1 V minimum supply voltage, were run from two AA-type NiMH cells in series, β would be 2.18. In this system, even a very low efficiency switching regulator is desirable: with as little as 46% efficiency, it would out perform an ideal linear regulator. With 90% efficiency, as is readily achieved using the design techniques presented in Section 5.4, the run-time would be 4.28 times longer than if the chipset were run directly from the battery source.

5.1.1 Converter Size vs. Extra Battery Size

While DC-DC conversion can significantly improve system run-time, this same enhancement of run-time may also be achieved by simply increasing the capacity of the battery source. Thus, from a system design standpoint, it is important to compare the volume required for the converter to the volume that would be required for this additional battery capacity.

The volume required by the DC-DC converter, $\Delta S_{DC\text{-}DC}$, is related to the load power, whereas the volume of the additional battery capacity, ΔS_B, is related to the integral of the load power—the energy consumed by the load. These two quantities can only be compared by specifying a run-time, t_A, giving:

$$\frac{\Delta S_B}{\Delta S_{DC-DC}} = \frac{D_{P(DC-DC)}}{D_{E(bat)}} \cdot t_A \cdot \left(K_0 - \frac{1}{\eta} \right) \tag{145}$$

Here, $D_{P(DC\text{-}DC)}$ is the power density of the DC-DC converter, $D_{E(bat)}$ is the volumetric energy density of the battery, and K_0 is defined as $K_0 \equiv K/\eta$. Since K_0 is equal to β or β^2 (see Table 5.1), the ratio (Equation 145) is seen to increase with increasing efficiency, as expected.

Small NiMH cells have an energy density of about 0.16 W-h/cm^3. Primarily because of packaging volume, smaller converters have somewhat lower power densities than standard commercial converters of 50-200 W, but ultra-low-power converters with power densities above 1 W/cm^3 can be achieved through the use of the techniques discussed in Section 5.3. Using these power and energy densities in conjunction with Equation 145, it is possible to evaluate the relative converter or battery volume required for an equal extension of system run-time.

For example, again consider the multimedia chipset described in Chapter 9. In the previous section, it was shown that a 90% efficient DC-DC converter can enhance the run-time from a two-cell NiMH source by a factor of 4.28. For an 8 h run-time, the volume required by 3.28 times more NiMH capacity is roughly 182 times greater than that required by the converter. As the run-time increases, the comparison is even more favorable for the converter. With shorter run-times, however, the ratio of battery volume to converter volume decreases. In this system, for run-times shorter than 2.5 minutes, the additional NiMH capacity requires less volume than the converter.

It may be concluded that, with the exception of systems designed for very short run-times, enhancing system run-time by adding a DC-DC converter will typically involve only a small increase in volume, much smaller than the increase in battery volume that would be needed for the same increase in run-time.

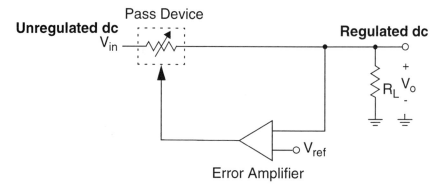

Figure 5.2: Block diagram of a linear (series-pass) regulator.

5.2 DC-DC Converter Topology Selection

At the power system level, there are two important design choices: battery voltage, and converter topology. In this section, three types of voltage converters are reviewed: linear regulators; switched capacitor converters; and switching regulators (true DC-DC converters), such as pulse-width modulated (PWM) converters. To design a power system with minimal size and loss, it is necessary to consider the interactions between battery voltage and topology, and these are discussed throughout the section.

5.2.1 Linear Regulators

Linear regulators, illustrated conceptually in Figure 5.2, are limited by two principle constraints. The output voltage, V_o, must be less than the input voltage, V_{in}, and the efficiency, η, can never be greater than V_o / V_{in}. However, linear regulators have the advantage of requiring few or no reactive components, and they can be very small and simple. This makes them attractive for portable applications.

A linear regulator can be efficient only in applications that require an output voltage just slightly below the input voltage. This requirement may be incompatible with other system design constraints, but in some systems it is practical, and, in this case, a linear regulator may be highly efficient. The achievable efficiency then depends on two parameters of the regulator: quiescent current and dropout voltage. The quiescent current determines the regulator's dissipation when the load is not drawing current, and in ultra-low-power applications, it may also contribute significantly to dissipation at full load.

If the input voltage of a linear regulator drops below a certain threshold, regulation is lost, and the output voltage will sag below the nominal regulation point. Dropout voltage is this minimum voltage difference between input and output required to maintain regulation. If it is not very low, it can conflict directly with the design requirement of having the output voltage only slightly less than the input voltage, and will therefore preclude high efficiency. This becomes especially important in low-voltage systems. With a 10 V output, a 2 V dropout voltage represents only a 20% increase in the minimum input power over what would be required with zero dropout voltage. However, with a 2 V output, a 2 V dropout voltage doubles the minimum input power.

Linear regulator circuits with low quiescent power, and PNP or MOSFET pass devices to allow low dropout voltage, are now commercially available. In the limited class of circuits that require a regulated voltage just below the input voltage of the regulator, these can provide a high-efficiency solution.

5.2.2 Switched Capacitor Converters

Switched-capacitor converters (also known as charge pumps) are widely used in ICs where a voltage higher than, or of opposite polarity to, the input voltage is needed. Unlike a PWM converter, a switched-capacitor converter requires no magnetic components. In addition, it is often possible to integrate the necessary capacitors, but applications are usually limited to those in which poor efficiency and very low output power are adequate. In this section, we examine the potential of switched-capacitor topologies for applications in which power density and efficiency are of prime importance.

Figure 5.3 illustrates the basic principle of operation of a switched-capacitor voltage doubler. The switches are closed in pairs, alternately. First the switches labeled ϕ_1 are closed, charging capacitor C_s to the input voltage, V_{in}. Then the ϕ_1 switches are opened, and the ϕ_2 switches are closed. This places C_s, which is now charged to V_{in}, in series with the input voltage, producing a voltage of $2V_{in}$ across the output. The cycle then repeats. The output capacitor maintains the output voltage near $2V_{in}$ during ϕ_1.

The same converter topology can be used as a step-down converter, producing an output voltage of half the input voltage, by exchanging the input and output terminals. By using more complex configurations, it is possible to produce any rational conversion ratio, for example by first stepping the voltage up by one

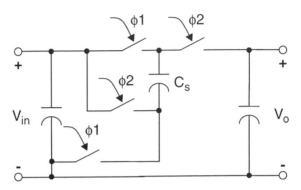

Figure 5.3: A switched-capacitor voltage doubler. Switches labeled ϕ_1 and ϕ_2 are closed alternately.

integer ratio, and then stepping down by another integer ratio. Some of the many possible topologies are discussed in [Oota90] and [Harada92].

Like a PWM DC-DC converter, a switched-capacitor converter may be built entirely of theoretically lossless elements—in this case, only switches and capacitors. However, a switched-capacitor converter is not ideally lossless. As the parasitic resistances in the capacitors and switches approach zero, the loss in the converter approaches a non-zero limit. This is in contrast to a PWM converter, in which the losses approach zero as parasitic effects are reduced.

The inherent losses in a switched-capacitor converter are due to unavoidable dissipation which occurs when a pair of capacitors, charged to different voltages, are shorted together through a switch. If two capacitors with values C_1 and C_2, initially charged to voltages V_{1o} and V_{2o}, respectively, are shorted together through a parasitic resistor R, the energy dissipated in the resistor will be

$$E_{diss} = \frac{1}{2} \cdot \frac{C_1 C_2}{C_1 + C_2} \cdot (V_{1o} - V_{2o})^2 \qquad (146)$$

Note that this is independent of the value of R.

To better understand these losses, consider the efficiency of the voltage doubler shown in Figure 5.3. During ϕ_2, the equivalent circuit is as shown in Fig-

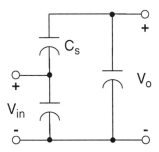

Figure 5.4: Equivalent voltage doubler circuit during ϕ_2.

ure 5.4. The charge flowing to the output is supplied by both the input and C_s. During ϕ_1, this same quantity of charge must be supplied from the input and stored on C_s for the next cycle. Since all the charge that flows out of the output must be supplied twice by the input, the average input current must equal twice the average output current, i.e., $I_{in} = 2I_o$. Thus, the efficiency is

$$\eta = \frac{V_o \cdot I_o}{V_{in} \cdot I_{in}} = \frac{V_o}{2V_{in}} \tag{147}$$

The efficiency would be 100% if V_o were in fact twice V_{in}. However, in order for a charge, Q, to flow into C_s during ϕ_1 and subsequently flow out of C_s during ϕ_2, the voltages applied across C_s during the two phases must differ by an amount $\Delta V = Q/C_s$. Assuming that the RC time constant determined by the parasitic resistance of the switches and C_s is small compared to the switching period so that the charge on C_s reaches its steady-state value before the end of each phase, and that the input and output capacitors are large enough to maintain constant V_{in} and V_o, the voltage drop is $\Delta V = 2V_{in} - V_o$. With a switching period of T_s, $Q = I_o \cdot T_s$, and so

$$2V_{in} - V_o = \frac{I_o \cdot T_s}{C_s} \tag{148}$$

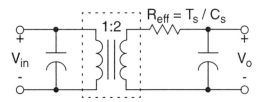

Figure 5.5: Equivalent circuit for the switched-capacitor voltage doubler.

The circuit may be modeled as shown in Figure 5.5, with an ideal doubler (shown as an ideal transformer) followed by an effective resistance

$$R_{eff} = T_s / C_s \qquad (149)$$

that accounts for the voltage drop ΔV. The effective resistance also accounts for the loss; calculating the dissipation in this resistor gives a result identical to that found from Equation 146.

In general, the model of a switched capacitor converter includes an ideal transformer with a fixed rational turns ratio, N, and an effective resistance. The ideal conversion ratio, N, can be chosen to bring V_o near the desired output voltage; to precisely regulate V_o, R_{eff} is varied through changes in the switching frequency. Using R_{eff} for regulation is undesirable, since increasing it to lower the output voltage produces additional power dissipation. However, N is fixed by the topology, and cannot be used to regulate the output.

This is the main limitation of switched-capacitor converters: they can efficiently *convert* voltages, but they cannot *regulate* these converted voltages any more efficiently than a linear regulator. Thus, their efficient application is limited to situations in which a voltage must be converted to another rationally related voltage, but regulation is not necessary, or to situations in which the regulation range is limited, and so the efficiency $\eta = V_o / (N \cdot V_{in})$ is adequate.

In practice, there are several other considerations that limit efficiency in a CMOS implementation of a switched-capacitor converter. In order for Equation 149 to hold, it is necessary for the time constant of the switched capacitor and the on-resistance of the switch to be much less than the switching period, i.e. $C_s \cdot R_{on} \ll T_s$. This requires the use of a large MOSFET to implement the switch, but the gate-drive for that device then requires substantial power, especially if a high switching frequency is used to minimize the required size of C_s. Thus, gate-drive loss must be considered in the design. The optimal FET size and switching frequency for one set of constraints are found in [Stratakos95].

If an on-chip capacitor is used for C_s, the stray capacitance from one of its plates to ground will be a substantial fraction of its terminal capacitance. This introduces $C_{stray} \cdot V^2 \cdot f_s$ loss, further hampering efficiency. Technologies for fabricating capacitors with low stray capacitance to ground, or off-chip capacitors are necessary to achieve high efficiency.

5.2.3 Switching Regulators

The switching regulator shown in Figure 5.6 converts an unregulated battery source voltage V_{in} to the desired regulated DC output voltage V_o. A single-throw, double-pole switch chops V_{in} producing a rectangular wave having an average voltage equal to the desired output voltage. A low-pass filter passes this DC voltage to the output while attenuating the AC ripple to an acceptable value. The output is regulated by comparing V_o to a reference voltage, V_{ref}, and adjusting the fraction of the cycle for which the switch is shorted to V_{in}. This pulse-width modulation (PWM) controls the average value of the chopped waveform, and thus controls the output voltage. Unlike a switched-capacitor converter, a switching regulator has an efficiency which approaches 100% as the components are made more ideal. In practice, efficiencies above 75% are typical, and efficiencies above 90% are attainable.

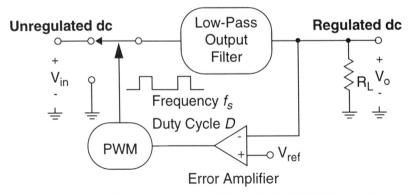

Figure 5.6: Block diagram of a pulse-width modulated (PWM) switching DC-DC converter.

There are several simple alternative arrangements of the switching and filter components that can be used to produce an output voltage larger or smaller
than the input voltage, with the same or opposite polarity. Some of these will be
discussed below. However, many of the design issues are similar, so first one
topology, the step-down (buck) converter, will be discussed in more detail.

Buck Converter

The buck circuit (Figure 5.7) can produce any arbitrary output voltage
$0 \leq V_o \leq V_{in}$. It works as follows: First the PMOS pass device, M_p, conducts, and
the inverter output node, v_x, is shorted to V_{in}. Then the pass device turns off, and
the NMOS rectifier device, M_n, turns on to pick up the inductor current, shorting
v_x to ground. The resulting waveform at the inverter output node is rectangular
with duty cycle D, period $T_s = f_s^{-1}$, and average value equal to the desired output
voltage. The second order low-pass filter (L_f and C_f) passes the desired DC, while
attenuating the AC component to an acceptable ripple value. In the ideal case, the
DC output voltage is given by the product of the input voltage and the duty cycle:

$$V_o = V_{in} \cdot D \tag{150}$$

The control of M_n and M_p is pulse-width modulated, adjusting the duty
cycle of the rectangular wave at v_x, and ultimately, the DC output voltage, to compensate for input and load variations. The pulse-width modulation is controlled

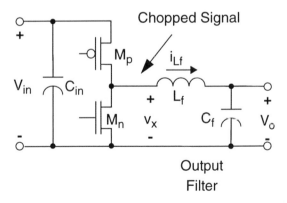

Figure 5.7: Low-voltage CMOS buck converter.

by a negative feedback loop, shown in the block diagram of Figure 5.6, but omitted from Figure 5.7 for simplicity.

Other Topologies

Two other basic configurations for PWM switching converters are the boost converter (Figure 5.8), and the buck-boost converter (Figure 5.10). All three basic topologies—buck, boost, and buck-boost—are similar in that they each have two complementary switches and one inductor. Their conversion ratios may all be adjusted by varying the duty cycle with frequency held constant. They can all be derived from the same basic switching cell [Kassakian91].

The boost converter produces output voltages $V_o \geq V_{in}$. A typical steady-state v_x (t) waveform is shown in Figure 5.9. In one portion of the cycle, (1-D), the NMOS device is on, and the input voltage is applied across the inductor, building up current and thus storing energy in the inductor. Next, the NMOS switch is turned off, and this attempt to interrupt the current in the inductor causes the voltage at node v_x to rise rapidly. The PMOS device is turned on at this point, limiting the voltage produced by this inductive kick to the voltage on the output capacitor. During the fraction of the cycle, D, that the PMOS device conducts,

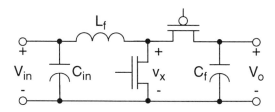

Figure 5.8: Low-voltage CMOS boost converter.

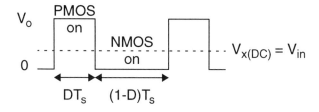

Figure 5.9: Nominal steady-state v_x (t) boost circuit waveform.

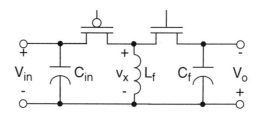

Figure 5.10: Low-voltage CMOS buck-boost circuit.

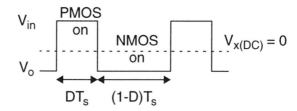

Figure 5.11: Nominal steady-state v_x *(t)* buck-boost circuit waveform.

some of the energy stored in the inductor is transferred to the output, along with additional energy flowing from the input. The cycle then repeats.

The boost converter may be considered a variation of the buck converter, but with power flow from the lower voltage side to the higher voltage side. The voltage at node v_x is a rectangular wave whose DC component is equal to the input voltage. (It must be equal, as the average voltage across the inductor must be zero for periodic steady state.) Thus, the input and output voltages are related by

$$V_{in} = V_o \cdot D \tag{151}$$

the same relation as for the buck converter, but with the input and output voltages reversed.

The operation of the buck-boost converter (Figure 5.10) is similar to that of the buck converter, in that the cycle starts with the input voltage applied across the inductor, in this case through the PMOS device for a duration, DT_s. However, when the PMOS device is turned off, the voltage at v_x heads downward, and the circuit produces an output voltage polarity opposite to that of the input (Figure 5.11). The energy transferred during this portion, *(1-D)*, of the cycle (while the

NMOS device conducts) is only the energy stored in the inductor, with none coming directly from the input. Setting the average voltage across the inductor equal to zero allows the conversion ratio to be found:

$$V_o = V_{in} \cdot \frac{D}{1 - D} \tag{152}$$

Note that this allows input voltages of smaller or larger magnitude than the input, hence the name "buck-boost."

These three basic topologies — buck, boost, and buck-boost — are a small subset of the many DC-DC converter topologies that have been proposed and that are used in practice. Other important classes of converter topologies include transformer-coupled circuits and soft-switching topologies, such as resonant converters. Although many of these topologies have important advantages in some applications, transformer coupling is usually unnecessary in portable systems, as discussed below, and soft switching can be achieved without the use of resonant techniques (Section 5.4.2). Thus the basic topologies are appropriate, perhaps optimal, for many low-power applications. The reader is referred to [Kassakian91] for more discussion of other topologies.

5.2.4 Effects of Conversion Ratio

The effect of conversion ratio on efficiency and component sizing can be an important factor in selecting the battery source voltage. While predetermined constraints may dictate the selection of battery voltage and converter output voltage and thus determine the required conversion ratio, in the design of a complete power system there is often a choice of battery voltage.

In general, a conversion ratio as close to 1:1 as possible minimizes the inductor size. For example, in Section 5.3.1, it will be shown that for a buck converter with a given output voltage, the required inductor value is proportional to the complement of the duty cycle, $(1-D)$. Thus, as the conversion ratio approaches 1:1, D approaches one, and the value and physical size of the inductor approach zero. Similarly, the inductor requirement in a boost converter approaches zero as the conversion ratio approaches 1:1. In a buck-boost converter, a 1:1 ratio still minimizes the inductor requirement, but because none of the energy is transferred directly—it is transferred from the input into the inductor, and then in a separate portion of the cycle, from the inductor to the output—a larger inductor is needed in this circuit, and the requirement does not approach zero as the conversion ratio approaches 1:1.

Thus to minimize inductor size, the preferred battery voltage is as close as possible to the desired output voltage, consistent with the constraint that, with a buck converter, the end-of-life battery voltage must be above the required output voltage. (For a boost converter, the constraint would be that the maximum battery voltage must be below the required output voltage.) Because of its more severe inductor requirements, a buck-boost topology should only be used for voltage polarity inversion or in other special applications.

Another important consideration for a CMOS converter implementation which includes complementary switches is that P-channel devices are inherently inferior to N-channel devices. On the basis of FET losses alone, it is desirable to choose a conversion ratio which ensures that current is carried by the NMOS device for a large fraction of the cycle. For example, consider the standard buck topology drawn in Figure 5.7. For a given output voltage and current, the losses in the power transistors are minimized if the NMOS device carries the inductor current for the majority of the cycle. This calls for a wide conversion ratio, as far from 1:1 as possible. With a 5:1 conversion ratio, for example, the PMOS device will conduct for only 20% of the cycle, and its losses can be made small.

However, for conversion ratios near 1:1, it is desirable to reconfigure the buck topology as shown in Figure 5.12. In this circuit, the NMOS device functions as the pass device, and, for conversion ratios near 1:1, it will have the longer conduction interval, while conduction in the PMOS device is minimized. Similar

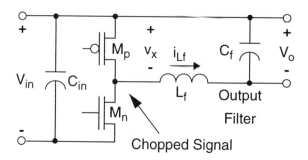

Figure 5.12: Alternative buck converter topology. For conversion ratios below 50%, M_n carries current for a larger fraction of the cycle than M_p.

reconfigurations of the boost and buck-boost topologies are possible to minimize losses at extreme duty cycles.

In discrete power conversion circuits, a transformer-coupled topology is often desirable to accomplish conversion over a wide voltage ratio, because the turns ratio in the transformer can produce most of the voltage ratio. This allows switching patterns similar to those in a 1:1 converter, minimizing inductor requirements (and relaxing the requirements for other components in a discrete implementation). However, in a small converter, the size of the transformer would probably outweigh any size reductions that would result from decreased inductor requirements. Thus, transformer-coupled circuits are likely to be useful in portable systems only for special applications, and will not be discussed further in this chapter. Special applications that could indicate the use of a transformer-coupled circuit could include high voltage requirements (e.g., for a display or backlight) and isolation. The reader is referred to [Kassakian91], and the references contained therein, for more details on these circuits.

In a system requiring many unique voltages for different sub-systems, the battery voltage should be selected as close as possible to the voltage at which the most power is required, minimizing the size and maximizing the efficiency of the converter supplying that voltage. The remaining converter topologies would then be chosen to accommodate that battery voltage, typically resulting in wider conversion ratios for the converters supplying the remaining sub-system voltages.

5.3 Converter Miniaturization

Since the portability requirement places severe constraints on physical size and mass, the volume and mass of a converter can be a critical design consideration. This section provides a summary of design techniques that may be used to reduce the size of a PWM DC-DC converter.

5.3.1 Output Filter Design

In Figure 5.13, the rectangular wave of the inverter output node is applied to the second order low-pass output filter of the buck circuit (L_f and C_f), which passes the desired DC component of v_x while attenuating the AC component to an acceptable ripple value. Load R_L draws a DC current I_o from the output of the filter. Figure 5.14 shows the nominal steady-state $i_{Lf}(t)$ and $v_o(t)$ waveforms for a rectangular input $v_x(t)$.

In order to achieve the large attenuation needed in a practical power circuit, $L_f C_f \gg \omega_s^{-2}$, where $\omega_s = 2\pi f_s$, and f_s is the switching frequency of the converter. In this case, the filter components may be sized independently, using time domain analysis, rather than frequency domain analysis. Neglecting the effects of output

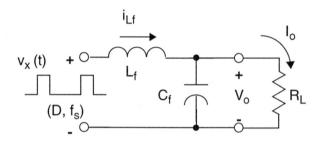

Figure 5.13: The output filter of the buck circuit (L_f and C_f) with load R_L.

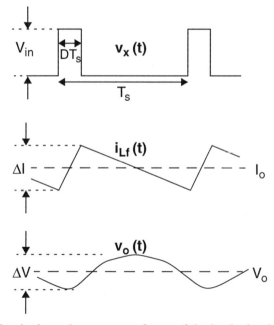

Figure 5.14: Nominal steady-state waveforms of the buck circuit output filter.

voltage ripple $(v_{o-AC} \ll v_{x-AC})$, for a rectangular input with period T_s, the AC inductor current waveform is triangular with period T_s and peak-to-peak ripple ΔI symmetric about the average load current I_o. The peak-to-peak current ripple may be found by integrating the AC component of the $v_x (t)$ waveform over a fraction, D, of one cycle, yielding:

$$\Delta I = \frac{V_{in} \cdot D \cdot (1 - D)}{L_f \cdot f_s} = \frac{V_o \cdot (1 - D)}{L_f \cdot f_s} \tag{153}$$

The output filter capacitor is selected to ensure that its impedance at the switching frequency, including its equivalent series resistance (ESR), is small relative to the load impedance. Thus, the AC component of the inductor current flows into the filter capacitor, rather than the load. For many capacitor technologies at frequencies above several hundred kilohertz, the resistive impedance dominates over the capacitive impedance. In high-current-ripple designs, a primary design goal is to minimize ESR to reduce both output voltage ripple and conduction loss (see Section 5.4). For this reason, a high-Q capacitor technology, such as multi-layer ceramic chip capacitors, is typically used, and even at high frequencies, ESR may be neglected in calculating output voltage ripple. Considering only capacitive impedance, the peak-to-peak output voltage ripple may be found through charge conservation. Assuming the AC inductor current flows only into the filter capacitor:

$$\Delta V = \frac{V_o \cdot (1 - D)}{8 \cdot L_f \cdot C_f \cdot f_s^2} \tag{154}$$

This output voltage ripple is symmetric about the desired DC output voltage V_o, and, for the $v_x (t)$ waveform shown in Figure 5.14, is piecewise quadratic with period T_s.

Equations 153 and 154 illustrate the two principle means of miniaturizing a DC-DC converter. First, it can be readily seen that the values of filter inductance and capacitance decrease as f_s^{-1}. Thus, a higher operating frequency typically results in a smaller converter. Second, because the requirement of interest is output voltage ripple, it is the $L_f C_f$ product, rather than the values of the individual components, that is important. Through choice of a higher current ripple, ΔI, a lower filter inductance solution may be obtained, often resulting in a smaller supply.

High Frequency Operation

As indicated by Equations 153 and 154, there are inherent size and cost advantages associated with higher frequency operation. The reactive filter components are likely to be the major contributors to the volume of a monolithic converter. For the same impedance, $j\omega_s L$ or $1/(j\omega_s C)$, a higher switching frequency, ω_s, enables the use of smaller value and smaller physical size reactive components. Ideally, the size of these components will decrease with f_s^{-1}. However, as will be described in Section 5.4.4, if the operating frequency of the circuit is increased, the sum of the losses in the power transistors and drive, if optimized, will increase roughly with $\sqrt{f_s}$. Thus, the general relationship between the size of a DC-DC converter and its losses is as illustrated in Figure 5.15. Here, operating frequency is used as a free-running variable, and the sum of the losses in the power transistors and drive is plotted against the volume of the converter.

If the cost and volume of the converter are decreased, additional space and resources are left for a larger or better battery, compensating for lower conversion efficiency. The system requirements and battery characteristics will help to determine which point on this curve is optimal for a specific application.

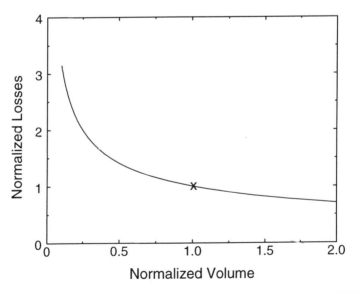

Figure 5.15: General trends in power transistor loss vs. the size of a DC-DC converter.

In practice, higher-frequency operation is limited mainly by two factors: frequency-dependent losses in the power train and controller, and diminishing returns in the miniaturization of the filter components. In Section 5.4, circuit-level optimizations are described which significantly reduce the frequency-dependent losses, yielding a class of miniature yet highly efficient converters that are well-suited for portable applications. Frequency limitations in inductive filter components will be addressed in Section 5.3.3.

High Current Ripple

Since the $L_f C_f$ product determines the output voltage ripple in Equation 154, the relative size and cost of inductance versus capacitance should be considered in the selection of these components. As the size, cost, and commercial availability of low-voltage ceramic chip capacitors are often superior to those of inductors, it is desirable to accomplish the majority of the filtering with the capacitor, allowing for the choice of a smaller-valued and smaller-sized inductor. This decision is restricted primarily by the increasing rms current in the inductor, which circulates throughout the power train, increasing conduction losses in proportion to $i_{Lf\text{-}rms}^2$.

In Figure 5.16, the peak-to-peak inductor current ripple, ΔI, is varied, and its effects on three key circuit parameters are shown. As illustrated by Equation 153, the value of filter inductance decreases with ΔI^{-1}. However, the physical size of L_f is roughly proportional to its peak energy storage, which in turn, is given by:

$$E_{L_f} = \frac{1}{2}L_f(I_o + \Delta I/2)^2 \tag{155}$$

and is minimized for $\Delta I = 2I_o$. The inductor current is approximated as a triangular AC waveform with peak-to-peak ripple ΔI superimposed on a DC component, I_o. Thus, the rms current is:

$$i_{L_f\text{-}rms} = \sqrt{I_o^2 + \frac{1}{3}\cdot\left(\frac{\Delta I}{2}\right)^2} \tag{156}$$

Although the preferred value of ΔI will depend slightly on the trade-off between size and loss in a particular application, it can be concluded that a peak-to-peak current ripple in the range $I_o < \Delta I < 2I_o$ is optimal for many applications. For $\Delta I < I_o$, the decrease in rms current due to ripple, and the resulting decrease in

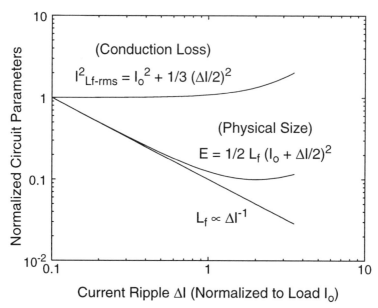

Figure 5.16: The effect of increased current ripple on the value of L_f, the physical size of L_f, and $i_{Lf\text{-}rms}{}^2$.

loss, are insignificant. There is no obvious benefit for $\Delta I > 2I_o$, but this is advantageous for one mode of operation (see Section 5.4.2).

5.3.2 Integration

A completely monolithic supply (active and passive elements) would meet the severe size and weight restrictions of a hand-held device. Because most portable applications call for low-voltage power transistors, their integration in a standard logic process is tractable. However, existing monolithic magnetics technology cannot provide inductors of suitable quality [Baringer93]. Emerging magnetics technology may allow completely monolithic supplies (see Section 5.3.3), but currently, magnetics, capacitors, and silicon circuitry are fabricated separately and assembled at the board level or in a multi-chip module (MCM). The extent of integration is the use of a monolithic silicon circuit, including all power transistors with their drive, and all control circuitry.

Such a monolithic solution not only results in a more compact design, it gives the designer more latitude in physical design and device sizing, allowing application-specific optimizations which are likely to yield a more efficient converter. In addition, parasitics from both the active devices and interconnect may be orders of magnitude lower on an IC than on a printed circuit board. Many of the frequency dependent losses in a power circuit increase in direct proportion to the energy storage of these parasitics; thus, integration enables higher efficiency at high operating frequencies than that obtained by a board-level solution.

5.3.3 Issues in Magnetic Component Design

Magnetic components (inductors and transformers) are essential for power electronics. With a few exceptions (such as switched-capacitor converters), switching power converters ubiquitously require inductors and/or transformers. Inductors are required as components of low-loss filters and resonant circuits. Transformers are used for isolation, converting voltage and current levels, and energy storage and transfer. Magnetic devices serve these and other functions in both traditional and newly introduced converter topologies.

Magnetic devices often are the physically largest components in a converter, and can be the most expensive. Typically, they must be custom designed for a particular circuit, either because suitable standard commercial parts do not exist, or because of the size, cost, and performance advantages possible in a custom design. For these reasons, it is important for a power converter designer to have some knowledge of magnetics design. A complete review of magnetics design is beyond the scope of this text. However, some of the special issues in magnetics for low-power portable applications will be addressed. A basic tutorial introduction to magnetics for power electronics can be found in [Kassakian91]. A more complete work on magnetics design is [Snelling88].

Magnetic Cores

In some cases, a coil (or pair of coils for a transformer) may have sufficient inductance for a power circuit application, even with no magnetic core material. Such an air-core coil is attractive because it can be fabricated on a PC board or a MCM substrate simply by patterning a spiral, ideally in multiple layers. However, there are many advantages to adding a core, and there are potential pitfalls in using an air-core coil.

An air-core coil is typically characterized at a given frequency by its inductance and its quality factor, $Q=\omega L/R$, where R and L are the effective values at a given frequency, ω. A high value of Q, in the range of 20 to 200, is required for efficient operation. The addition of a magnetic material can, ideally, increase the inductance of a coil by a factor equal to the relative permeability of the material, μ_r, without affecting the resistance of the coil. Since practical magnetic materials for power applications often have permeabilities in the thousands, the value of Q can be greatly increased by the addition of a magnetic core. However, the increase is not as large as might be expected from this naive analysis. The core dissipates power. These losses contribute to the effective value of R. In order to combat the resulting lowering of Q, and/or to prevent magnetic saturation of the core, an air gap is often introduced in the magnetic path through the core. This decreases the gain in L. Nonetheless, the increase in Q is substantial, and it is rarely possible to make practical magnetic components for power circuits operating below 1 MHz without the use of a magnetic core.

An additional advantage of a magnetic core is that it may serve to contain the flux. The external high-frequency flux from an air-core coil may cause RFI problems in nearby circuits, and will induce eddy currents in any nearby metal objects. The resulting losses can severely lower the Q of the coil, and even decrease its inductance. It may be possible to place the coil in a volume empty of metal components to avoid eddy current problems, but this additional volume becomes another reason to prefer using a component with a core, which is typically smaller than an air-core coil with the same inductance, even without this extra volume.

The most commonly used magnetic material for power applications is MnZn ferrite. Although it has a lower saturation flux density than typical magnetic metal alloys, it has much higher resistivity ($\approx 10\ \Omega$-cm, vs. $20 \times 10^{-8}\ \Omega$-cm for NiFe alloy). This allows operation at high frequencies (between 20 kHz and 1 MHz) without the severe eddy current losses that would result from using a lower-resistivity material.

Despite the high resistivity of ferrites, significant losses do occur in high frequency operation due to hysteresis. Since the hysteresis losses increase rapidly with frequency and flux level, it is often necessary to reduce flux level at high frequency in order to keep losses under control. This typically limits the flux density to well below the saturation flux density, and thus hysteresis loss is the most important parameter in determining the power handling capability, as well as the efficiency, of a magnetic component based on a ferrite core.

Although power handling density of magnetic components should ideally increase proportionally to frequency, the losses in ferrites increase faster than this and thus, the power handling is a weaker function of frequency, typically reaching a maximum around 500 kHz to 1 MHz for MnZn ferrites. To further improve power density, different materials are needed. NiZn ferrites may give a factor of two or three improvement in power density at frequencies in the range of 10 to 30 MHz [Stijntjes89]. More substantial improvements require a superior technology.

One emerging technology is microfabrication using thin-film magnetic metal alloy core materials. A process similar to that used for IC manufacture (more closely related to the process used for thin-film magnetic recording heads) is used to deposit and pattern thin films of magnetic alloys and copper coils. These magnetic materials can have very low hysteresis loss, and can be operated near their saturation flux density ($\approx$ 1 T) without disproportionate losses. The use of thin films and fine patterning can control eddy current losses in both the magnetic material and the coil. Experimental devices have been demonstrated in principle [Yamasawa90] [Yamaguchi93a] [Yamaguchi93b] [Yamaguchi93c] [Yachi91] [Yachi92], and calculations show that much higher power density will be possible [Sullivan93]. Although fabricating the magnetics on a separate substrate from the silicon circuitry and combining the two in a multi-chip module (MCM) is more likely to be economically viable, it is also possible to fabricate the magnetics and silicon circuitry on a single substrate, as has been experimentally demonstrated in [Mino93].

Either because of the use of planar microfabrication techniques, or for packaging and thermal dissipation considerations, magnetics for low-power converters are often designed in a planar configuration. While this presents no particular difficulties for transformer design, it can be a problem for inductors that use a gap in the magnetic path. The resulting magnetic field distribution can introduce severe eddy current losses in the conductor. For a description of this problem, and some remedies, see [Chew91] [Sullivan93].

5.4 Circuit Optimizations for High Efficiency

The power train of a low-output-voltage buck circuit is shown in Figure 5.17. Listed below are the chief sources of dissipation that cause the conversion efficiency of this circuit to be less than unity. In this section, methods which reduce these losses are described.

Conduction Loss:

Current flow through non-ideal power transistors, filter elements, and interconnections results in dissipation in each component:

$$P_q = i_{rms}^2 \cdot R \tag{157}$$

where i_{rms} is the root mean squared current through the component, and R is the resistance of the component.

Gate-Drive Loss:

Raising and lowering the gate of a power transistor each cycle dissipates an average power:

$$P_g = E_g \cdot f_s \tag{158}$$

where E_g is directly proportional to the gate energy transferred per cycle (which can include some energy due to Miller effect), and includes dissipation in the drive circuitry.

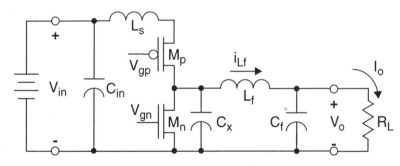

Figure 5.17: Low-output-voltage buck circuit, including parasitic inductance L_s.

Capacitive Switching Loss:

In a hard-switched converter, MOSFET M_p charges parasitic capacitance C_x to V_{in} each cycle, dissipating an average power:

$$P_{C_x} = \frac{1}{2} \cdot C_x \cdot V_{in}^2 \cdot f_s \tag{159}$$

where C_x includes reverse-biased drain-body junction diffusion capacitance C_{db} and some or all of the gate-drain overlap (Miller) capacitance C_{gd} of the power transistors, wiring capacitance from their interconnection, and stray capacitance associated with L_f. In ultra-low-power monolithic converters, C_x may be dominated by parasitics associated with the connection of an off-chip filter inductor, which include a bonding pad, bond wire, pin, and board interconnect capacitance. In circuit topologies which do not recover the energy stored on C_x each cycle through the inductor, the factor of 1/2 is removed from Equation 159.

Short-Circuit Loss:

A short-circuit path may exist temporarily between the input rails during transitions. To avoid potentially large short-circuit losses, it is necessary to provide dead-times in the conduction of the MOSFETs to ensure that the two devices never conduct simultaneously.

Body Diode Reverse Recovery:

If the durations of the dead-times are too long, the body diode of the NMOS power transistor may be forced to pick up the inductor current for a fraction of each cycle. When the PMOS device is turned on, it must remove the excess minority carrier charge from the body diode, dissipating an energy bounded by:

$$E_{RR} = Q_{rr} \cdot V_{in} \tag{160}$$

where Q_{rr} is the stored charge in the body diode.

Stray Inductive Switching Loss:

Energy storage by the stray inductance L_s in the loop formed by the input decoupling capacitor C_{in} and the power transistors causes $\frac{1}{2} L_s I^2 f_s$ dissipation.

Quiescent Operating Power:

The PWM and other control circuitry consume power. In low-power applications, this control power may contribute substantially to the total losses, even at full-load.

5.4.1 Synchronous Rectification

The focus of this chapter is low-output-voltage CMOS DC-DC converters, in which the switching elements, modelled by the single-throw double-pole switch in the block diagram of Figure 5.6, are implemented by two MOSFETS. The more conventional implementation consists of one controlled switch and one uncontrolled switch (a diode). The pure CMOS implementation allows an important advantage.

Consider the conventional buck circuit of Figure 5.18. Even if all other losses in the circuit are made negligible, the maximum efficiency is limited by the forward bias diode voltage, V_{diode}. Since the diode conducts for a fraction $(1-D)$ of the switching period, the maximum efficiency this circuit can obtain is given by:

$$\eta_{max} = \frac{V_o}{V_o + (1-D) \cdot V_{diode}} \tag{161}$$

Figure 5.19 shows η_{max} versus V_o for $V_{in} = 6$ V, and forward diode drops of 0.7 V and 0.3 V.

If the diode in Figure 5.18 is replaced by an NMOS device which is gated when the diode would have conducted (M_n in Figure 5.17), the forward drop can be made arbitrarily small by making the device sufficiently large. In this way, the gated NMOS device, used as a synchronous rectifier, can perform the same function as the diode more efficiently. Assuming all other losses, including the gate-drive for the synchronous rectifier, are still negligible, the maximum efficiency of the low-voltage buck converter approaches unity.

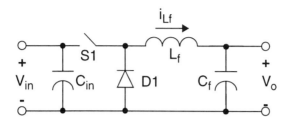

Figure 5.18: Conventional buck circuit with pass device S_1 and diode.

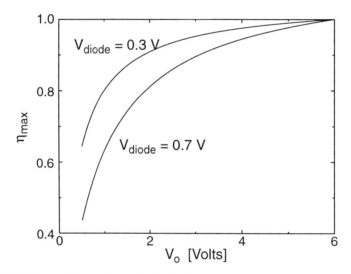

Figure 5.19: The maximum theoretical efficiency of the conventional buck circuit as a function of V_o for $V_{in} = 6$ V.

5.4.1.1 Synchronous Rectifier Control

Although the synchronous rectifier may reduce conduction loss at low output voltage levels, it comes at the expense of an additional drive signal and increased gate-drive losses. In addition, as mentioned above, without proper control of the rectifier, a short-circuit path may exist temporarily between the input rails during transients. In the rectifier control scheme described in Section 5.4.3, the dead-times, which ensure that M_p and M_n never conduct simultaneously, are

adapted in a negative feedback loop to achieve nearly ideal zero-voltage switched
turn-on transitions of both MOSFETs (see below).

5.4.2 Zero-Voltage Switching

When the low-voltage buck circuit of Figure 5.17 is hard-switched, it dissipates power in proportion to $C_x V_{in}^2 f_s$ as a result of the step charging of parasitic
capacitance C_x through a resistive path (M_p). In addition, it is likely to exhibit
either substantial short-circuit loss (if no dead-time is provided), or reverse recovery loss (if a dead-time is provided). In a soft-switched circuit, the filter inductor
is used as a current source to charge and discharge this capacitance in an ideally
lossless manner, allowing additional capacitance to be shunted across C_x, slowing
the inverter output node transitions. In this way, appropriate dead-times may be
set such that the power transistors are switched with $v_{ds} = 0$, eliminating all associated switching loss.

Figure 5.17 and Figure 5.20 show the low-voltage buck circuit and associated steady-state waveforms for ideal zero-voltage switching operation. The

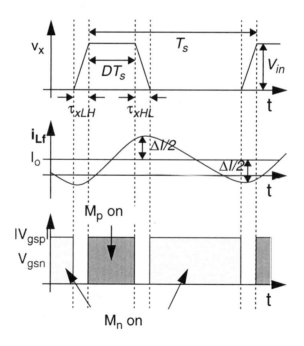

Figure 5.20: Nominal steady-state ZVS waveforms.

soft-switching behavior is similar to that described in [Maksimovic93] and by other authors.

Assume that at a given time (the origin in Figure 5.20), the rectifier (M_n) is on, shorting the inverter output node to ground. Since by design, the output is DC and greater than zero, a constant negative potential is applied across L_f, and i_{Lf} is linearly decreasing. If the value of filter inductance is small enough, the current ripple exceeds the average load ($\Delta I > 2I_o$), and i_{Lf} ripples below zero. As illustrated in Section 5.3.1, for such a value of current ripple, the physical size of the inductor is close to minimum.

If the rectifier is turned off after the current reverses (and the PMOS device, M_p, remains off), L_f acts approximately as a current source, charging the inverter output node. To achieve a lossless low-to-high transition at the inverter output node, the PMOS device is turned on when $v_x = V_{in}$. In this scheme, a pass device gate transition occurs exactly when $v_{dsp} = 0$.

With the PMOS device on, the inverter output node is shorted to V_{in}. Thus, a constant positive voltage is applied across L_f, and i_{Lf} linearly increases, until the high-to-low transition at v_x is initiated by turning M_p off. As indicated by Figure 5.20, at this time, the sign of current i_{Lf} is positive. Again, L_f acts as a current source, this time discharging C_x. If the NMOS device is turned on with $v_x = 0$, a lossless high-to-low transition of the inverter output node is achieved, and M_n is switched at $v_{dsn} = 0$.

In this scheme, a form of soft-switching, the filter inductor is used to charge and discharge all capacitance at the inverter output node (and supply all Miller charge) in a lossless manner, allowing the addition of a shunt capacitor at v_x to slow these transitions. Since the power transistors are switched at zero drain-source potential, this technique is known as zero-voltage switching (ZVS), and essentially eliminates capacitive switching loss. Furthermore, because the inductor current in a ZVS circuit reverses, the body diode turns off through a short circuit (rather than through a potential change of V_{in}), nearly eliminating the dissipation associated with reverse recovery, a factor which might otherwise dominate switching loss, particularly in low-voltage converters.

5.4.2.1 Design of a ZVS Buck Circuit

As discussed in Section 5.3.1, the inductor current waveform in a buck circuit is triangular with maximum and minimum values $I_o + (\Delta I)/2$ and

$I_o - (\Delta I)/2$ which are relatively constant over the entire dead-time. In a ZVS circuit, $\Delta I > 2I_0$, allowing L_f to charge and discharge the inverter output node. The ratio of theses soft-switched transition times is given by the ratio of currents available for each commutation:

$$\frac{\tau_{xLH}}{\tau_{xHL}} = \frac{\Delta I/2 + I_o}{\Delta I/2 - I_o} \qquad (162)$$

and approaches unity for large inductor current ripple. Here, τ_x indicates a soft-switched inverter output node transition interval, with subscripts LH and HL denoting low-to-high and high-to-low transitions, respectively. Using Equation 162, the inductor current ripple, ΔI, is chosen to limit the maximum asymmetry in the durations of the soft-switched transitions to a reasonable value at full load. From this value of ΔI, L_f is selected according to Equation 153. Given specifications on the maximum tolerable output voltage ripple, ΔV, the value of C_f is then chosen using Equation 154.

To slow the soft-switched transitions to durations for which dead-times may be programmed or adjusted, extra capacitance is added to the inverter output node. The total capacitance required to achieve a given low-to-high transition time is approximately

$$C_x = \frac{\tau_{xLH} \cdot (\Delta I/2 - I_o)}{V_{in}} \qquad (163)$$

5.4.3 Adaptive Dead-Time Control

To ensure ideal ZVS of the power transistors, the periods when neither conducts (the dead-times) must exactly equal the inverter output node transition times:

$$\tau_{DLH} = \tau_{xLH} \qquad (164)$$

$$\tau_{DHL} = \tau_{xHL} \qquad (165)$$

Here, τ_D indicates a dead-time and τ_x indicates an inverter node transition time, with subscripts LH and HL denoting low-to-high and high-to-low transitions respectively.

In practice, it is difficult to maintain the relationships of Equations 164 and 165. As indicated by Figure 5.20, the inductor current ripple is symmetric about

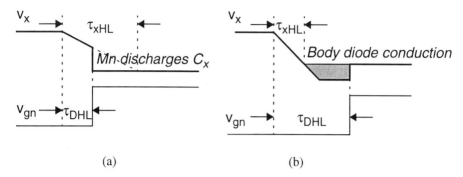

Figure 5.21: Non-ideal ZVS and its impact on conversion efficiency.

the average load current. As the average load varies, the DC component of the i_{Lf} waveform is shifted, and the current available for commutating the inverter output node is modified. Thus, the soft-switched transition times are load dependent.

In one approach to soft-switching, a value of average load may be assumed, yielding estimates of the inverter output node transition times. Fixed dead-times are based on these estimates. In this way, losses are reduced, yet perhaps not to negligible levels. To keep losses small in these circuits, they may still need to be operated at lower frequencies, thereby increasing the size of the passive filter components, and decreasing the power density.

In portable battery-operated applications where volume and efficiency are at a premium, this approach to soft-switching may not be adequate. To illustrate the potential hazards of fixed dead-time operation, Figure 5.21 shows the impact of non-ideal ZVS on conversion efficiency through reference to a high-to-low transition at the inverter output node. In Figure 5.21a, the dead-time is too short, causing the NMOS device to turn on at $v_{dsn} > 0$, partially discharging C_x through a resistive path and introducing losses. Since, as indicated by Equation 163, shunt capacitance with a value much larger than the intrinsic parasitics is typically added to slow the soft-switched transitions in a ZVS circuit, this loss may be substantial. In Figure 5.21b, the dead-time is too long, and the inverter output node continues to fall below zero until the drain-body junction of M_n becomes forward biased. In low-voltage applications, the forward-bias body diode voltage is a significant fraction of the output voltage; thus, body diode conduction must be avoided for efficient operation.

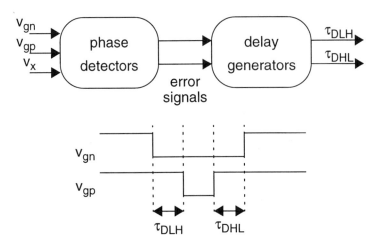

Figure 5.22: A conceptual representation of the adaptive dead-time control scheme described in [Stratakos94].

To provide effective ZVS over a wide range of loads, an adaptive dead-time control scheme has been outlined in [Stratakos94]. Figure 5.22 shows a block diagram of the approach. A phase detector generates an error signal based on the relative timing of v_x and the gate-drive signals. A delay generator adjusts the dead-times based on these error signals. Using this technique, effective ZVS is ensured over a wide range of operating conditions and process variations.

5.4.4 Power Transistor Sizing

Through use of ZVS with an adaptive dead-time control scheme, switching loss is essentially eliminated. If the filter components in the buck circuit of Figure 5.17 are ideal, and series resistance and stray inductance in the power train are made negligible, the fundamental mechanisms of power dissipation will include on-state conduction loss and gate-drive loss in the power transistors. When sizing a MOSFET for a particular power application, the principal objective is to minimize the sum of the dissipation due to these mechanisms.

During their conduction intervals, the power transistors operate exclusively in the ohmic region, where $r_{ds} = R_o \cdot W^{-1}$ (the channel resistance is inversely proportional to gate-width with constant of proportionality R_o). Thus, the on-state conduction loss in a FET is given by:

$$P_q = \frac{i^2_{ds-rms} \cdot R_o}{W} \tag{166}$$

Since the device parasitics generally increase linearly with increasing gate-width, the gate-drive loss can be expressed as a linear function of gate-width W:

$$P_g = E_{go} \cdot f_s \cdot W \tag{167}$$

where E_{go} is the total gate-drive energy consumed in a single off-to-on-to-off gate transition cycle and f_s is the switching frequency of the converter. In a ZVS circuit, the filter inductor supplies all the Miller charge, so E_{go} contains no dissipation due to Miller effect.

Using an algebraic minimization, the optimal gate-width of the power transistor,

$$W_{opt} = \sqrt{\frac{i^2_{ds-rms} \cdot R_o}{E_{go} \cdot f_s}} \tag{168}$$

is found to balance on-state conduction and gate-drive losses, where

$$P_{q-opt} = P_{g-opt} = \sqrt{i^2_{ds-rms} \cdot R_o \cdot E_{go} \cdot f_s} \tag{169}$$

and $P_t = P_g + P_q$ is at its minimum value, P_{t-min}. Figure 5.23 illustrates normalized power transistor losses as a function of gate-width.

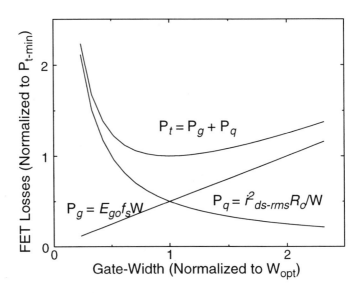

Figure 5.23: Normalized power transistor losses vs. gate-width.

5.4.5 Reduced-Swing Gate-Drive

To ensure that the duration of the low-to-high soft-switched transition is kept reasonably short in a ZVS buck circuit, the inductor current ripple must be made substantial. This gives rise to large circulating currents in the power train, and therefore, when the power transistors are sized according to Equation 168, increased gate-drive loss. Since gate-drive losses increase in direct proportion to f_s, this proves to be the limiting factor to higher-frequency operation of soft-switched converters. To reduce gate-drive losses, a number of resonant gate-drives have been proposed [Maksimovic90] [Theron92][Weinberg92]. While several such techniques have demonstrated the ability to recover a significant fraction of the gate energy at lower frequencies, due to the resistance of the poly-silicon gate of a power transistor, none are likely to be practical in the 1 MHz frequency range. Furthermore, each requires additional off-chip reactive components and is therefore not well suited to portable applications.

Rather than attempting to recover gate energy in a resonant circuit, another approach to reducing gate-drive dissipation is to reduce the gate energy consumed

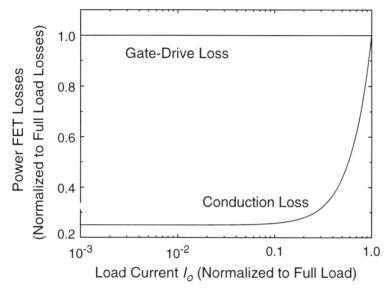

Figure 5.24: Power transistor losses versus a one thousand to one load variation for a ZVS PWM buck converter.

per cycle. By decreasing the gate-source voltage swing between off-state conduction $(|V_{GS}| = 0)$ and on-state conduction $(|V_{GS}| = V_g)$, gate energy may be, to the first order, quadratically reduced. However, because the channel resistance of the device increases with $(V_g - |V_t|)^{-1}$, where $|V_t|$ is the magnitude of the device threshold voltage, gate-swing cannot be arbitrarily reduced, implying the existence of an optimum V_g [Stratakos95].

5.4.6 PWM-PFM Control for Improved Light-Load Efficiency

While a PWM DC-DC converter can be made to be highly efficient at full load, many of its losses are independent of load current, and it may, therefore, dissipate a significant amount of power relative to the output power at light loads. Figure 5.24 plots power transistor losses (conduction and gate-drive) versus a one thousand to one load variation for a ZVS PWM buck converter with a peak-to-peak inductor current of two times the full load current. The power transistors are optimized for full load operation. From this plot, it may be concluded that a PWM converter which is 95% efficient at full load is roughly 3% efficient at one thousandth full load.

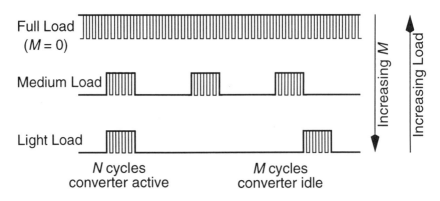

Figure 5.25: A conceptual illustration of PWM-PFM control.

One control scheme which achieves high efficiency at light loads is a hybrid of pulse-width modulation and pulse-frequency modulation (PFM), commonly referred to as "burst mode". In this scheme, conceptually illustrated in Figure 5.25, the converter is operated in PWM mode only in short bursts of N cycles each. Between bursts, both power FETs are turned off, and the circuit idles with zero inductor current for M cycles. During this period, the output filter capacitor sources the load current. When the output is discharged to a certain threshold below V_o, the converter is operated as a PWM converter for another N cycles, returning charge to C_f. Thus, the load-independent losses in the circuit are reduced by the ratio $N/(N+M)$. As the load current decreases, the number of off cycles, M, increases.

During the N cycles that the converter is active, it may be zero-voltage switched. However, the transitions between idle and active modes require an additional energy overhead. When idle mode is initiated by turning both FETs off, the body diode of M_n must pick up the inductor current until $i_{Lf} = 0$. In low-voltage applications, body diode conduction is highly dissipative. When the converter is reactivated, M_p must charge C_x to V_{in}, introducing additional losses. The minimum number of off cycles, M_{min}, can be chosen to ensure that the energy saved by idling is substantially larger than this loss overhead.

Circuit implementations of PWM-PFM control may be found in [Williams88] and [Locascio93].

REFERENCES:

[Baringer93] W. Baringer and R. Brodersen, "MCMs for Portable Applications," *IEEE Multi-Chip Module Conf.*, 1993.

[Caruthers94] F. Caruthers, "Battery Technology Charges Ahead," *Computer Design's OEM Integrations*, pages 10-14, May 1994.

[Chew91] W. Chew and P. Evans, "High frequency inductor design concepts," *22nd Annual Power Electronics Specialists Conference*, pages 673-678, June 1991.

[Harada92] I. Harada, F. Ueno, T. Inoue, and I. Oota, "Characteristics analysis of fibonacci type sc transformer," *IEICE Transactions on Fundamentals of Electronics, Communications and Computer Sciences*, E75-A(6):655-62, June 1992.

[Kassakian91] J. Kassakian, M. Schlecht, and G. Verghese, *Principles of Power Electronics*, Addison-Wesley, 1991.

[Locascio93] J. Locascio and W. Cho, "New controllers for battery systems increase systems efficiency," in *Power Quality USA*, 1993.

[Maksimovic90] D. Maksimovic, "A MOS Gate Drive with Resonant Transitions," *IEEE Power Electronics Specialists Conference*, pages 96-105, 1990.

[Maksimovic93] D. Maksimovic, "Design of the Zero-Voltage Switching Quasi-Square-Wave Resonant Switch," *Proc. IEEE Power Electronics Specialists Conference*, 1993.

[Mino93] M. Mino, T. Yachi, A. Tago, K. Yanagisawa, and K. Sakakibara, "Microtransformer with monolithically integrated rectifier diodes for micro-switching converters," *24nd Annual Power Electronics Specialists Conference*, pages 503-508, June 1993.

[Oota90] I. Oota, F. Ueno, and T. Inoue, "Analysis of a switched-capacitor transformer with a large voltage-transformer-ratio and its applications," *Electronics and Communications in Japan, Part 2 (Electronics)*, 73(1):85-96, January 1990.

[Snelling88] E. Snelling, *Soft Ferrites, Properties and Applications*, Butterworths, second edition, 1988.

[Stijntjes89] T. Stijntjes, "Power ferrites; performance and microstructure," *Crystal Properties and Preparation*, 27-30(1):587-94, 1989.

[Stratakos94] A. Stratakos, S. Sanders, and R. Brodersen, "A Low-Voltage CMOS DC-DC Converter for a Portable Battery-Operated System," *IEEE Power Electronics Specialists Conference.*, pages 619-626, 1994.

[Stratakos95] A. Stratakos, C. Sullivan, S. Sanders, and R. Brodersen, "DC Power Supply Design in Portable Systems," UC Berkeley ERL document, memorandum number M95/4, 1995.

[Sullivan93] C. Sullivan and S. Sanders, "Microfabrication of transformers and inductors for high frequency power conversion," *24nd Annual Power Electronics Specialists Conference*, pages 33-40, June 1993.

[Theron92] P. Theron, P. Swanepoel, J. Schoeman, J. Ferreira, and J. van Wyk, "Soft Switching Self-Oscillating FET-Based DC-DC Converters," *IEEE Power Electronics Specialists Conference*, volume 1, pages 641-648, 1992.

[Weinberg92] S. Weinberg, "A Novel Lossless Resonant MOSFET Driver," *IEEE Power Electronics Specialists Conference*, volume 2, pages 1002-1010, 1992.

[Williams88] J. Williams, "Achieving Microamp Quiescent Current in Switching Regulators," *Linear Technology Design Notes*, DN 11(11), June 1988.

[Yachi91] T. Yachi, M. Mino, A. Tago, and K. Yanagisawa, "A new planar microtransformer for use in micro-switching-converters," *22nd Annual Power Electronics Specialists Conference*, pages 1003-1010, June 1991.

[Yachi92] T. Yachi, M. Mino, A. Tago, and K. Yanagisawa, "A new planar microtransformer for use in micro-switching-converters," *IEEE Trans. on Magnetics*, 28(4):1969-73, 1992.

[Yamaguchi93a] K. Yamaguchi, E. Sugawara, O. Nakajima, and H. Matsuki, "Load characteristics of a spiral coil type thin film microtransformer," *IEEE Trans. on Magnetics*, 29(6):3207-3209, 1993.

[Yamaguchi93b] K. Yamaguchi, S. Ohnuma, T. Imagawa, J. Toriu, H. Matsuki, and K. Murakami, "Characteristics of a thin film microtransformer with spiral coils," *IEEE Trans. on Magnetics*, 29(5):2232-2237, 1993.

[Yamaguchi93c] M. Yamaguchi, S. Arakawa, H. Ohzeki, Y. Hayashi, and K. Arai, "Characteristics and analysis for a thin film inductor with closed magnetic circuit structure," *IEEE Trans. on Magnetics*, 29(5), 1993.

[Yamasawa90] K. Yamasawa, K. Maruyama, I. Hirohama, and P. Biringer, "High-frequency operation of a planar-type microtransformer and its application to multilayered switching regulators," *IEEE Trans. on Magnetics*, 26(3):1204-1209, May 1990.

6

Adiabatic Switching

Lars Svensson

University of Southern California, Information Sciences Institute

The fundamental cause of CMOS dynamic power dissipation is the organization of the energy transport in the circuit. Charging a node with a node capacitance, C, to a voltage, V, means storing a signal energy, $E_{sig} = CV^2/2$, on the node. In a level-restoring CMOS circuit with rail-to-rail swing[1], the signal charge, $Q = CV$, is drawn from the power supply at a constant voltage, V. Thus, as pointed out in Chapter 3, an energy $E_{inj} = QV = CV^2$ is injected into the circuit from the power supply. The injected energy is twice the signal energy; half of it is dissipated for the other half to be delivered to its destination. When the node is pulled low, the charge is drained from the node to ground, and the other half of the injected energy is consequently dissipated. Thus, all energy drawn from the supply is used only once before being discarded.

To decrease the dissipation, the designer must minimize the switching events, reduce the node capacitance, decrease the voltage swing, or apply some combination of these methods. Depending on the throughput requirements of the circuit and on the power-delay product, or switching energy, of the available logic style and MOS devices, the energy to carry out a certain operation in a certain time will vary, but the CV^2 barrier is impossible to circumvent as long as the overall approach of using energy only once is retained.

---

[1] In this chapter, such logic styles are collectively called "conventional" styles.

Adiabatic switching allows the recycling of energy to reduce the total energy drawn from the power supply. This chapter discusses the limits to energy recycling achievable with a given silicon technology and a given switching speed and voltage swing. Logic circuits have to be reorganized, sometimes radically, to make signal energy recycling possible. The efficiency of some example circuits of different styles is analyzed in this chapter.

The term "adiabatic" is usually used to describe thermodynamic processes that exchange no heat with the environment. Here, the "process" is the transfer of electric charge between nodes in a circuit. Any energy dissipated as a result of the transfer is spread into the environment as heat. As is the typical use of the term in thermodynamics, fully adiabatic (and thus completely dissipation-less) operation is an ideal condition that may be asymptotically approached as the process is slowed down. A dissipation decrease with increased switching time is therefore the defining property of adiabatic switching. In many practical cases, the dissipation for a certain charge transfer is composed of an adiabatic component and a non-adiabatic one, and only the former will decrease when the process is slowed down. Such a partially-adiabatic system will not operate completely without dissipation, regardless of how slowly it is operated.

When adiabatic switching is compared to the voltage-scaling approach described in Chapter 4, both similarities and differences come to light:

- Both approaches address mainly the dynamic dissipation, so reducing the switching frequency and the driven capacitance is beneficial whether the driving circuit uses conventional or adiabatic switching.

- Adiabatic switching allows dissipation to be reduced further without changing the voltage swing, something which cannot be done with conventional circuits.

- In both cases, the dissipation improvement comes at the cost of slower switching speed, so the amount of hardware has to be increased for constant throughput, as described in Chapter 4.

- The "exchange rate" between dissipation and switching speed is different for the two approaches. Typically, voltage scaling initially gives more dissipation improvement for a certain speed reduction. Adiabatic charging is, however, more scalable and may be able to ultimately offer less dissipation at low speeds.

• Voltage scaling is conceptually simpler, requiring no radical re-thinking of the way the logic circuits are organized. There is no consensus yet on the best ways to construct logic systems that use adiabatic switching.

The choice between adiabatic switching and voltage scaling is determined by the circumstances. The conceptual simplicity and high payback of voltage scaling make it preferable in many cases. In situations when the voltage swing cannot be scaled down, adiabatic switching is the only known way to trade speed for power dissipation. Most commonly, these situations arise in interfaces between different parts of a system, where the logic operations performed are rarely more complex than simple power amplification or buffering. Examples include off-chip bus drivers, where the voltage swing is set by industry standards (such as 0–5 or 0–3.3 volts), and drivers for transducers and micromechanical devices, where the necessary swing is determined by some physical phenomenon employed by the device. Additionally, noise immunity may set the lower limit on practical signal energies, especially in environments rich in noise sources (such as commutators, electromagnetic actuators, cosmic rays, and radio transmitters).

When no "external" limits such as those mentioned in the previous paragraph must be taken into account, the voltage swing may be decreased further, but not indefinitely. The switching speed of conventional restoring CMOS gates quickly decreases when the supply voltage approaches the sum of the threshold voltages for the PMOS and NMOS devices, as described in Chapter 4. Gates in other styles of logic may be functional down to one threshold voltage. Beyond this, any decrease in the supply voltage will cause a disproportionally large increase in switching time. Further voltage scaling is practical only if the threshold voltages are scaled as well (cf. Section 4.5). The threshold voltages, in turn, are limited by the desire to keep a sufficient ratio of the on- and off-resistances of the devices. Estimates of the minimum practically useful threshold voltages vary from 50 mV to 200 mV [Mead] [Burr91] [Liu93]. It appears that a radical reorganization of the logic or new kinds of switching elements will be necessary to decrease the dissipation beyond what is possible with voltage scaling methods only. In this chapter, we address the former of these alternatives.

It must finally be noted that the bulk of the research into adiabatically-switching digital circuits has taken place in the last two years. The ideas and techniques described in this chapter are thus necessarily less mature than the voltage-scaling approach that is the main theme of this book. The field is, however, very active, and new results are published frequently.

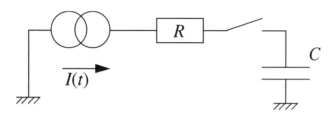

Figure 6.1: Current source charging a capacitance through a switch with a certain on-resistance.

6.1 Adiabatic Charging

We first consider a very simple circuit, illustrated in Figure 6.1. A time-dependent current source, $I(t)$, is used to charge a capacitance, C, through a switch with a resistance, R. This circuit is used as a model of a CMOS circuit with a certain output resistance driving a capacitive load. Note the similarity to the conventional model, where a load capacitance is charged through a switch resistance from a *voltage* source. As in the conventional case, the source models the power supply.

The load capacitance is discharged at time 0. The capacitance voltage as a function of time, $V_C(t)$, is then given by:

$$V_C(t) = \frac{1}{C}\int_0^t I(\theta)d\theta = \frac{1}{C}\tilde{I}(t)t \tag{170}$$

$\tilde{I}(t)$ is the average current from 0 to t:

$$\tilde{I}(t) = \frac{C \cdot V_C(t)}{t} \tag{171}$$

The energy dissipation in R from 0 to $t = T$ is then given by:

$$E_{diss} = R\int_0^T I(\theta)^2 d\theta \geq R\int_0^T \tilde{I}(T)^2 d\theta = R\tilde{I}(T)^2 T = \frac{RC}{T}CV_C(T)^2 \tag{172}$$

with equality when $I(t) \equiv \tilde{I}(T)$, that is, when the current is constant. Other distributions of the current over time give higher dissipation. The influence of the current waveform may be quantified in a shape factor, ξ:

$$\xi = \frac{\displaystyle\int_0^T I(\theta)^2 d\theta}{\tilde{I}(T)^2 T} \geq 1 \qquad (173)$$

Then:

$$E_{diss} = \xi \frac{RC}{T} CV_C(T)^2 \qquad (174)$$

This expression was introduced by Seitz and co-workers [Seitz85] (who apparently assumed constant-current charging and therefore omitted ξ).

The voltage on the output of the current generator in Figure 6.1, V_I, is connected to I (and thereby to V_C) by the following differential equation:

$$V_I = RI + V_C = RC\frac{d}{dt}V_C + V_C \qquad (175)$$

A constant V_I (the "conventional" charging case described in Chapter 3) corresponds to an exponential current waveform with the time constant RC. A constant current corresponds to a linear voltage ramp.

We may now make several observations:

- Of all possible distributions of charging current over time, a constant current causes the least dissipation. In this sense, constant-current charging is the most efficient way to charge a capacitance through a resistance to a certain voltage in a certain time.

- When the exponential current waveform of the conventional case is substituted in Equation 172, R and T cancel out, and the dissipation is once again given by $E_{diss} = (1/2) CV_C(T)^2$.

- The dissipation is lower than for the conventional case if the current is constant and $T > 2RC$.

- The dissipation may be made *arbitrarily small* by further extending the charging time: $E_{diss} \sim T^{-1}$.

- A smaller R also brings a lower dissipation. Again, this is in contrast to the conventional case, where dissipation depends only on the capacitance and the voltage swing.

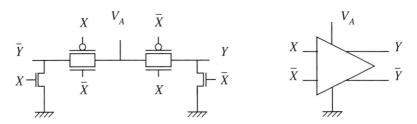

Figure 6.2: Adiabatic amplifier; circuit schematic and logic symbol.

We have efficiently moved energy from the power supply onto a load capacitance by using slow, constant-current charging. Reversing the current source will cause the energy to flow from the load capacitance back into the supply. Thus, in marked contrast to the conventional case, energy is not discarded (dissipated) after being used only once. The power supply must be designed to be able to retrieve the energy fed back to it; otherwise, only half of the potential benefit is realized.

It is clear from Equation 175 that to achieve the non-exponential current waveform desirable for low dissipation, it is necessary to provide a non-constant supply voltage. Adiabatic-switching circuits thus require non-standard power supplies with time-varying voltage and current. We will sometimes refer to these as "pulsed-power supplies." A low overall dissipation can be achieved only if the voltage and current waveforms of the supply both allow power-frugal logic circuits to be built and are possible to generate efficiently. A compromise between the conventional constant-voltage case and the ideal constant-current case may be reached by using sinusoidal waveforms, which can be efficiently generated with inductor-based pulsed-power supplies (as described in Section 6.5).

6.2 Adiabatic Amplification

We will now investigate a simple adiabatic-switching buffer or amplifier circuit. We will analyze its efficiency for different charging times and voltage swings and compare it to that of conventional buffers. The exposition is similar to that of Athas *et al.* [Athas94a]. The example amplifier circuit analyzed here has also been implemented; it should be noted that other circuit topologies are possible. The analyses for other cases will differ in the details, but the overall approach should be generally applicable.

Figure 6.2 shows a simple adiabatic amplifier for capacitive loads. It consists of two transmission gates (T-gates) and two NMOS clamps. The input is dual-rail encoded, since both signal polarities are needed to control the T-gates. The output is also dual-rail encoded, which is required when other T-gates (such as those in other amplifiers) are to be controlled by the output signal. Also, dual-rail signalling keeps the capacitive load on the power supply data-independent, which simplifies the power supply design (cf. Section 6.5).

The operation of the amplifier is straightforward. First, the input is set to a valid value: X and $\bar{X}$ cannot be equal. Next, the amplifier is "energized" by applying to V_A a slow voltage ramp from 0 to V_{dd}. The load capacitance connected to one of the outputs is adiabatically charged to V_{dd} through one of the T-gates, while the other output is clamped to ground. When charging is complete, the output signal pair is valid and can be used as an input to other circuits. Next, the amplifier is de-energized by ramping the voltage on V_A back to 0. The signal energy that was stored on the load capacitance flows back into the power supply connected to V_A. The input signal pair must be stable throughout the process.

The dissipation caused by the operation of the amplifier is easy to determine. As shown in Section 6.1, the dissipation caused by adiabatically charging and discharging a capacitance depends on the on-resistance of the switch. The analysis assumed that the resistance was linear. MOS devices are highly non-linear, but a T-gate can be linearized to a first approximation by carefully selecting the device widths, as is shown next.

A T-gate is turned on with minimal on-resistance when the gate of the PMOS device is grounded and the gate of the NMOS device is connected to V_{dd}. Both devices are in the triode region when the voltage drop across the T-gate is small, which is the intended region for adiabatic circuits. Following Mead and Conway [Mead80], we model the conductance of the NMOS device, G_n, as:

$$G_n = \frac{C_n}{K_n}(V_{dd} - V_{ch} - V_{th}) \qquad (176)$$

$$K_n = \frac{L^2}{\mu_n} \qquad (177)$$

V_{ch} is the average channel voltage, V_{th} is the threshold voltage, and C_n is the gate capacitance of the device. K_n is a process constant that combines mobility,

μ_n, and channel length, L (the minimum channel length allowed in the process is used for all devices). Likewise, for the PMOS device, we get:

$$G_p = \frac{C_p}{K_p}(V_{ch} - V_{th}) \qquad (178)$$

These equations do not take into account body effects nor the difference in threshold voltage between NMOS and PMOS devices. Accuracy is therefore limited to within a factor of two.

The sum of the two conductances may be simplified by selecting the widths of the two MOS devices such that $C_n/K_n = C_p/K_p$:

$$G_p + G_n = \frac{C_n}{K_n}(V_{dd} - V_{ch} - V_{th} + V_{ch} - V_{th}) = \frac{C_n}{K_n}(V_{dd} - 2V_{th}) \qquad (179)$$

The on-resistance of the T-gate is then independent of the channel voltage:

$$R_{TG} = \frac{K_n}{C_n(V_{dd} - 2V_{th})} \qquad (180)$$

Equations 179 and 180 are valid only when both devices are conducting, which is not the case when V_{ch} is within one threshold voltage of either supply rail. At these extremes, the on-resistance will be less than the value given by Equation 180, so the formulation can be used as an estimate of the upper bound on the resistance. In practice, the body effect increases the resistance further.

The energy efficiency of the amplifier may now be analyzed with the help of Equations 174 and 180. The dissipation in the amplifier caused by first charging and then discharging one load capacitance, C_L, is:

$$E_{load} = 2\xi\frac{R_{TG}C_L}{T}C_LV_{dd}^2 = \frac{2\xi}{T}\frac{K_n}{C_n(V_{dd} - 2V_{th})}C_L^2V_{dd}^2 \qquad (181)$$

Additionally, parasitic effects such as diffusion capacitance of the T-gates and the clamp NMOS devices will contribute to the total load capacitance. Terms modelling these effects may be easily added to Equation 181. For simplicity, parasitics are neglected in this analysis.

Let $V_{dd} = m \cdot V_{th}$, and collect all process constants in one parameter, $\tau_n = K_n / V_{th}$. Then:

$$E_{load} = \left(2\xi \frac{\tau_n}{T} \frac{1}{(m-2)} \frac{C_L}{C_n} \right) C_L V_{dd}^2 = \left(2\xi \frac{\tau_n}{T} \frac{m^2}{(m-2)} \frac{C_L}{C_n} \right) C_L V_{th}^2 \qquad (182)$$

We see that the dissipation decreases linearly with increasing T. Also, since $V_{dd} = m \cdot V_{th}$, the dissipation increases only linearly with the voltage swing, as opposed to the V_{dd}^2 dependence of the conventional case.

Energy is also dissipated to drive the input capacitances. When this energy is taken into account, the dependence on T and V_{dd} is affected, as is shown in the following sections.

6.2.1 One-Stage Adiabatic Buffer in Conventional System

We first analyze the case when the inputs are driven conventionally to V_{dd}, so that all the input energy, E_{in}, is dissipated. We assume that the total input capacitance of each of the input lines is proportional to the gate capacitance of the T-gate NMOS device, with a constant of proportionality, α.[2] As an example, if no parasitic capacitances are considered and the gate capacitances of the clamp devices are neglected, $\alpha = (C_n + C_p)/C_n$. The dissipation caused by driving the inputs of one of the T-gates is given by:

$$E_{in} = \alpha C_n V_{dd}^2 \qquad (183)$$

The total energy dissipated per cycle for one driven output is then:

$$E_{total} = E_{in} + E_{load} = \alpha C_n V_{dd}^2 + \left(2\xi \frac{\tau_n}{T} \frac{1}{(m-2)} \frac{C_L}{C_n} \right) C_L V_{dd}^2 \qquad (184)$$

Equation 184 defines an important trade-off in adiabatic CMOS circuits with conventionally-driven inputs. When the channel width, and thereby the input capacitance, is increased, the energy dissipated in charging the load decreases, but the energy dissipated in charging the inputs increases proportionately. The

---

[2.] This α should not be confused with the activity factor of the previous chapters.

minimal energy dissipation is achieved by choosing the device sizes such that the two terms of Equation 184 are equal. The optimum NMOS gate capacitance is:

$$C_{n_{opt}} = \sqrt{\frac{2\xi}{\alpha}\frac{\tau_n}{T}\frac{1}{(m-2)}}C_L \tag{185}$$

Inserting Equation 185 into Equation 184 yields the following expression for the minimum dissipation:

$$E_{total_{min}} = \sqrt{8\xi\alpha\frac{\tau_n}{T}\frac{1}{(m-2)}}C_L V_{dd}^2 = \sqrt{8\xi\alpha\frac{\tau_n}{T}\frac{m^4}{(m-2)}}C_L V_{th}^2 \tag{186}$$

Equation 186 illustrates an important limitation for adiabatic charging with conventionally-driven control signals: because of the complete dissipation of the controlling gate energy, the total switching energy will only scale as $T^{-1/2}$, as opposed to the T^{-1} scaling of Equation 174.

6.2.2 Two-Stage Adiabatic Buffer in Conventional System

A variation on the solution described above uses two amplifier stages, where a smaller adiabatic amplifier is used to drive the input signals of the final stage, as shown in Figure 6.3. Let C_{n1} and C_{n2} be the gate capacitances of the NMOS devices of the T-gates of the first and second amplifier, respectively. The input capacitance of the second amplifier, αC_{n2}, then constitutes the load of the first amplifier. Allotting half of the total charging time to each stage, we get:

$$E_{total} = \alpha C_{n1} V_{dd}^2 + 4\xi\frac{\tau_n}{T}\frac{1}{(m-2)}\left(\frac{\alpha^2 C_{n2}^2}{C_{n1}} + \frac{C_L^2}{C_{n2}}\right)V_{dd}^2 \tag{187}$$

With both C_{n1} and C_{n2} as free variables, the minimum total energy dissipation is:

$$E_{total_{min}} = 8\left(\alpha\xi\frac{\tau_n}{T}\frac{1}{(m-2)}\right)^{3/4}C_L V_{dd}^2 = 8m^2\left(\alpha\xi\frac{\tau_n}{T}\frac{1}{(m-2)}\right)^{3/4}C_L V_{th}^2 \tag{188}$$

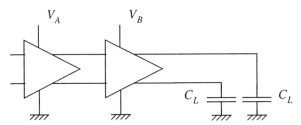

Figure 6.3: Two cascaded adiabatic amplifiers driving capacitive loads. V_A and V_B are two pulsed-power supply voltages, chosen to guarantee that the output of the first amplifier is valid when the second amplifier is energized. The input of the first amplifier is driven conventionally.

From Equation 186 to Equation 188, the energy scaling improves from $T^{-1/2}$ to $T^{-3/4}$. It can be shown that for a cascade of n such amplifiers, the energy dissipation scales as:

$$E_{total_{min}} \sim T^{2^{-n}-1} \tag{189}$$

This result is not very useful in practice, since parasitics can no longer be neglected when the capacitances of the final gate-drive MOS devices become comparable to that of the load. Also, if the switching is to be performed within a constant time interval, the ramp time for each of the steps must decrease at least linearly with the number of cascaded amplifier stages.

6.2.3 Fully-Adiabatic System

As shown above, conventionally driven control signals cause a sub-linear dependence of energy on time. True T^{-1} scaling can be achieved only if the system uses adiabatic charging throughout. It is not straightforward to build such a system (some issues are outlined in Section 6.3), but the resulting dissipation for charging a certain capacitance is easy to calculate. Assume a cascaded arrangement like that shown in Figure 6.3, but with the inputs of the first amplifier adiabatically driven, its device dimensions fixed, and a charging time T allowable for each amplifier. Then, the dissipation for driving the load and the inputs of the second amplifier is:

$$E_{total_{min}} = \left(2\xi \frac{\tau_n}{T} \frac{1}{(m-2)} \right) \left(\frac{\alpha^2 C_{n2}^2}{C_{n1}} + \frac{C_L^2}{C_{n2}} \right) V_{dd}^2 \qquad (190)$$

The C_{n2} that minimizes the dissipation is independent of m and of τ_n/T, and the dissipation is linear in T:

$$C_{n2_{opt}} = \sqrt[3]{\frac{C_L^2 C_{n1}}{2}} \qquad (191)$$

$$E_{total} = \left(2\xi \frac{\tau_n}{T} \frac{1}{(m-2)} \left(\sqrt[3]{2} + \frac{1}{\sqrt[3]{4}} \right) \sqrt[3]{\frac{C_L}{C_{n1}}} \right) C_L V_{dd}^2 \qquad (192)$$

6.2.4 Comparison with Conventional Buffer

We can now compare the dissipations of the adiabatic buffers with that of a conventional one, for different speeds and voltage swings. To drive the input and load of a conventional buffer with mobility-scaled device sizes (giving the same total input capacitance as for the T-gate) causes a dissipation given by:

$$E_{total} = (\alpha C_n + C_L) V_{dd}^2 \qquad (193)$$

By rewriting Equation 90 of Chapter 3, the charging time of a conventional buffer may be approximated as:

$$T = \frac{2K_n V_{dd}}{(V_{dd} - V_{th})^2} \cdot \frac{C_L}{C_n} = 2\tau_n \frac{m}{(m-1)^2} \cdot \frac{C_L}{C_n} \qquad (194)$$

Thus:

$$E_{total} = \left(1 + 2\alpha \frac{\tau_n}{T} \frac{m}{(m-1)^2} \right) C_L V_{th}^2 \qquad (195)$$

We now plot Equations 186, 188, 192, and 195 as functions of the switching speed for different values of m. These plots are shown in Figures 6.4 and 6.5. For each plot, the dissipation is normalized to $C_L V_{dd}^2$. The time scale is normalized to the delay time of a conventional buffer when $C_L = 20C_n$, as given by Equation 194. The variants of the adiabatic buffer improve with increased supply voltage and charging time, whereas the conventional buffer gets little dissipation benefit from ncreased switching time. With a lower supply voltage, the

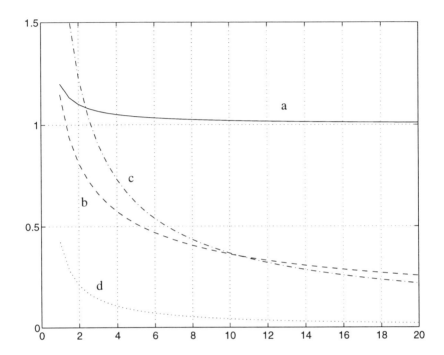

Figure 6.4: Comparison of dissipation as a function of the transition time for a) conventional driver, b) one-stage adiabatic driver, c) two-stage adiabatic driver, and d) driver in fully adiabatic system. The voltage swing is fixed to three times the threshold voltage ($m = 3$). Dissipation is normalized to $C_L V_{dd}^2$. Transition time is normalized to that of a conventional driver when the load capacitance is 20 times the gate capacitance of the NMOS device of the driver.

cross-over point where the adiabatic approach is preferable occurs at a longer switching time.

Care must be taken when these diagrams are interpreted. First, in all cases the dissipation for driving *one* line conventionally is compared to that of driving *one* line adiabatically; but the design of the adiabatic amplifier presupposes dual-rail signalling. Second, the dissipation in the power supply (which generates the voltage ramps) is not taken into account; it is non-trivial to minimize this dissipation. Third, the use of T-gates in the buffers causes a lower limit on the usable supply voltage: if $m < 2$, there will be a region in which neither device is conducting, and thus the resistance (and, according to the model, the dissipation)

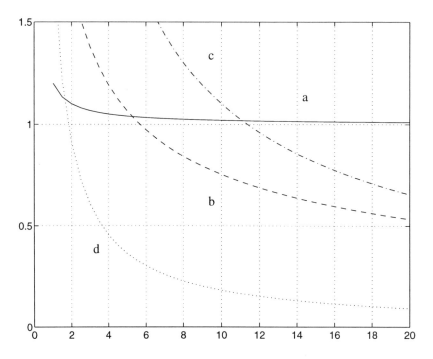

Figure 6.5: As Figure 6.4, but for a lower supply voltage: $m = 2.1$. With the lower supply voltage, the transition time has to be increased more over that of the conventional buffer for the adiabatic approach to yield lower dissipation.

goes to infinity. The latter limitation may be avoided with other types of adiabatic circuits (cf. Section 6.3).

To investigate how well the theory reflects reality, an adiabatic amplifier for stand-alone use as a line driver for the address bus of a memory board was designed, implemented, and tested [Athas94b]. The most significant deviation from the theoretical results was caused by dissipation in the power supply (which was similar to those described in Section 6.2.5). The non-unity shape factor and additional parasitics further reduced the dissipation gain from a theoretical factor of 20 (according to Equation 181) to a factor of 6.3 (when the dissipation in the power supply is included).

6.2.5 Supply Voltage Influence

In Section 6.2.4, we compared the dissipations of the different adiabatic amplifiers with that of a conventional buffer at several fixed supply voltages. If we are free to choose the voltage swing for minimum dissipation, we might try to scale down signal energies (by decreasing V_{dd}) as far as the application permits and then apply adiabatic-charging techniques to reduce switching energies further. Such an approach does, however, not necessarily lead to minimal dissipation for adiabatic circuits, since the reduced gate drive at lower supply voltages increases the on-resistance of the switches. For a T-gate-based driver, the resistance increase counteracts the dissipation gain caused by the reduced signal energies, leading to a process-dependent optimal voltage swing which yields the lowest overall dissipation.

It is easy to derive the optimal voltage swing for the T-gate-based buffers analyzed above. For the one-stage buffer, we find the voltage by rewriting Equation 186:

$$E_{total_{min}} = \sqrt{8\xi\alpha\frac{\tau_n}{T}C_L V_{th}^2 \frac{m^2}{\sqrt{m-2}}} \tag{196}$$

The energy is minimized when $m = 8/3$, and thus $V_{dd} = (8/3)\,V_{th}$. Similar derivations for the two-stage buffer (described by Equation 188) and the fully adiabatic system (described by Equation 192) give the values $m = 16/5$ and $m = 4$, respectively. Note that these optima are based on a simple model for the on-resistance of the T-gate. The body effect and the regions where only one FET device is conducting are not taken into account. SPICE simulations and more detailed analyses (that take body effects, subthreshold regions, and waveform shapes into account) all show that the minima are shallow and that the values given above yield close to the minimal dissipation.

6.3 Adiabatic Logic Gates

A straightforward extension of the adiabatic amplifier of Section 6.2 allows the implementation of arbitrary combinational logic functions that use adiabatic

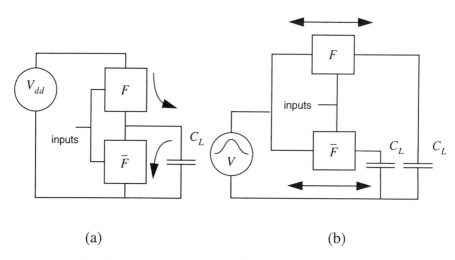

(a)   (b)

Figure 6.6: (a) A conventional CMOS logic gate. (b) A corresponding adiabatic gate may be constructed by reorganizing the switch networks as shown, replacing the PMOS and NMOS devices of the conventional gate with T-gates. The DC supply voltage must be replaced by a pulsed-power source. The arrows indicate the charge flow when the load capacitances are charged and discharged.

switching. As shown in Figure 6.6, a conventional logic gate may be transformed into an adiabatically-switched counterpart by replacing each of the PMOS and NMOS devices in the pull-up and pull-down networks with T-gates, and by using the expanded pull-up network to drive the true output and the expanded pull-down network for the complementary output. Both networks in the transformed circuit are used both to charge and discharge the output capacitances. The DC V_{dd} source of the original circuit must be replaced by a pulsed-power source with varying voltage to allow adiabatic operation. The optimal sizes of the T-gates of the networks can be determined from the equations of Section 6.2. The resulting combinational circuits can switch fully adiabatically, but their device count is twice that of the conventional counterpart, and their input capacitance is relatively large because of the T-gates.

As with conventional CMOS logic, it is desirable for reasons of performance and complexity management to partition a large block of logic into smaller ones and then compose them to implement the original larger function. However, non-adiabatic flow of energy will occur if values are allowed to ripple through a

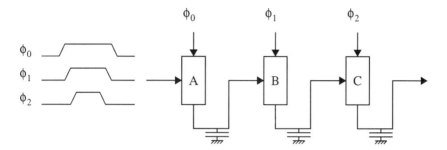

Figure 6.7: Retractile cascade of adiabatic-switching logic gates. The supply voltages are ramped up in order, energizing the gates and charging the load capacitances. The voltages must then be ramped down in reverse order to ensure that the energy stored on the capacitances is recovered. The pulsed-power voltage source waveforms are shown on the left.

chain of cascaded adiabatic logic gates. Adiabatic operation is possible only if the inputs of every gate are held stable while the gate is energized.

Hall observed that a cascade of initially de-energized circuits may be adiabatically energized in succession [Hall92]. The gates must then be de-energized in reverse order before the input values to the cascade may change (Figure 6.7). These "retractile cascades" are impractical for several reasons: they require a large and possibly indeterminate number of supply voltage waveforms; these waveforms all have different pulse widths; and since an N-stage cascade requires time proportional to N to produce each result, the latency is proportional to N and the throughput is proportional to $1/N$.

Pipelining can be used to improve the throughput of the system. By using latches to hold the inputs of a retractile stage, we may circumvent the requirement that the preceding stage stays energized until the current stage has been de-energized. Conventional latches, however, cause dissipation independent of the charging time and therefore the unfavorable sub-linear dependence on charging time described in Sections 6.2.1 and 6.2.2. It is possible to avoid using conventional latches, but at a considerable complexity cost, as will be seen in Section 6.3.1 below. An alternative is to use latches with a small, fixed dissipation and allow the dissipation of the combinational parts to scale as $T^{-1/2}$, which allows the use of logic styles with less overhead. One such style is described in Section 6.3.2.

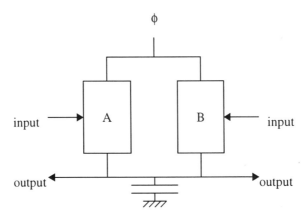

Figure 6.8: Pipelinable adiabatic gate. The load capacitance may be charged through one function network, A, and discharged through another, B. The inputs to the first network must be valid only during the charging phase, and the preceding stage can compute new values during the discharge of the load capacitance.

6.3.1 Fully-Adiabatic Sequential Circuits

The requirement for latches to hold the inputs of a function network arises from the need to provide stable control signals throughout the full charge-discharge cycle of the output capacitances. If the retractile stage is expanded to include an explicit discharge path, as shown in Figure 6.8, the controlling inputs of the energizing path need not be kept stable to handle de-energizing. A means for pipelining that does not use latches then exists, and T^{-1} scaling can be retained.

The control signals for the de-energizing path must be stable throughout the discharge to ensure adiabatic operation. Therefore, they cannot be derived directly from the outputs of the stage being de-energized [Koller93]. Any attempt at schemes based on the same idea is bound to fail for thermodynamic reasons [Landauer61]. The de-energizing path must instead be controlled by signals derived from the output of the *following* stage in the pipeline. A pure copy of the signal is sufficient, but this method only defers the problem and is not a practical solution when resources are limited. The ability to derive control signals for the de-energizing path is guaranteed if all logic blocks implement functions that are invertible, so that their inputs may be recomputed from their output.

It is possible to assemble a fully adiabatic pipeline by constructing all of the logic stages according to Figure 6.8 and restricting the function blocks to invertible functions only [Athas94a][Younis93]. The exact logic style chosen for the pipeline determines the timing and sequence requirements for the supply-voltage waveforms. Since these not only power the computations but also pace them, they may equally well be thought of as clock signals, similar to the "hot clocks" of Seitz [Seitz85]. The use of exclusively invertible functions makes the pipeline reversible: if the clocks are reversed in time, the pipeline runs backwards. Reversible logic has been studied in theoretical physics since Bennett showed that computations need not destroy information [Bennett73]. Since then, a large body of theory has been developed [Bennett85][Bennett89][Merkle93].

The complexities of two mutually inverse function blocks are usually similar, so the typical circuit overhead for the separate discharge path is a factor of two, not counting the cost associated with the restriction to invertible functions. The latter cost can be quite prohibitive [Athas94c] and is likely to limit the use of fully-adiabatic solutions to very small circuits with simple functionality, and cases where very slow switching is acceptable as long as the energy dissipation per transition is low enough.

6.3.2 Partially-Adiabatic Sequential Circuits

It is possible to enjoy the benefits of adiabatic switching without the complications of reversible logic. Less complex circuitry may compensate for the unfavorable dissipation scaling with time. Several such schemes have been suggested[3] [Seitz85] [Kramer94] [Denker94] [Gabara94].

In this section, we will describe a partially-adiabatic circuit style recently proposed by Denker [Denker94], based on an idea by Koller [Koller93]. We will refer to it as the "flip-flop" style. This scheme has some desirable properties: its device count is low, approaching that of precharged dynamic logic; and a pipeline of these gates can be operated with a four-phase clock, a relatively simple requirement compared to some other suggestions. It is quite different from the adiabatic amplifiers described in Section 6.2, in that no T-gates are used. The analysis in Section 6.2 is therefore not immediately applicable.

[3.] Other styles that achieve sub-$C_L V_{dd}^2$ dissipation but get no benefit from slower switching have also been described [Hinman93]; regardless of their other qualities, such schemes cannot really be considered adiabatic, based on the definition on page 182.

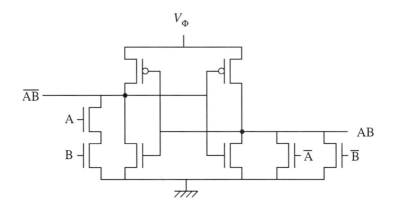

Figure 6.9: Flip-flop-style adiabatic charging logic gate. The example shows a gate implementing an AND/NAND function. Both inputs and outputs are dual-rail-encoded. Load capacitances are not shown. V_Φ is connected to the pulsed-power supply.

Figure 6.9 shows an example of a flip-flop gate. It consists of two inverters, cross-coupled in a flip-flop configuration, and two pull-down networks, which implement two complementary logic functions. Both inputs and outputs are dual-rail-encoded.

The gate works as follows. Assume that the load capacitances (which represent inputs of other gates and wiring capacitances) and the supply/clock line are all initially at 0 V. First, the inputs are given a set of valid values. This will connect exactly one of the output lines to ground, since the networks implement complementary functions. Next, the supply line is ramped from 0 to V_{dd}. Initially, no significant current flows, since both the PMOS devices are turned off. When the supply voltage reaches V_{th} (the PMOS threshold voltage), the device whose gate is grounded by a pull-down network will turn on, and the corresponding load capacitance will begin to charge up. This causes an initial, non-adiabatic dissipation of approximately $(1/2)\,C_L V_{th}^2$. The remainder of the charging, up to V_{dd}, of the chosen output capacitance will be adiabatic, and its dissipation will decrease with increased charging time. When the charging is complete, the output values are valid and the high inputs are allowed to go low. This disconnects all pull-down networks, but the feedback through the cross-coupled inverters will preserve the output values. Other gates, which use the outputs of the current gate as inputs, may be energized while the inputs of the current gate are ramped down.

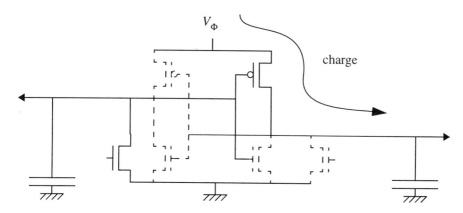

Figure 6.10: Charging an output capacitance of a buffer, implemented
in the style of Figure 6.9. The pull-down device on the left grounds the
gate of the PMOS device, creating a path from the source to the load.
Devices drawn dashed are not conducting.

When the output values have been sampled, the gate may be de-energized
by ramping down the supply voltage. The charge and energy on the load capacitance flows back into the power supply through the PMOS device. When the supply voltage reaches V_{th}, the PMOS device is turned off, and no further energy
recovery will take place.

The charge left on the load capacitance at the end of the discharge will leak
away slowly through the turned-off devices. If, however, the supply voltage is
ramped up again before much charge has leaked away, and the same pull-down
network is tied, the charging will resume close to V_{th}. Thus, the non-adiabatic
dissipation will be less than $(1/2)\,C_L V_{th}^2$; ideally, with very low leakage, it will
be close to 0. In case the other pull-down network is tied, both load capacitances
will change values non-adiabatically, resulting in an energy dissipation of
$2 \cdot (1/2) \cdot C_L V_{th}^2$. Non-adiabatic dynamic dissipation therefore occurs mainly
when the output value changes, which is also the case for conventional logic.

Using the same style of analysis as in Section 6.2, we now calculate the
efficiency of a buffer implemented in this style when charging a capacitive load.
Figure 6.10 shows the active parts of such a buffer during the charging phase. Initially, assume that the load capacitance is charged up to V_{th} (the PMOS threshold
voltage). The supply voltage is ramped up from 0 to V_{dd}; the charging device will
turn on when the supply voltage passes V_{th}, and the load will be charged from

V_{th} to V_{dd}. According to Equation 178, the resistance of the charging device can be modelled as:

$$R_p(V_{ch}) = \frac{K_p}{C_p(V_{ch} - V_{th})} \tag{197}$$

Again, this does not take the body effect into account. The resistance depends on the channel voltage in a highly non-linear fashion (it grows without bounds when V_{ch} approaches V_{th}). It is therefore not possible to break out the dependence on the current waveform shape into a ξ factor, as in Equation 174. Instead, we calculate the dissipation for one specific case: that of a sinusoidal current, as would be generated by an inductive power supply with very low output impedance (cf. Section 6.5). Charging from V_{th} to V_{dd} will then require a current given by:

$$I(t) = \frac{\pi}{2} \frac{V_{th}(m-1)C_L}{T} \cdot \sin\left(\frac{\pi t}{T}\right) \tag{198}$$

where $V_{dd} = m \cdot V_{th}$. The voltage drop across the device is small when the charging time is long, so the channel voltage is approximately the same as the voltage on the load capacitance:

$$V_{ch} \approx V_C = V_{th} + \int_0^t I(\theta)d\theta = V_{th}\left(1 + (m-1)\sin^2\left(\frac{\pi t}{2T}\right)\right) \tag{199}$$

The resistance as a function of time is then:

$$R_p(t) = \frac{\tau_p}{C_p(m-1)}\csc^2\left(\frac{\pi t}{2T}\right) \tag{200}$$

where $\tau_p = K_p/V_{th}$. The instantaneous power dissipation during the charging is given by:

$$P(t) = I(t)^2 R_p(t) = \frac{\pi^2}{4}(m-1)\frac{\tau_p}{T^2}\frac{V_{th}^2 C_L^2}{C_p} \cdot \sin^2\left(\frac{\pi t}{T}\right)\csc^2\left(\frac{\pi t}{2T}\right) \tag{201}$$

The total adiabatic energy dissipation is:

$$E_{ff} = \int_0^T P(t)dt = 4(m-1)\left(\frac{\pi^2 \tau_p}{8}\frac{C_L^2}{T C_p}V_{th}^2\right) \tag{202}$$

The adiabatic part of the dissipation for the flip-flop driver grows linearly with the voltage swing above the threshold voltage. There is no "optimum" voltage, as for the charging through a T-gate (described by Equation 182); a lower swing yields a lower dissipation. The price paid is the non-adiabatic $C_L V_{th}^2$ dissipated when the output value changes. In the T-gate driver, this dissipation is avoided through the use of an NMOS device which ensures low resistance and adiabatic behavior throughout the charging, regardless of the previous output value.

According to Equation 202, the dissipation is proportional to τ_p. This is since the flip-flop driver, as described here, charges and discharges its load capacitances through PMOS devices. Since the carrier mobility is larger for NMOS devices (and therefore $K_n > K_p$ and $\tau_p > \tau_n$), the dissipation may be decreased by "turning the circuit upside down" so that the load capacitances are charged through NMOS devices instead.

The assumptions used for the derivation of Equation 202 are largely the same as those used in Section 6.2 (the load capacitance is large compared to the gate capacitance of the driver; the voltage drop across the driver device is small; the body effect is neglected; etc.). An additional assumption is that the current will stay sinusoidal despite the severe resistance variations. This is an obvious oversimplification: with an inductive supply, the real current waveform depends on the behavior of the whole ensemble of circuits driven by the same voltage phase.

It must finally be stressed that the comparison of the dissipation of two buffers does not tell the whole story about the total dissipation of systems using these buffers. The simplicity and low device count of the gate style outlined here makes it possible to build small and appealingly simple circuits. This in turn lowers the overall driven capacitances and contributes to low dissipation even for relatively large values of m and V_{dd}.

6.4 Stepwise Charging

As we have seen, the key to low dissipation in an adiabatic-switching circuit is to lower the average voltage drop traversed by the charge that flows onto the load capacitance. The ideal case, yielding the smallest dissipation for a given charging time, is when the charging current can be kept constant. This requires that the power supply be able to generate linear voltage ramps. Practical supplies have

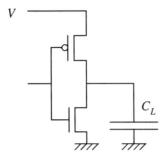

Figure 6.11: Conventional CMOS signal driver with capacitive load.

used resonant inductor circuits (as outlined in Section 6.5) to approximate the constant currents and linear ramps with sinusoidal signals.

Inductor circuits have several practical drawbacks. High-Q inductors cannot be integrated on a chip together with other circuits and thus complicate the board-level design. The efficiency of the inductive solutions may rely on accurately timed control pulses. While small timing errors can cause ringing and efficiency loss, larger errors may cause damaging voltage spikes. The problem of designing an efficient power supply for variable capacitive loads has led some workers towards dual-rail logic, which approximately doubles the amount of circuitry.

This section describes *stepwise charging*, an alternative way to generate approximate voltage ramps [Svensson94]. Being inductor-less, it suffers none of the drawbacks mentioned above. In addition, it requires no changes in the power supply and may therefore be included in existing designs with relative ease. On the other hand, the circuit solutions involved are more complex and require more silicon area. The method is likely to be useful to drive those few nodes in a circuit that cause a large part of the dissipation, such as output pads and busses. It is not readily extensible to logic gates.

Again, consider the conventional CMOS driver, repeated in Figure 6.11. All charge delivered to the load is injected at the supply voltage, and thus the dissipation caused by charging C_L to V is $(1/2) C_L V^2$, regardless of the speed with which the charging takes place. A full cycle causes twice this dissipation.

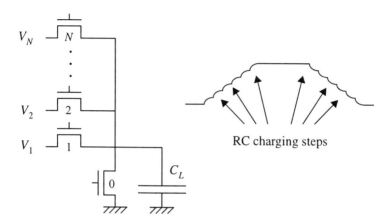

Figure 6.12: A stepwise driver for a capacitive load. Voltages are given by the formula $V_i = (i/N) \ V$. Switches are shown as NMOS devices, but some may be PMOS. The voltage waveform on the load capacitance is also shown.

If several voltage sources are available, we may save energy by carrying out the charging in several steps. Figure 6.12 shows a basic stepwise driver for a capacitive load, implemented with NMOS devices. A bank of N voltage supplies with evenly distributed voltages is used. The load is charged by connecting V_1 through V_N to the load in succession (by closing switch 1, opening switch 1 and closing switch 2, etc.). To discharge the load, V_{N-1} through V_1 are switched in in the same way, and then switch 0 is closed, connecting the output to ground.

For each step, the dissipation is again given by the transferred charge and the average voltage drop across the switch resistance:

$$E_{\text{step}} = Q\bar{V} = C_L \frac{V}{N} \cdot \frac{V}{2N} = \frac{1}{2} C_L \frac{V^2}{N^2} \tag{203}$$

N steps are used to charge C_L all the way to the supply voltage V, so the total energy dissipation is:

$$E_{\text{stepwise}} = N \cdot E_{\text{step}} = N \cdot \frac{1}{2} C_L \frac{V^2}{N^2} = \frac{1}{2} C_L \frac{V^2}{N} \tag{204}$$

Again, a full charge-discharge cycle will cause twice the dissipation of the charging only. Thus, according to this simplified analysis, charging by N steps reduces

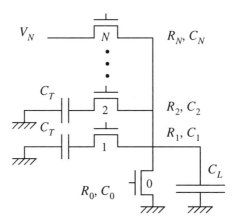

Figure 6.13: A stepwise driver with all but one of the necessary voltages supplied by tank capacitors, C_T. The R_i, C_i pairs represent the on-resistance and the gate capacitance of the ith switch.

the energy dissipation per charge-discharge cycle, and thereby the total power dissipation, by a factor of N. The decrease comes from a reduction of the voltage drop across the switch through which the charge flows, which is now on average a factor N smaller than in the conventional case.

While avoiding the inductor problems, this scheme suffers from a few other drawbacks. Most notably, multiple supply voltages are needed. Not only is the overhead of routing many supply lines to each circuit in a system undesirable; more importantly, a power supply able to efficiently generate the voltages would be complex and expensive.

These problems can be circumvented by using the circuit in Figure 6.13. It shows the same circuit as Figure 6.12, but with all supplies except one replaced with large "tank" capacitors. The tank capacitor voltages will change little during the entire charging of the load capacitance, provided that the capacitors are large enough and initially charged to the evenly distributed voltages of Figure 6.12. The behavior and dissipation will then be the same as for the circuit in Figure 6.12. No extra circuitry is required to maintain the voltages on the tank capacitor bank: it can be shown that the tank capacitor voltages converge to the desired, evenly distributed voltages. Thus, only one supply line must be routed to the chip, and the power supply is no more complicated than in the conventional case. The main cost is a larger chip area for the switches and the package pins set aside for the tank capacitors (these would most often be kept off-chip, since they must be large compared to the driven capacitance). Note, however, that several drivers may share the capacitor bank, and hence the drivers for an entire bus may need only a few extra package pins.

Equation 204 indicates that dissipation decreases linearly with increasing N. N cannot, however, usefully be made arbitrarily large, because each step requires that a switch be turned on and off, which itself causes dissipation. Also, the energy used to control each switch depends on the width of the device, which should be just enough to allow the charging to "complete" before the next step commences. Thus, for a given total allowable charging time, there is an optimal number of steps and a set of optimal device sizes which lead to minimal total dissipation. We find these by deriving an expression for the gate drive energy as a function of N, and then minimize the sum of the gate-drive energy and the dissipation caused by the charge flowing onto the load capacitance.

In the derivations that follow, we assume that the voltage swing is identical in all parts of the circuit, so that the gate of a switch device is tied either to ground or to V. We model the channel of switch i as a linear resistance, R_i, and its gate as a linear capacitance, C_i, connected to ground. All switch devices are assumed to be minimum-length; only the width will be varied. We introduce the switch quality measure, given by:

$$\rho_i = R_i C_i \tag{205}$$

ρ_i is independent of the switch width. However, it varies with i, since the channel voltage is different for different i; the voltage dependence is shown in Figure 6.14. It is also different for PMOS and NMOS devices. Finally, ρ_i is obviously process-dependent. It is similar to the τ of Mead and Conway [Mead80], in that it quantifies the maximum speed attainable in the process, but it takes the behavior across the entire voltage range into account. Referring again to Figure 6.14, it is clear that a PMOS device is a better switch at high channel voltages, while an NMOS device is preferable at lower voltages.

Assume that the gates of all the switches are driven conventionally, so that the dissipation is given by CV^2. The total gate energy spent to charge and discharge the load capacitance once is then:

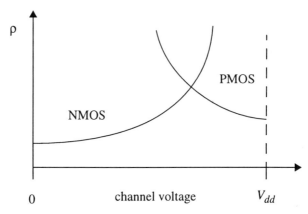

Figure 6.14: Conceptual dependence of the switch quality, ρ, on the channel voltage. For high channel voltages, a PMOS device with its gate tied to ground conducts better than an equally-sized NMOS device with its gate tied to V_{dd}. A lower value means a better switch.

$$E_{\text{gates}} = \left(\sum_{i=1}^{N} C_i + \sum_{i=0}^{N-1} C_i \right) V^2 \tag{206}$$

Next, we relate the switch gate capacitances, C_i, to N. Allot each step one Nth of the total available charging time, T. Then:

$$\frac{T}{N} = jR_iC_L \tag{207}$$

Here, j is a new parameter: the number of RC time constants spent waiting for each charging step to "complete." If its value is chosen too small, there will still be a significant voltage across a switch when the next switch is to take over. Hence, there is an increase in the average voltage across each switch, and therefore a dissipation increase. If on the other hand j is chosen unnecessarily large, time is wasted that could have been used to increase the number of steps. Suitable values range from 2 to 4. We see from Equation 207 that if j is the same for all steps, all the switch devices should have equal on-resistance. The required gate capacitance for switch i is then given by:

$$C_i = \frac{\rho_i}{R_i} = \frac{Nj\rho_i}{T}C_L \tag{208}$$

Substitute Equation 208 into Equation 206:

$$E_{\text{gate}} = \frac{Nj}{T}\left(\sum_{i=1}^{N}\rho_i + \sum_{i=0}^{N-1}\rho_i\right)C_L V^2 \tag{209}$$

Equation 209 still cannot be used immediately for optimization, since the dependence on N is complicated and non-intuitive. We therefore introduce $\bar{\rho}$, a weighted average of ρ_i for the different switches:

$$\bar{\rho} = \frac{1}{2N}\left(\sum_{i=1}^{N}\rho_i + \sum_{i=0}^{N-1}\rho_i\right) \tag{210}$$

$\bar{\rho}$ is close to the unweighted average of ρ over the entire voltage range if N is sufficiently large. The unweighted average is independent of N and may be used in place of the weighted average when finding the best N. We get:

$$E_{\text{gate}} = \frac{Nj}{T}\cdot 2N\bar{\rho}\cdot C_L V^2 \tag{211}$$

Combining Equations 211 and 204, we get the following expression for the total dissipation for a full charge-discharge cycle:

$$E_{\text{tot}} = \left(\frac{1}{N} + 2N^2 j\frac{\bar{\rho}}{T}\right)C_L V^2 \tag{212}$$

The N that minimizes E_{tot} is given by:

$$N_{\text{opt}} = \sqrt[3]{\frac{T}{4j\bar{\rho}}} \tag{213}$$

The corresponding energy dissipation is:

$$E_{\text{opt}} = \frac{3}{2}\sqrt[3]{\frac{4j\bar{\rho}}{T}}C_L V^2 \tag{214}$$

By using the number of stages given by Equation 214, the designer can minimize the power dissipation of the driver. The minimum is rather shallow, however, so a lower N (as would most often be dictated by practical consider-ations) will still give a considerable improvement over the conventional case; $N = 2$ already gives almost 50% reduction unless T is very small. Once N and j have been selected, the gate capacitance, and thereby the width, of each switch is given by Equation 208. The values of ρ for a certain process can be found by circuit simulation or by measuring the on-resistances of test devices of known width.

The accuracy of this design procedure is surprisingly good, given the many simplifications that have been made. A test driver with six steps (2μ CMOS, $T = 500$ns) was predicted by the equations to reduce the dissipation by 80%. Circuit simulation of the extracted layout resulted in 77% improvement. Measurements on the fabricated chip yielded 73% improvement [Svensson94].

The improvement figures do not take the energy needed to generate the sequence of control signals into account. In the test chip, this was done with a state machine that had not been optimized for low power. In a practical chip, part of this overhead may be amortized over several drivers that share part of the control logic (for example, all signals on a bus would switch in synchrony).

A stepwise-charging driver may save an order of magnitude in dissipation over the conventional counterpart when the switching may be slow compared to what would be achievable with a conventional driver. It uses only capacitors and switches and may therefore be integrated (but the tank capacitances would mostly be kept off-chip). Finally, it does not require a special power supply, and is therefore considerably easier to design into a conventional system.

6.5 Pulsed-Power Supplies

The preceding sections have discussed the design of adiabatic-switching gates and circuits. The pulsed-power supply waveforms were assumed to be given. This section provides an introduction to the problems of designing a resonant inductive power supply, and gives an analysis of the attainable efficiency for a simple example.

Figure 6.15 depicts a very simple inductive pulsed-power supply. It consists of an inductor, L, and three switches labelled A, B, and C. The load presented to the power supply by the logic circuit is mainly capacitive (the total capacitance of all the driven MOS gates, plus wiring capacitance and other stray capacitances). Most of the capacitance is charged through device channels, which adds a resistive component to the load. In this section, we will assume that the load can be represented by a linear resistance in series with a linear capacitance.

The circuit works as follows. Assume that the load is initially fully discharged, and that only switch B is tied. To charge the load, B is cut and A is tied. A and L now form an RLC circuit together with the load. Charge will flow

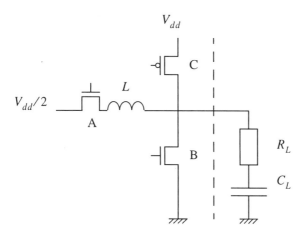

Figure 6.15: A simple inductive power supply driving a load with a resistive and a capacitive component, representing an adiabatically switching logic circuit.

through L onto C_L. Provided that the RLC circuit is underdamped, the load voltage will swing past $V_{dd}/2$; if the circuit is highly underdamped, the voltage will peak at close to V_{dd} when the inductor current goes to 0. At this instant, A must be cut and C tied to pull the output all the way to V_{dd}. Similarly, discharging the load is accomplished by cutting C, tying A, waiting for the load voltage to reach its lowest value, and cutting A and tying B.

The power-supply circuit is a second-order system; its dissipation during a full charge-discharge cycle is easily found using standard time-domain circuit analysis. The state of the circuit at the end of a full cycle is identical to that at the beginning, and for reasons of symmetry, the amount of charge pulled from the $V_{dd}/2$ connection during the charge phase is equal to that pushed back during the discharge phase. Thus, the dissipated energy is equal to the energy injected from V_{dd} through switch C. The waveforms are given by the values of L, C_L, and R_{total}, which consists of R_L and R_A, the on-resistance of switch A. Assuming a heavily underdamped circuit, we get:

$$I(t) = \frac{V_{dd}}{2\omega L} e^{-\frac{\alpha t}{2}} \sin \omega t \qquad (215)$$

$$\alpha = \frac{R_{total}}{L} = \frac{R_L + R_A}{L} \qquad (216)$$

$$\omega = \sqrt{\frac{1}{LC_L} - \frac{\alpha^2}{4}} \approx \frac{1}{\sqrt{LC_L}} \qquad (217)$$

The maximum voltage reached at the load (at time $T = \dfrac{\pi}{\omega} \approx \pi\sqrt{LC_L}$) is:

$$V_{max} = V_{dd}(1 - \varepsilon) \qquad (218)$$

$$\varepsilon = \frac{1}{2}\left(1 - e^{-\frac{\alpha\pi}{2\omega}}\right) \approx \frac{\alpha\pi}{4\omega} \approx \frac{\pi^2}{4} \cdot \frac{R_{total}C_L}{T} \qquad (219)$$

The injected energy, and therefore the dissipated energy, is given by:

$$E_{inj} = V_{dd}(V_{dd} - V_{max})C_L = \varepsilon \cdot C_L V_{dd}^2 \qquad (220)$$

We make the following observations:

- Comparing Equations 174, 219, and 220, we see that $\xi = \pi^2/8 \approx 1.234$ for sinusoidal currents.

- Unless C_L is constant from cycle to cycle, the charging time T will have to vary (assuming that L is held constant).

- The $V_{dd}/2$ source may be replaced with a large tank capacitance, C_T, charged to $V_{dd}/2$. As long as $C_T \gg C_L$, the behavior change is negligible.

- The dissipated energy may be separated into two terms, one depending on R_A and the other on R_L:

$$E_{inj} = \frac{\pi^2}{4} \cdot \frac{R_A C_L}{T} \cdot C_L V_{dd}^2 + \frac{\pi^2}{4} \cdot \frac{R_L C_L}{T} \cdot C_L V_{dd}^2 \qquad (221)$$

The latter term depends only on the load and is independent of the power supply, as long as the voltage and current levels stay the same.

We will now size switch A to minimize the overall dissipation, including that of the circuit driving the gate of switch A. R_A is given by Equation 176, with $V_{ch} = V_{dd}/2$:

$$R_A = \frac{2K_n}{C_A(V_{dd} - 2V_{th})} = \frac{2\tau_n}{C_A(m-2)} \qquad (222)$$

Since τ_n is process dependent, it may be different for the power-supply switch than for the logic circuits. Substitute Equation 222 into the first term of

Equation 221, and assume that the gate of switch A is driven conventionally. We must minimize this expression with respect to C_A:

$$E_{supply} = C_A V_{dd}^2 + \frac{\pi^2 \tau_n}{4} \frac{2}{T} \frac{C_L}{(m-2)} \cdot \frac{C_L}{C_A} C_L V_{dd}^2 \tag{223}$$

This expression is similar to Equation 184. The optimal value for C_A is:

$$C_{A_{opt}} = \sqrt{\frac{\pi^2 \tau_n}{2} \frac{1}{T} \frac{1}{(m-2)}} C_L \tag{224}$$

The total energy dissipation is given by:

$$E_{tot} = \left(\sqrt{2\pi^2 \frac{\tau_n}{T} \frac{1}{(m-2)}} + \frac{\pi^2 R_L C_L}{4} \frac{C_L}{T} \right) C_L V_{dd}^2 \tag{225}$$

When the charging time is increased, the last term of Equation 225 becomes negligible due to its T^{-1} dependence, and the $T^{-1/2}$ term dominates. We recognize the $\sqrt{1/T}$ dependence from Section 6.2.1. Again, the situation may be improved by driving the gate of the switch device adiabatically, which would allow us to use a wider switch. However, a very large gate capacitance also means that the drain of switch A (the right-hand side in Figure 6.15) has a large stray capacitance. When switch A is cut, this stray capacitance forms an RLC circuit with L and the load. Ringing will ensue due to the voltage present across L when switch A is cut. The ringing not only produces high voltages at the drain of switch A, but also causes dissipation proportional to the drain stray capacitance.

The ringing problem can be circumvented by removing the controlling switch device from the main charge path, as shown in Figure 6.16. The RLC circuit is now free-running, and the efficiency is limited mainly by the load resistance, but one degree of freedom has been lost, in that the supply voltage waveforms provided to the load are now close to sine waves. The energy dissipation per cycle is reduced if the charging time is chosen to be very long (by increasing L). Eventually, the limit on energy dissipation per operation will be set by the parasitic series resistances of the available and affordable inductors.

Most published papers on power supplies for adiabatic circuits describe variants on this theme [Hinman93][Younis93][Athas94b] [Athas94c] [Gabara94]. As an additional variation, only the timing of the switches must be changed to operate the circuit of Figure 6.16 as an edge-resonant circuit [Maksimovic91]

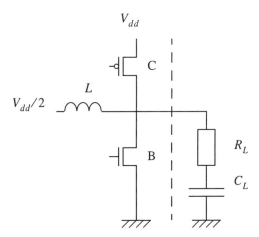

Figure 6.16: Inductive power supply like the one in Figure 6.15, but lacking switch A, which causes most of the dissipation.

[Wik94]. When the load is discharged, switch B is tied long enough to let a significant current build up in the inductor. When B is cut, the current keeps flowing and charges the load almost like a constant-current source. Switch C is tied when the load reaches V_{dd} and held closed for long enough for the current to reverse.

In another variation, if several symmetrical clock waveforms are used, C_T may be replaced by the "opposite" load capacitance with no significant change in efficiency, as shown in Figure 6.17. Such a circuit can only be used when the capacitances to be driven by the two phases are equal. Again, the switch in the main path of the charge may be removed for improved efficiency, but at the price of less controllability.

Other types of power supplies are conceivable. In addition to other kinds of inductive supplies, it would be possible to use a switched-capacitance circuit like those used as signal drivers in Section 6.4, since the load presented to the power supply is mainly capacitive; the latter approach would appear to give a rather low overall efficiency. It has also been speculated that high-Q circuits of other kinds, such as piezo-electric resonators, might become useful [Solomon94].

We finally note that until now, the power supplies required for adiabatic circuits have received considerably less attention than the logic circuits themselves. More work is clearly needed in this area for adiabatic switching to become useful at very small switching energies.

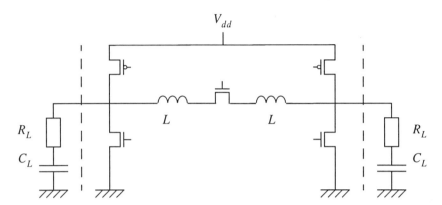

Figure 6.17: Symmetrical clock waveform generator; used to generate two clock waveforms with opposite phase. During the transition, charge flows from one load capacitance to the opposite one.

6.6 Summary

Most approaches to the reduction of dynamic power dissipation in electronic switching circuits start from the ubiquitous equation $P = fCV^2$. The threefold path to lower dissipation is indicated by the three factors in the expression: to reduce power, we must reduce the switching frequency, the driven capacitance, or the supply voltage, or (preferably) some combination of these. When, for whatever reason, these factors cannot be further reduced, adiabatic switching is the only known approach to further reduce dynamic dissipation.

Adiabatic circuits can clearly be used to great benefit when a small number of circuit nodes of significant capacitance must be driven to a high voltage (compared to the threshold voltage of the process). Such situations arise wherever the swing is fixed by some physical phenomenon. Examples include capacitive transducers, micromechanical devices, and LCD panels. In addition, voltage swing is sometimes determined by industry signalling standards. Then, adiabatic switching may allow a designer to reduce the interface part of the power budget by circumventing the $P = fCV^2$ barrier. The benefits increase with the voltage swing and with the switching time and are relatively easy to quantitatively assess.

Logic circuits of higher complexity than mere buffers may also benefit from adiabatic switching. The range of viable circuit styles is larger than in the driver case, and there is no consensus yet as to what schemes would be preferable in terms of circuit complexity and overall power dissipation. In part, this is since the complexity and efficiency of the required pulsed-power supplies is difficult to factor into the equations. While buffer circuits can use relatively simple power waveforms, many of the proposed logic schemes rely on intricate timing and level relationships between several supply voltages.

The design of the pulsed-power supplies may turn out to be the most critical element to the ultimate practicality of ultra-low-dissipation adiabatic circuits. The efficiency of most pulsed-power supplies presented so far is limited by the on-resistance of a small number of switches, through which the supply currents to the logic circuit passes. Also, many current low-power systems make heavy use of gated-clock design styles, where the clock signals of an idle block are stopped to avoid unnecessary logic transitions inside the circuit. The equivalent state in an adiabatic circuit would be one where the pulsed-power supply signals are held constant for an inactive block. A fine-grain control of the clock signals would be necessary to do this selectively for the different parts of a system. It is presently unclear how to efficiently achieve this.

In adiabatic switching circuits, the on-resistance of switches determines the dissipation but has no influence on the charging times. Broadly speaking, a process improvement (increased transconductance, reduced parasitic capacitances) that increases the possible speed in a conventional circuit allows lower dissipation in an adiabatic one. Thus, the qualitative comparisons described in this chapter should stay valid while semiconductor processes evolve.

The full impact of adiabatic switching on the design of low-power switching circuits is yet to be determined. Its utility in interface circuits seems evident, but it is at this point unclear whether superior overall dissipation can be achieved for switching circuits with more complex logic functions. New logic styles which offer substantially better power-delay products may still emerge.

6.7 Acknowledgments

This chapter is founded on the adiabatic-switching work carried out at the Information Sciences Institute (ISI) of the University of Southern California, together with W.C. Athas, J.G. Koller, N. Tzartzanis, and E. Chou. In addition, discussions with J.S. Denker, D.J. Frank, T. Gabara, J.S. Hall, T.F. Knight, S. Mattisson, L. Peterson, P.M. Solomon, and T.R. Wik have been very helpful.

The work on adiabatic switching at ISI is supported by ARPA under contract no. DABT63-92-C0052.

REFERENCES:

[Athas94a] W.C. Athas, L."J." Svensson, J.G. Koller, N. Tzartzanis, and E. Chou, "Low-power digital systems based on adiabatic-switching principles," *IEEE Transaction on VLSI Systems*, pp. 398–407, Dec 1994.

[Athas94b] W.C. Athas, J.G. Koller, and L."J." Svensson, "An energy-efficient CMOS line driver using adiabatic switching," *Fourth Great Lakes Symposium on VLSI Design*, pp. 196–199, March 1994.

[Athas94c] W.C. Athas and L."J." Svensson, "Reversible logic issues in adiabatic computing," *IEEE Workshop on Physics and Computation*, PhysComp '94, pp 111–118, IEEE Press, Nov 1994.

[Bennett73] C.H. Bennett, "Logical reversibility of computation," *IBM J. Res. Dev.*, vol 17, pp. 525–532 (1973).

[Bennett85] C.H. Bennett and R. Landauer, "The fundamental physical limits of computation," *Scientific American*, pp. 48–56, July 1985.

[Bennett89] C.H. Bennett, "Time/space trade-offs for reversible computation," *SIAM J. Computing*, vol. 18, pages 766–776, 1989.

[Burr91] J. B. Burr and A.M. Peterson, "Energy considerations in multichip-module based multiprocessors," *IEEE International Conference on Computer Design*, pp. 593–600, 1991.

[Denker94] J. S. Denker, "A review of adiabatic computing," *IEEE Symposium on Low Power Design*, pp 94–95, 1994.

[Gabara94] T. Gabara, "Pulsed power supply CMOS - PPS CMOS," *IEEE Symposium on Low Power Design*, pp 98–99, 1994.

[Hall92] J.S. Hall, "An electroid switching model for reversible computer architectures," *Proc. ICCI'92, 4th International Conf. on Computing and Information*, 1992.

[Hinman93] R. T. Hinman and M. F. Schlecht, "Recovered energy logic: a highly efficient alternative to today's logic circuits," *IEEE Power Electronics Specialists Conference Record*, pp. 17–26, 1993.

[Koller93] J.G. Koller and W.C. Athas, "Adiabatic switching, low energy computing, and the physics of storing and erasing information," *Workshop on Physics and Computation*, PhysCmp '92, Oct. 1992; IEEE Press, 1993.

[Kramer94] A. Kramer, J.S. Denker, S.C. Avery, A.G. Dickinson, and T.R. Wik, "Adiabatic computing with the 2N-2N2D logic family," *1994 Symposium on VLSI Circuits: Digest of technical papers*, IEEE Press, June 1994.

[Landauer61] R. Landauer, "Irreversibility and heat generation in the computing process," IBM J. Res. Dev., vol 5, pp. 183–191 (1961).

[Liu93] D. Liu and C. Svensson, "Trading speed for low power by choice of supply and threshold voltages," *IEEE JSSC*, vol 28, no. 1, Jan 1993.

[Maksimovic91] D. Maksimovic, "A MOS gate drive with resonant transitions," *IEEE Power Electronics Specialists Conference*, pages 527–532. IEEE Press, 1991.

[Mead80] C.A. Mead and L. Conway, *Introduction to VLSI systems*, Addison-Wesley, Reading, 1980.

[Mead] C. Mead, "Scaling of MOS technology to submicrometer feature sizes," *Journal of VLSI Signal Processing*, 8, 9–25 (1994).

[Merkle93] R. Merkle, "Reversible electronic logic using switches," *Nanotechnology*, vol. 4, pages 21–40, 1993.

[Seitz85] C. L. Seitz, A.H. Frey, S. Mattisson, S.D. Rabin, D.A. Speck, and J.L.A. van de Snepscheut, "Hot-clock NMOS," *1985 Chapel Hill Conference on VLSI,* pp. 1–17, 1985.

[Solomon94] P.M. Solomon and D.J. Frank, "The case for reversible computation," *1994 International Workshop on Low-Power Design.*

[Svensson94] L."J." Svensson and J.G. Koller, "Driving a capacitive load without dissipating fCV^2," *IEEE Symposium on Low Power Design,* pp. 100–101, 1994.

[Wik94] T.R. Wik, Private communication, October 1994.

[Younis93] S.G. Younis and T.F. Knight, "Practical implementation of charge recovery asymptotically zero power CMOS," *1993 Symposium on Integrated Systems*, pp. 234–250. MIT Press, 1993.

Minimizing Switched Capacitance

In the previous chapter, power dissipation was minimized in CMOS circuits by aggressive supply voltage scaling. Since CMOS circuits do not dissipate power if they are not switching, another approach to low power design is to reduce the switching activity to the minimal level required to perform the computation. This can range from simply powering down the complete circuit or portions of it, to more sophisticated schemes in which the clocks are gated or optimized circuit architectures are used which minimize the number of transitions. The focus of this chapter is on minimizing the switched capacitance at all levels of the design. The following sections describe a system level approach to minimize the switched capacitance which involves optimizing algorithms, architectures, logic design, circuit design, and physical design.

7.1 Algorithmic Optimization

The choice of algorithm is the most highly leveraged decision in meeting the power constraints. The ability for an algorithm to be parallelized will be critical and the basic complexity of the computation must be highly optimized.

7.1.1 Minimizing the Number of Operations

Minimizing the number of operations to perform a given function is critical to reducing the overall switching activity. To illustrate the power trade-offs that can be made at the algorithmic level, consider the problem of compressing a video data stream using the vector quantization algorithm. Vector quantization

(VQ) is a lossy compression technique which exploits the correlation that exists between neighboring samples and quantizes samples together rather than individually. Detailed description of vector quantization can be found in [Gersho92].

In this section, the focus will be on evaluating the computational complexity of encoding algorithms for VQ. Typically, the distortion metric used to compare the input image with the codebook entries is the mean square error. The basic calculation performed is:

$$D_i = \sum_{j=0}^{15} \left(X_j - C_{ij} \right)^2 \tag{226}$$

where X_j are the elements of the input vector (in this example, each vector contains 16 pixels) and C_{ij} are the elements of the codebook vector. Three VQ encoding algorithms will be evaluated: full search, tree search, and differential codebook tree-search. A codebook size of 256 is assumed.

7.1.1.1 Full Search Vector Quantization

Full search is a brute-force VQ in which the distortion between the input vector and every entry in the codebook is computed. Here the distortion as shown in Equation 226 is computed 256 times (for **C0** through **C255**) and the codeindex corresponding to minimum distortion is determined and sent over to the decoder. For each distortion computation, there are 16 memory accesses (to fetch the entries in the codeword), 16 subtractions, and 16 multiplications, and 15 additions. In addition to this, the minimum of 256 distortion values, which involves 255 comparison operations must be determined.

7.1.1.2 Tree-structured Vector Quantization

In order to reduce the computational complexity required by an exhaustive full-search vector quantization scheme, a binary tree-search is typically used. The basic idea is to perform a sequence of binary searches instead of one large search. As a result, the computational complexity increases as log N instead of N, where N is the number of nodes at the bottom of the tree. Figure 7.1 shows the structure for the tree search. At each level of the tree, the input vector is compared against two codebook entries. If for example at level 1, the input vector is closer to the left entry, then the right portion of the tree is never compared below level 2 and

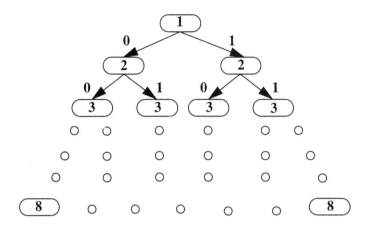

Figure 7.1: Tree structured Vector Quantization.

an index bit 0 is transmitted. This process is repeated till the leaf of the tree is reached. The TSVQ will in general have some degradation over the performance of the full search VQ due to the constraint on the search. However, this is typically not very noticeable [Gersho92] and the power reduction relative to the full search VQ is very significant. Here only 2 x $\log_2$ 256 = 16 distortion calculations have to be made compared to 256 distortion calculations in the full search VQ. The number of comparison operations is also reduced to 8.

7.1.1.3 Differential Codebook Tree-structure Vector Quantization

Another option is to use the same search pattern as the previous scheme but perform computational transformations to minimize the number of switching events [Fang92]. In the above scheme, at each level of the tree, the distortion difference between the left node and right node needs to be computed. This is summarized by the following equation:

$$D_{left-right} = \sum_{j=0}^{15} \left(X_j - C_{leftj} \right)^2 - \sum_{j=0}^{15} \left(X_j - C_{rightj} \right)^2 \qquad (227)$$

This equation can be manipulated to reduce the number of operations.

$$D_{left-right} = \sum_{j=0}^{15} \left(\left(X_j - C_{leftj} \right)^2 - \left(X_j - C_{rightj} \right)^2 \right) \qquad (228)$$

$$D_{left-right} = \sum_{j=0}^{15} \left(X_j^2 + C_{leftj}^2 - 2X_j C_{leftj} - X_j^2 - C_{rightj}^2 + 2X_j C_{rightj} \right)(229)$$

$$D_{left-right} = \sum_{j=0}^{15} \left(C_{leftj}^2 - C_{rightj}^2 \right) + \sum_{j=0}^{15} 2X_j \left(C_{rightj} - C_{leftj} \right) \qquad (230)$$

The first term in Equation 230 can be precomputed for each level and stored. By storing and accessing $2 \times (C_{rightj} - C_{leftj})$, the number of memory access operations can be reduced; that is, by changing the contents of the code-book through computational transformations, the number of switching events - number of multiplications, additions/subtractions and memory accesses- can be reduced. Table 7.1 shows a summary of the computational complexity per input vector (16 pixels). This example clearly demonstrates the dramatic reduction in computational complexity that can achieved through algorithmic optimizations.

Table 7.1 Computational complexity of VQ encoding algorithms.

Algorithm	# of Memory Accesses	# of Multiplications	# of Adds	# of Subs
Full Search	4096	4096	3840	4096
Tree Search	256	256	240	264
Differential Tree Search	136	128	128	0

7.1.2 Minimizing Temporal Bit Transition Activity by Choice of Data Representation

Communicating data bits in an appropriately coded form can reduce the switching activity. Coding has long been used in communication systems to control the statistics of the transmitted data symbols, or in other words to control the spectrum of the transmitted signal. The goal of coding has been to either remove undesired correlation among information bits (scrambling, e.g. for encryption) or to introduce controlled correlation (redundancy, e.g. for spectrum shaping of transmitted signal, timing recovery, error correction in unreliable channels). Coding for reduced switching activity falls under the second category - introducing sample to sample correlation such that the total number of bit transitions is reduced.

Consider the simple case where one chip needs to send n-bit data words to another chip, and where all the 2^n possible n-bit data words are possible as would be the case in general. The common way to do this inter-chip communication, without using any coding, is to transmit the data bits over a set of n wires. The idea behind using coding to reduce the switching activity is shown in Figure 7.2 where a reduction in the switching activity is obtained at the cost of extra hardware in the form of an encoder on the sender chip, a decoder on the receiver chip, and possibly larger number of wires m (where $m >= n$) between the two chips. A coding scheme with simple encoder and decoder, and a small number of wires m ($>= n$) is obviously desirable.

A coding scheme is non-redundant if $m = n$ and the 2^n elements of the set of n-bit data words will be mapped among themselves. It is obvious that one cannot devise a non-redundant coding scheme to reduce the switching activity without knowing the statistics of the sequence of words that is being communicated.

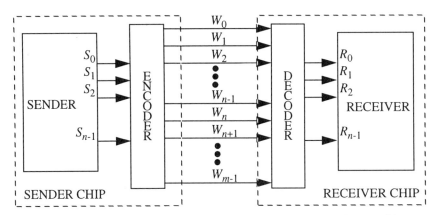

Figure 7.2: Data coding to reduce the switching activity on highly capacitive inter-chip I/O Lines. Note that $m >= n$

When $m > n$ the coding scheme has redundancy because the 2^n unique data words to be sent get mapped to a larger set of 2^m data words that are transmitted over the wires. This mapping may be done in many different ways. One simple way is to use a one-to-one mapping so that 2^m-2^n of the m-bit patterns are never transmitted. An advantage of such a scheme is that both the encoder and the decoder are memoryless. Another way to do the mapping is to use a static one-to-many mapping of the 2^n unique data words to be sent to the 2^m data words that are transmitted. The current state is then used to select a specific m-bit data word to transmit out of the various choices corresponding to a n-bit data word that needs to be sent. This scheme would require an encoder with memory, but the decoder is still memoryless. More elaborate schemes where both the encoder and decoder have memory are also possible.

7.1.2.1 Example #1: One-Hot Coding

Of the many coding schemes that exist in the literature, one which has relevance to switching activity reduction is the *one-hot* coding which is sometimes used in state assignment of finite state machines. In the context of the inter-chip communication problem, one-hot coding is a redundant coding scheme with a one-to-one mapping between the n-bit data words to be sent and the m-bit data words that are transmitted. The two chips are connected using $m = 2^n$ wires, and a n-bit data word is encoded for transmission by placing a '1' on the i-th wire where $0 <= i <= 2^n$-1 is the binary value corresponding to the bit pattern, and '0' on the remaining m-1 wires. Obviously both the encoder and the decoder are memoryless.

One-hot encoding requires an exponential number (2^n) of wires, but guarantees that precisely one 0→1 and one 1→0 bit transition occurs when a different data word is sent. Assuming the n-bit data words to be sent are independent and uniformly randomly distributed, the reduction in dynamic power consumption due to one-hot coding will be:

$$\frac{P_{\text{one-hot}}}{P_{\text{no coding}}} = \frac{\alpha_{\text{one-hot}}CV^2f}{\alpha_{\text{no coding}}CV^2f} = \frac{\alpha_{\text{one-hot}}}{\alpha_{\text{no coding}}} = \begin{cases} 1 & \text{for } n = 1 \\ \frac{4(1-2^{-n})}{n} & \text{for } n \geq 2 \end{cases}$$

$$(231)$$

While the one-hot coding gives a large reduction in switching activity, the exponential number of wires required often makes it an impractical scheme. For example, if $n=8$, the one-hot coding gives a 75% reduction in switching activity but requires $m=256$ wires.

7.1.2.2 Example #2: Gray-coding

Another encoding strategy which has been used for low-power is gray-coding [Su94]. A gray code sequence is a set of numbers in which adjacent numbers only have one bit difference. Gray coding is most useful when the data being transmitted over a bus is sequential and highly correlated. In this case, the number transitions for binary representation over gray-coding approaches 2 since for large values of n, the number of transitions for binary representation will approach 2 while gray-code will always have one transition. Table 7.2 shows the binary representation and gray-code representation for decimal numbers 0 through 15.

Table 7.2 Binary and Gray-code representation.

Decimal Value	Binary Code	Gray Code
0	0000	0000
1	0001	0001
2	0010	0011
3	0011	0010
4	0100	0110
5	0101	0111
6	0110	0101
7	0111	0100
8	1000	1100
9	1001	1101
10	1010	1111
11	1011	1110

Table 7.2 Binary and Gray-code representation.

Decimal Value	Binary Code	Gray Code
12	1100	1010
13	1101	1011
14	1110	1001
15	1111	1000

The conversion from binary code to gray code representation is defined as follows. Let $B = <b_{n-1}, b_{n-2},...,b_1,b_0>$ be the binary representation and $G = <g_{n-1},g_{n-2},...,g_1,g_0>$ be the gray code representation for the number. Note that gray code is a non-redundant representation.

$$g_{n-1} = b_{n-1} \qquad\qquad (232)$$

$$g_i = b_{i+1} \oplus b_i \, (i=n\text{-}2,0) \qquad\qquad (233)$$

For example, a binary code representation $B= <1,1,0,1>$ is equivalent to a gray code representation $<b_3, b_3 \oplus b_2, b_2 \oplus b_1, b_1 \oplus b_0>$ which is equal to $<1,0,1,1>$.

The conversion from gray code to binary also uses XOR operation, but is slightly more complex. For the MSB, once again:

$$b_{n-1} = g_{n-1} \qquad\qquad (234)$$

For the rest of the bits, b_i is equal to the XOR of g_i and all the bits of G preceding g_i.

$$b_i = b_{i+1} \oplus g_i \, (i=n\text{-}2,0) \qquad\qquad (235)$$

For example, a gray code representation $G = <1,1,0,1>$ is equivalent to a binary representation $<g_3, g_3 \oplus g_2, g_3 \oplus g_2 \oplus g_1, g_3 \oplus g_2 \oplus g_1 \oplus g_0>$ which is equal to $<1,0,0,1>$.

Gray coding has been applied to code the address lines for both instruction access and data access to reduce the number of transitions [Su94]. For instruction accesses, it was found that gray coding significantly reduced the switching activity since the temporal transitions were typically sequential. When a branch occurs, the temporal transition is not sequential and therefore there will be more than one bit transition for the gray coding approach. For data access, it was found that the transitions were approximately equal for both representations since the data accesses were not as sequential. For random data patterns, it was found that binary and gray coding have approximately equal transition activity. Figures 7.3 and 7.4 show the reduction in switching activity for instruction address coding and data address coding for a set of benchmark programs. *BPI* is the number of bit transitions per instruction executed.

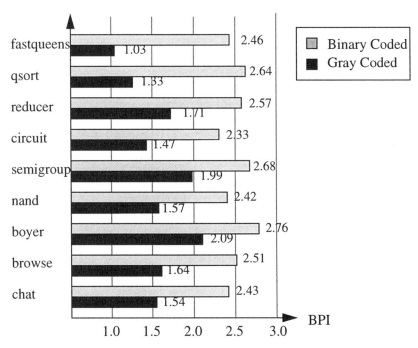

Figure 7.3: Temporal transition activity comparison for instruction addresses [Su94]. (© 1994 IEEE)

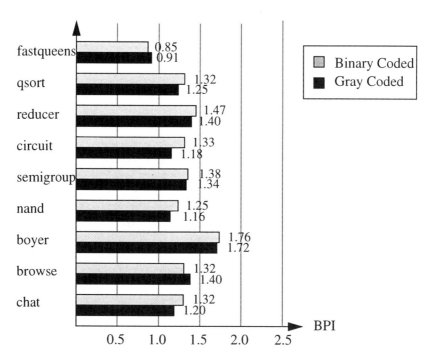

Figure 7.4: Temporal transition activity comparison for data addresses [Su94].
(© 1994 IEEE)

7.1.2.3 Example #3: Bus Inversion Coding

Another coding scheme, proposed in [Stan94], is a redundant coding scheme with a one-to-many mapping between the n-bit data words that need to be sent and the m-bit (where $m=n+1$) data words that are transmitted:

If the i-th data word to be transmitted over the wires is S_i, then either S_i or $\overline{S}_i$ is transmitted depending on which would result in fewer number of bit transitions. Note that $\overline{S}_i$ is the bit-wise inverse of S_i. Further, to tell the receiver as to what is being transmitted - S_i or $\overline{S}_i$ - an extra wire P is used to carry this information. This extra wire thus encodes the polarity of the data word. The possibility of a transition on the polarity wire is taken into account when deciding whether to transmit S_i or $\overline{S}_i$.

The scheme is simple: the decoder is memoryless, the encoder only uses the current state of the wires for its memory, and both the decoder and the encoder can be implemented with little area and power overhead. The number of extra wires required is one (i.e., $m=n+1$), which is the smallest possible number of any redundant coding scheme.

Analysis shows that for uniformly distributed n-bit data words the reduction in dynamic power consumption due to the above coding scheme is:

$$\frac{P_{coding}}{P_{no\ coding}} = \frac{\alpha_{coding}CV^2f}{\alpha_{no\ coding}CV^2f} = \frac{\alpha_{coding}}{\alpha_{no\ coding}} = \left(1+\frac{1}{n}\right)\left(1-\frac{{}^nC_{\lceil n/2\rceil}}{2^n}\right) \quad (236)$$

Figure 7.5 is a plot of the above ratio as a function of the number of bits n. As the plot shows, the maximum reduction in switching activity is obtained for $n=2$ for which the reduction is 25%. Even for 32 bit busses, a reduction of 11% is obtained by using the coding scheme at the cost of one extra wire. It is easy to see from the analytic expression for $\alpha_{coding}/\alpha_{no\ coding}$ as well as from the plot in Fig-

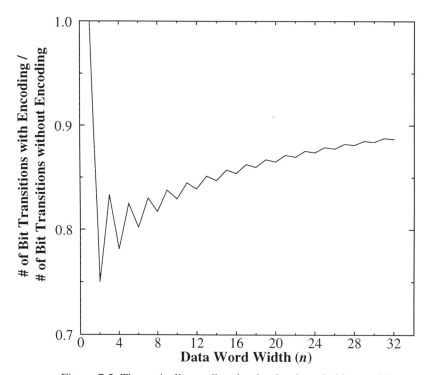

Figure 7.5: Theoretically predicted reduction in switching activity for a stream of uniformly distributed data words.

ure 7.5 that the efficacy of the coding technique disappears for large values of n, as the ratio $\alpha_{coding}/\alpha_{no\ coding}$ converges to 1.

Since the coding technique works better for smaller values of n, one can use the following extension which uses a larger number of extra wires to obtain larger reductions in switching activity: divide the n-bit data bus into smaller groups (of 2 or more bits) and code each group independently by associating a polarity bit with each of the group. For example, a 32-bit bus can be divided into 4 groups of 8-bits which are then coded independently using a total of 4 extra wires, thus giving 18.3% reduction in switching activity as opposed to the 11.3% reduction that is obtained by coding the entire bus together using only one extra wire. In the extreme, by using 16 extra wires, and coding the original 32 bits in 16 groups of 2 bits, one can get a reduction in switching activity of 25%. Since groups have at least 2 bits (coding 1 bit is useless), this extended approach allows one to make use of up to $n/2$ extra wires.

7.1.2.4 Example #4: Two's Complement vs. Sign Magnitude

Another choice of data representation is two's complement vs. sign magnitude. In most signal processing applications, two's complement is typically chosen to represent numbers since arithmetic operations (addition and subtraction) are easy to perform. One of the problems with two's complement representation is sign-extension, which causes the MSB sign-bits to switch when a signal transitions from positive to negative or vice-versa (for example, going from -1 to 0 will result in all of the bits toggling). Therefore using a two's complement representation can result in significant switching activity when the signals being processed switch frequently around zero and when they do not utilize the entire bit-width (i.e., the dynamic range is much smaller than the maximum possible value determined from the bit-width) since a lot of the MSB bits will perform sign-extension. Even if a signal utilizes the entire bit-width, arithmetic operations such as scaling can reduce the signal dynamic range.

One approach to minimizing the switching in the MSB's is to use a sign-magnitude representation, in which only one bit is allocated for the sign and the rest for the magnitude. In this case, if the dynamic range of a signal does not span the entire bitwidth, only one bit will toggle when the signal switches sign, as opposed to the two's complement representation where due to sign extension several of the bits will switch. To illustrate this, consider gaussian data applied to a 16-bit data-bus, and the let the signal have a mean of 0 and a $3\sigma = 2^{11}$. Figure

7.6a shows the transition probabilities vs. bit position number for two's complement representation as a function of the first order correlation coefficient ρ =
$cov(X_n,X_{n+1})/\sigma^2$. A ρ = 0 indicates that the data is uncorrelated and therefore the
transition probability for the MSB's is 1/2 (i.e., there is a 50% chance that the output will change sign). A large positive correlation coefficient (e.g. +0.99) implies
that the signal changes very slowly and therefore switches sign very infrequently
(i.e., the transition probability is close to zero). Similarly, a negative correlation
coefficient with a large magnitude (e.g. -0.99) implies that the signal changes frequently from positive to negative.

The upper breakpoint (which represents the dynamic range of signal) lies at
$\log_2(3\sigma)$ = 11bits in this case. Therefore, for the two's complement representation, the bits from 11 to 15 indicate the activity of the sign-bit. If a sign-magnitude representation is used (shown in Figure 7.6b), the bits 11-14 will have a low
transition activity, and bit 15, whose transitions represent the sign transition probability, will have the same activity as the two's complement representation. Also,
the transition region has lower activity for the sign-magnitude representation.
Clearly, there is an advantage in using sign-magnitude representation over two's
complement to reduce the switching activity.

From the above example, it is clear that the reduction in the number of
transitions for sign-magnitude over two's complement is a function of both the
dynamic range of the signal and the signal correlation coefficient. Consider evaluating the reduction in the number of transitions as a function of dynamic range
and the correlation coefficient. Let the normalized dynamic range of a signal be
defined as 3σ / Maximum Amplitude. For a 16 bit example, this turns out to be:
$3\sigma / 2^{16}$-1. Let the number of transitions represent the sum of the transitions per
clock cycle for all the bits,

$$Number\ of\ Transitions\ =\ \sum_{i=1}^{16} \alpha_i \qquad (237)$$

Figure 7.7 shows a plot of the total number of transition as a function of the
normalized dynamic range and the correlation factor for two's complement representation. For a correlation factor of ρ = 0, the number of transitions is equal to 8
and is independent of the normalized dynamic range since the data is random and
each bit has a 50% probability of transitioning (=> 8 transitions for a 16-bit bus).
For very correlated data (e.g. ρ = 0.99), the MSB's don't switch very often while

the LSB's switch 50% of the time. Therefore, for a small normalized dynamic range, the average number of transitions is very small and for a large dynamic range it approaches the value of 8 (since the lower region with activity of 50% extends out to more bits). Similarly, for very anti-correlated data (e.g. $\rho = -0.99$), the MSB switch very frequently. Therefore, a small normalized dynamic range will result in a lot of transitions (since many bits are allocated to sign-extension) while the number of transition approaches 8 for high values of the normalized dynamic range.

Figure 7.8 shows the ratio of the number of transitions required in the two's complement representation to the number of transitions in the sign-magnitude representation as a functions of signal correlation and normalized dynamic range (plotted on a log axis). From this plot, it is clear that the biggest win for sign-magnitude representation is when the dynamic range is smallest and the signal is very anti-correlated.

The above analysis suggests that sign-magnitude has some advantages in terms of the number of the transitions on busses. However, addition and subtraction computation are difficult to implement in sign-magnitude representation. Sign-magnitude is therefore most useful for cases where large busses have to be driven (for example external memory access), where the overhead for converting back to two's complement is quite insignificant compared to the reduction in capacitance switched in the large busses. That is, a small overhead capacitance is added to the system to reduce the *overall* switched capacitance.

Two's Complement Representation

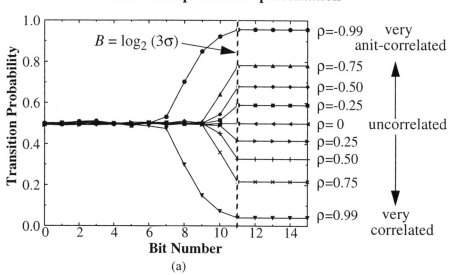

(a)

Sign-magnitude Representation

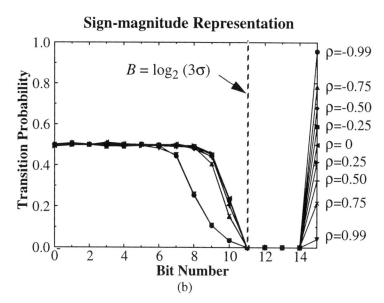

(b)

Figure 7.6: Transition activity for different number representations.

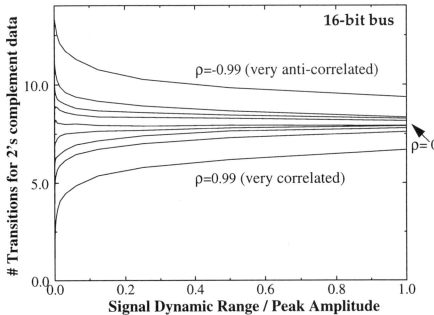

Figure 7.7: Transition activity for two's complement representation.

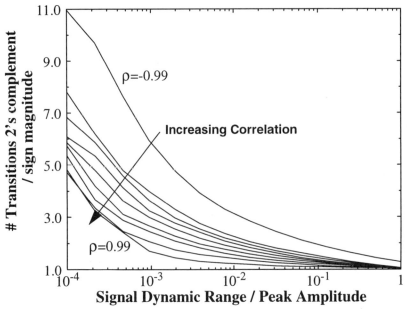

Figure 7.8: Activity reduction for sign-magnitude over two's complement.

7.2 Architecture Optimization

Architecture optimization can also be used to significantly reduce the switching activity by optimizing the number representation, optimizing the ordering of operations, optimizing resource utilization, and minimizing glitching activity.

7.2.1 Optimizing Data Representation for Arithmetic Computation

As mentioned in the previous section, sign-magnitude has some advantages in terms of reducing the number of the transitions on busses, but addition/subtraction operations are difficult to implement. However, in several applications that require multiple additions inside a sample period (e.g. multiply-accumulate units of DSP filters), an architecture optimization can be used that trades silicon area for lower switching activity without altering the throughput. To illustrate this consider a correlator example in which the correlation length is 1024; the samples, whose values range from -7 to +7, are accumulated at 64MHz.

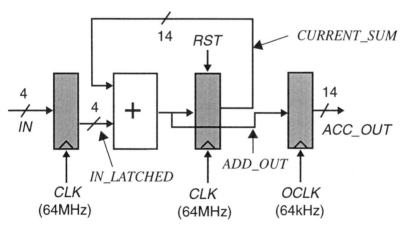

Figure 7.9: Two's complement implementation of an accumulator.

Figure 7.9 shows a conventional architecture for an accumulator that uses two's complement representation. The accumulated result (adding 1024, 4-bit numbers) is transferred to an accumulator register at 64kHz. In this architecture, the MSB of the input, bit 3 (assuming that the LSB is bit 0), is tied to bits 4-13 of the adder input for sign-extension and therefore anytime the input switches sign, the MSB bit (which indicates the sign) will switch, resulting in all of the higher order input bits to the adder switching.

Figure 7.10 shows the transition probabilities for three signals in the two's complement datapath assuming uniformly distributed inputs. For this distribution, the input is equally likely to be positive or negative, and therefore the sign-extension bits 3-13 have a transition probability close to 1/2. Since the accumulator acts as a low-pass filter (i.e., $CURRENT\_SUM_N = CURRENT\_SUM_{N-1} + IN_N$), the higher order bits have little switching activity even when the input is rapidly varying; that is, the accumulator smoothes the input signal. This is shown in Figure 7.11 which shows the $CURRENT\_SUM$ output and the input value for 1024 samples (here the input is random and varying from -7 to +7). Although the $CURRENT\_SUM$ has low switching activity, the adder output (before the latch) has significant switching activity due to glitching, as seen from Figure 7.10. The glitching activity arises since all of the input bits to the adder switch each time the input changes sign, and this results in high switching activity of the adder (even though the final adder output at the end of the cycle does not change relative to its value at the beginning of the cycle).

Another approach for implementing the accumulator is to use a sign-magnitude representation whose datapath is shown in Figure 7.12. Here two accumu-

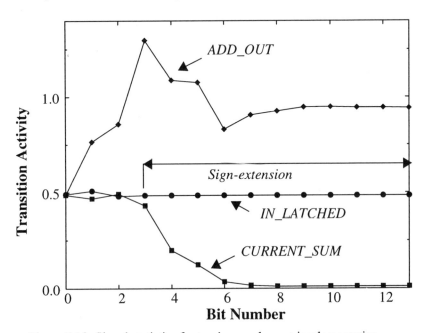

Figure 7.10: Signal statistics for two's complement implementation
of the accumulator datapath assuming random inputs.

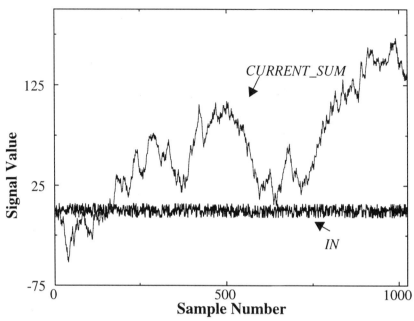

Figure 7.11: Signal Value for two's complement implementation for random inputs.

lator datapaths are used, one that sums all positive numbers and one that sums all
negative numbers. The latched sign-bit from the input register is used to generate
gated clocks that enables the positive datapath latch or the negative datapath
latch; this ensures that this scheme does not increase effective clock load. At the
end of 1024 cycles, the positive and negative accumulated values are transferred
to separate registers. A subtract operation is then performed at the lower frequency and therefore is quite negligible in terms of capacitance overhead. The
key to low-power is that there is no sign-extension is being performed and therefore the adder has a low switching activity in the higher order bits. In fact, the
higher order bits only need an incrementer (as opposed to a full-adder required by
the two's complement implementation), reducing the number of gates that are
switched in the accumulator. Figure 7.13 shows the transition probabilities for the
output of the adders in both datapaths. Also shown in the figure is the transition
activity for the two's complement implementation and the sum of the transition
activities for the sign-magnitude implementation; we can see that the glitching
activity is significantly reduced in the sign-magnitude implementation. The
sign-magnitude implementation, however, requires control circuitry (overhead

capacitance) to generate the timing signals for the various latches in the implementation.

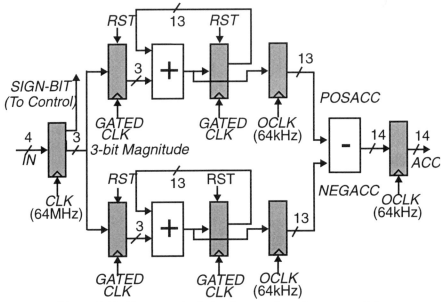

Figure 7.12: Sign magnitude implementation of an accumulator.

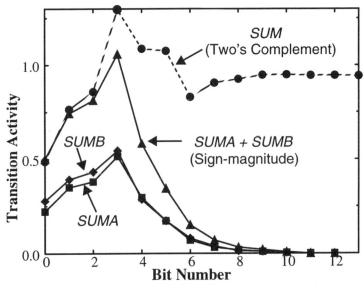

Figure 7.13: Signal statistics for Sign Magnitude implementation
of the accumulator datapath assuming random inputs.

By keeping the computation for positive data and negative data separate, the power is not very sensitive to rapid fluctuations in the input data. Table 7.3 shows the power estimates for various input patterns. Again, the biggest advantage is when the sign toggles frequently. For the case when the input changes very slowly, the sign-magnitude implementation consumes more power (15%) due to the capacitance switched by the control overhead circuitry.

Table 7.3 Number representation trade-off for arithmetic.

Input Pattern (1024 cycles)	Two's Complement Power, 3V	Sign Magnitude Power, 3V
Constant (IN= 7)	1.97 mW	2.25 mW
Ramp (-7,-6,...7,-7)	2.13 mW	2.43 mW
Random	3.42 mW	2.51 mW
Min -> Max -> Min (-7, +7,-7,+7,...)	5.28 mW	2.46 mW

7.2.2 Ordering of Input Signals

The switching activity can be reduced by optimizing the ordering of operations in a design. To illustrate this, consider the problem of multiplying a signal with a constant coefficient, which is a very common operation in signal processing applications. Multiplications with constant coefficients are often optimized by decomposing the multiplication into shift-add operations and using the canonical sign digit representation. Consider the example in which a multiplication with a constant is decomposed into $IN + IN >> 7 + IN >> 8$. The shift operations represent a scaling operation, which has the effect of reducing the dynamic range of the signal. This can be seen from Figure 7.14 which shows the transition probability for the 3 signals IN, $IN >> 7$ and $IN >> 8$. In this example, IN has a large variance and almost occupies the entire bit-width. The shifted signals are scaled and have a lower dynamic range.

Consider the two alternate topologies for implementing the two required additions. In the first implementation, IN and $IN >> 7$ are added in the first adder and the sum (SUM_1) is added to $IN >> 8$ in the second adder. In this case, the

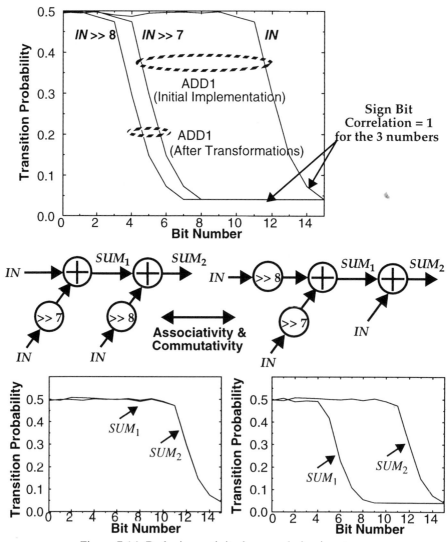

Figure 7.14: Reducing activity by re-ordering inputs.

SUM_1 transition characteristics are very similar to the characteristics of the input *IN* since the amplitude of *IN* >>7 is much smaller than *IN* and the 2 inputs have identical sign-bits (since a shift operation does not change the sign). Similarly SUM_2 is very similar to SUM_1 since SUM_1 is much larger in magnitude than *IN* >>8. In the second implementation (obtained by applying associativity and com-mutativity), the two small number *IN*>>7 and IN>>8 are summed in the first adder and the output is added to *IN* in the second adder. In this case, the output of the first adder has a small amplitude (since we are adding 2 scaled numbers of the same sign) and therefore lower switching activity. The second implementation switched 30% less capacitance than the first implementation. This example demonstrates that the ordering of operations can result in reduced switching activity.

7.2.3 Reducing Glitching Activity

As described in Chapter 3, static designs can exhibit spurious transitions due to finite propagation delays from one logic block to the next (also called critical races and dynamic hazards); i.e., a node can have multiple transitions in a single clock cycle before settling to the correct logic level. To minimize the "extra" transitions and power in a design, it is important to balance all signal paths and reduce the logic depth. For example, consider the two implementations for adding four numbers shown in Figure 7.15 (assuming a cascaded or non-pipelined implementation). Assume that all primary inputs arrive at the same time. Since there is a finite propagation delay through the first adder for the chained case, the second adder is computing with the new *C* input and the previous output of *A* + *B*. When the correct value of *A* + *B* finally propagates, the second adder recomputes the sum. Similarly, the third adder computes three times per cycle. In the tree implementation, however, the signal paths are more balanced and the amount of extra transitions is reduced. The capacitance switched for a chained implementation is a factor of 1.5 larger than the tree implementation for a four input addition and 2.5 larger for an eight input addition. The above simulations were done on layouts generated by the LagerIV silicon compiler using the IRSIM switch-level simulator over 1000 uncorrelated random input patterns.

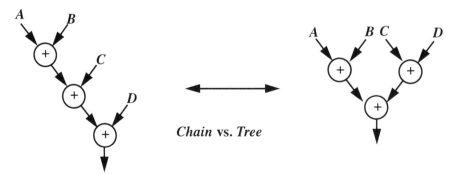

Chain vs. Tree

Figure 7.15: Reducing the glitching activity.

The results presented above indicate that increasing the logic depth (through more cascading) will increase the capacitance due to glitching while reducing the logic depth will increase register power. Hence the decision to increase or decrease logic depth is based on a trade-off between glitching capacitance vs. register capacitance. Also note that reducing the logic depth can reduce the supply voltage while keeping throughput fixed.

7.2.4 Degree of Resource Sharing

A signal processing algorithm has a required number of operations that have to be performed within the given sample period. One strategy for implementing signal processing algorithms is the direct mapping approach, where there is a one to one correspondence between the operations on the signal flow graph and operators in the final implementation. Such an architecture style is conceptually simple and requires a small or no controller. Often, however, due to area constraints or if very high-throughput is not the goal (e.g. speech filtering), time-multiplexed architectures are utilized in which multiple operations on a signal flowgraph can be mapped onto the same functional hardware unit.

Given that there is a choice between time-multiplexed and fully-parallel architectures, as is the case in low to medium throughput applications, an important question which arises is what architecture will result in the lowest switching activity. To first order, it would seem that the degree of time-multiplexing would not affect the capacitance switched by the logic elements or interconnect. For example, if a data flowgraph has five additions to be performed inside the sample period, it would seem that there is no difference between an implementation in which one physical adder performs all five additions or an implementation in which there are five adders each performing one addition per sample period. It would seem that the capacitance switched should only be proportional to the number of times the additions were performed but this is not the case since resource sharing can destroy signal correlations and increase switching activity. To demonstrate this trade-off, consider two examples of resource sharing: sharing busses and sharing execution units.

7.2.4.1 Example 1: Time-sharing Busses (Output of two counters)

Consider an example of two counters whose outputs are sent over parallel and time-multiplexing busses as shown in Figure 7.16. In case 1, we have two separate busses running at some frequency f while case 2 has one bus running twice as fast. Therefore for a fixed voltage, the power consumption for the parallel implementation is given by:

$$P_{no\text{-}bussharing} = 1/2\left(\Sigma\alpha_i\right) C_{nbs} \, V^2 f + 1/2\left(\Sigma\alpha_i\right) C_{nbs} \, V^2 f$$
$$= \left(\Sigma\alpha_i\right) C_{nbs} \, V^2 f \qquad (238)$$

where C_{nbs} is the physical capacitance per bit of BUS1 and BUS2 (assume for simplicity that all bits have the same physical capacitance) and α_i is the transition activity for bit i (for both 0->1 and 1->0 transitions and therefore there is a factor of 1/2 in Equation 238). In this case, since the output is coming from a counter, the activity is easy to compute. $\Sigma\alpha_i = 1 + 1/2 + 1/4 + \ldots + 1/128$ since the LSB switches every cycle, the 2nd LSB switches every other cycle, etc. This means that each bus will have on the average a total of approximately two transitions per clock cycle.

For the time-multiplexed implementation, there is a single bus, C_{bs}, which is switched at twice the frequency. Typically C_{bs} will be smaller than C_{nbs}. The goal here is to show that the activity α' is modified due to time-multiplexing and the power consumption for this implementation is given by:

$$P_{bussharing} = 1/2 \, \Sigma\alpha' \, C_{bs} \, V^2 \, 2f = \Sigma\alpha' \, C_{bs} \, V^2 f \qquad (239)$$

Figure 7.16 shows the plot of total number of transitions per cycle (i.e. $\Sigma\alpha'$) for the time-shared case as a function of the skew between the counter outputs. The figure also shows the number of transitions for the non-time-shared implementation which is independent of skew and is equal to 4 (for 2 busses). As seen from this figure, except for one value of the counter skew (when the skew is $= 0$), the "parallel" implementation has the lower switching activity. This example shows that time-sharing can significantly modify the signal characteristics and cause an increase in switching activity.

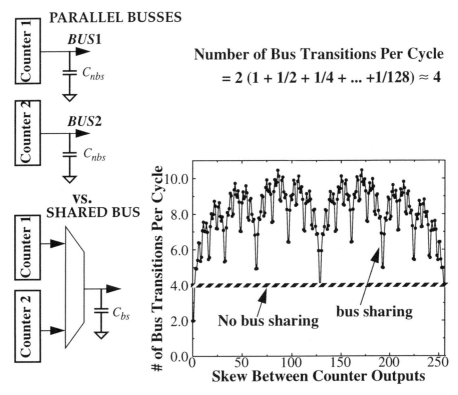

Figure 7.16: Activity trade-off for time-multiplexed hardware:
bus-sharing example.

7.2.4.2 Example 2: Time-sharing Execution Units

The second example is a simple second order FIR filter which will demonstrate that time-multiplexing of execution units can increase the switched capacitance. The FIR filter is described as follows:

$$Y = a_0 \bullet X + a_1 \bullet X@1 + a_2 \bullet X@2 = A + B + C \tag{240}$$

where @ represents the delay operator and $A = a_0 \bullet X$, $B = a_1 \bullet X@1$ and $C = a_2 \bullet X@2$. Also let $O1 = A + B$. The value of the coefficients are: $a_0 = 0.15625$, $a_1 = 0.015625$, and $a_2 = -0.046875$.

One possible implementation might be to have two physical adders, one performing $A + B = O1$ and the other performing $O1 + C$. An alternate implementation might be to have a single time-multiplexed adder performing both additions. So, in cycle 1, $A + B$ is performed, and in cycle 2, $O1 + C$ is performed. The topologies for the two cases are shown in Figure 7.17. In the first implementation, the adders can be chained and therefore the effective critical path can be reduced (chaining two ripple carry adders of N bits has a delay = $(N + 1)$ • Delay of one bit); it is clear that this will result in extra glitching activity but since the objective here is to isolate and illustrate the activity modification only due to time-multiplexing, the first implementation is assumed to be registered. The transition activities (obtained using IRSIM assuming speech input data) for the two adders for the parallel implementation and the average transition activity per addition for the time-multiplexed implementation are shown in Figure 7.17.

The results once again indicate that time-multiplexing can increase the overall switching activity. The basic idea is that in the fully-parallel implementation, the inputs to the adders change only once every sample period and the source of inputs to the adders are fixed (for example, $IN1$ of adder1 always comes from the output of the multiplier $a_0 * X$). Therefore, if the input changes very slowly (i.e the input data is very correlated), the activity on the adders become very low. However, in the time multiplexed implementation, the inputs to the adder change twice during the sample period and more importantly arrive from different sources. For example, during the first cycle, $IN2$ of the time-multiplexed adder is set to B and during the second cycle, $IN2$ of the time-multiplexed adder is set to C. Thus, even if the input is constant (which implies that the sign of X, $X@1$ and $X@2$ is the same) or slowly varying, the sign of B and C will be different since the sign of the coefficients a_1 and a_2 are different. This will result in the input to adder switching more often and will therefore result in higher switching activity. That is, even if the input to the filter does not change, the adder can still be switching. For this example, the time multiplexed adder switched 50% extra capacitance per addition compared to the parallel case (even without including the input changes to adders or the multiplexor overhead).

7.3 Logic Optimization

7.3.1 Logic Minimization and Technology Mapping

Changing the optimization criteria for synthesizing logic structures (both combinational and sequential) to one that addresses low power can yield reductions up to 25%. Techniques have been proposed which choose logic to minimize switching [Shen92], position registers through re-timing to reduce glitching activ-

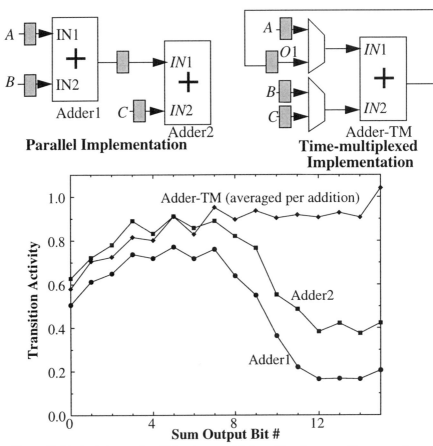

Figure 7.17: Activity trade-off for time-multiplexed hardware: adder example.

ity [Monteiro93] and to decrease power during the technology mapping phase [Tiwari93], and choosing gates from a library which reduces switching [Tsui93] - for example a three input AND gate can be implemented as a single 3-input AND gate or two 2-input AND gates with different power results.

7.3.2 Activity Trade-off for Various Logic Structures

It has been shown in Chapter 3 that the choice of logic topology can have a strong influence on the total transition activity, which will directly impact the switching activity. A logic function such as an adder can be implemented using different approaches. For example, an adder can be implemented using several topologies including ripple-carry, carry-select, carry-lookahead, conditional sum, and carry skip. Various circuit topologies for adders and multipliers have been investigated for impact on the transition activity [Callaway92]. Table 7.4 summarizes the results of the number of transitions per addition for different topologies assuming a randomly distributed input pattern.

Table 7.4 Average number of gate transitions per addition [Callaway92].

Adder Type	16 bits	32 bits	64 bits
Ripple Carry	90	182	366
Carry Lookahead	100	202	405
Carry Skip	108	220	437
Carry Select	161	344	711
Conditional Sum	218	543	1323

Though these simulation results were not obtained from layout and thus the capacitance switched is not known, it gives insight into the switching activity for various topologies. They used a metric of 1/(Number of Transitions * Delay) to evaluate the various adders and concluded that a lookahead topology was the best based on this metric. Figure 7.18 shows the probability distributions of the number of transitions for various adders obtained from gate level simulations.

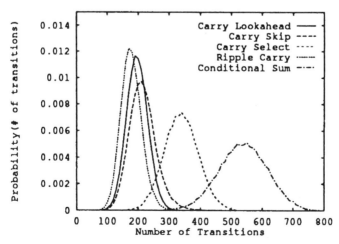

Figure 7.18: Histogram of transition activity for various 32-bit adder topologies [Callaway92]. (© 1992 IEEE)

7.3.3 Logic Level Power Down

Powering down has traditionally been applied only at the chip and module levels, however, the application to the logic level can also be very beneficial to reduce the switching activity at the expense of some additional control circuitry [Alidina94]. Assume a pipelined system for comparing the output of two numbers from a block of combinational logic as shown in Figure 7.19; the first pipeline stage is a combinational block and the next pipeline stage is a comparator which performs the function $A > B$, where A and B are generated in the first stage (i.e., from the combinational block). If the most significant bits, $A[N\text{-}1]$ and $B[N\text{-}1]$, are different then the computation of $A{>}B$ can be performed strictly from the MSB's and therefore the comparator logic for bits $A[N\text{-}1{:}0]$ and $B[N\text{-}2{:}0]$ is not required (and hence the logic can be powered down). If the data is assumed to be random (i.e there is a 50% chance that $A[N\text{-}1]$ and $B[N\text{-}1]$ are different), the power savings can be quite significant. One approach to accomplish this is to gate the clocks as shown in Figure 7.19.

The XNOR output of the $A[N\text{-}1]$ and $B[N\text{-}1]$ is latched by a special register to generate a gated clock. This gated clock is then used to clock the lower order registers. Since the latch to the lower bits is conditional, they must be made static. The standard TSPC register [Yuan89] is shown in Figure 7.20 as modified to sup-

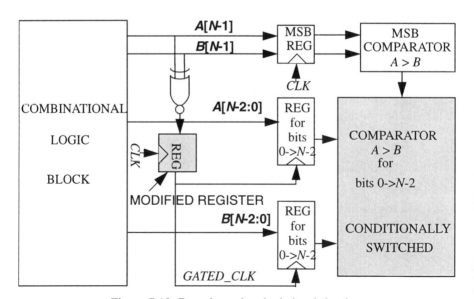

Figure 7.19: Data dependent logic level shutdown.

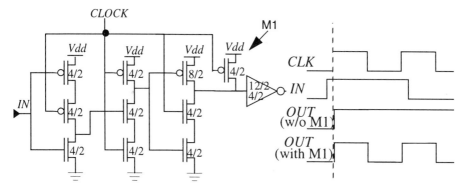

Figure 7.20: Schematic of a modified TSPC latch that is used to generate gated clocks.

port clock gating. As shown in the timing on the right side of Figure 7.20, with the extra PMOS device M1, the output can be forced to a ZERO during the low phase of the clock. Without this device, it will not be possible to generate rising edges for the gated clock on two consecutive rising edges of the system clock. Gated clocks have been used extensively in the system presented in Chapter 9.

7.4 Circuit Optimization

There are a number of options available in choosing the basic circuit approach and topology for implementing various logic and arithmetic functions. Choices between static vs. dynamic implementations, passgate vs. conventional CMOS logic styles, and synchronous vs. asynchronous timing are just some of the options open to the system designer. At another level, as mentioned previously, there are also various structural choices for implementing a given logic function; for example, to implement an adder module one can utilize a ripple-carry, carry-select, or carry-lookahead topology. In this section, the trade-offs with respect to low-power design between a selected set of circuit approaches will be discussed.

7.4.1 Dynamic Logic vs. Static Logic

The choice of using static or dynamic logic is dependent on many criteria other than just its low power performance, e.g. testability and ease of design. However, if only the low power performance is analyzed it would appear that dynamic logic has some inherent advantages in a number of areas including reduced switching activity due to hazards, elimination of short-circuit dissipation and reduced parasitic node capacitances. Static logic has advantages since there is no pre-charge operation and charge-sharing does not exist. Below, each of these considerations will be discussed in more detail.

7.4.1.1 Spurious Transitions

As mentioned earlier, static designs can exhibit spurious transitions due to finite propagation delays from one logic block to the next; i.e., a node can have multiple transitions in a single clock cycle before settling to the correct logic level. These spurious transitions dissipate extra power over that strictly required to perform the computation. The number of these extra transitions is a function of input patterns, internal state assignment in the logic design, delay skew, and logic depth. To be specific about the magnitude of this problem, an 8-bit ripple-carry adder with an uniformly distributed set of random input patterns, will typically consume an extra 30% in energy. Though it is possible with careful logic design to eliminate these transitions, dynamic logic intrinsically does not have this problem, since any node can undergo at most one power-consuming transition per clock cycle.

7.4.1.2 Short-circuit currents

Short circuit (direct-path) currents are found in static CMOS circuits. However, by sizing transistors for equal rise and fall times, the short-circuit component of the total power dissipated can be kept to less than 20% [Veendrick84] (typically < 5-10%) of the dynamic switching component. Dynamic logic does not exhibit this problem, except for those cases in which static pull-up devices are used to control charge sharing or when clock skew is significant.

7.4.1.3 Parasitic capacitance

Dynamic logic typically uses fewer transistors to implement a given logic function, which directly reduces the amount of capacitance being switched and thus has a direct impact on the power-delay product [Hodges88][Shoji88]. However, extra transistors may be required to insure that charge-sharing does not result in incorrect evaluation.

7.4.1.4 Switching activity

The one area in which dynamic logic is at a distinct disadvantage is in its necessity for a precharge operation. Since in dynamic logic every node must be precharged every clock cycle, this means that some nodes are precharged only to be immediately discharged again as the node is evaluated, leading to a higher activity factor. As described in Chapter 3, a dynamic NOR gate has an activity of 0.75 assuming uniform inputs, while the activity factor for the static NOR counterpart will be only 3/16, excluding the component due to glitching. In general, gate activities will be different for static and dynamic logic and will depend on the type of operation being performed and the input signal probabilities. In addition, the clock buffers to drive the precharge transistors will also require power that is not needed in a static implementation.

7.4.1.5 Power-down modes

Lastly, power-down techniques achieved by disabling the clock signal have been used effectively in static circuits, but are not as well-suited for dynamic techniques. If the logic state is to be preserved during shut-down, a relatively small amount of extra circuitry must be added to the dynamic circuits to preserve the state, resulting in a slight increase in parasitic capacitance and slower speeds.

7.4.2 Pass Transistor Logic vs. Conventional CMOS Logic

The type of logic style used (static vs. dynamic, pass gate vs. conventional CMOS) affects the physical capacitance in the circuit. The physical capacitance is a function of the number of transistors that are required to implement a given function. For example, one approach to reduce the physical capacitance is to use transfer gates over conventional CMOS gates to implement logic functions, as is used in the CPL (Complementary Passgate Logic) family.

In Figure 7.21, the schematic of a typical static CMOS logic circuit for a full adder is shown along with a static CPL version [Yano90]. The passgate design uses only a single transmission NMOS gate, instead of a full complementary passgate to reduce node capacitance. Passgate logic is attractive as fewer transistors are required to implement important logic functions, such as XOR's which only require 2 pass transistors in a CPL implementation. This particularly efficient implementation of an XOR is important since it is key to most arithmetic functions, permitting adders and multipliers to be created using a minimal number of devices. Likewise, multiplexors, registers, and other key building blocks are simplified using passgate designs.

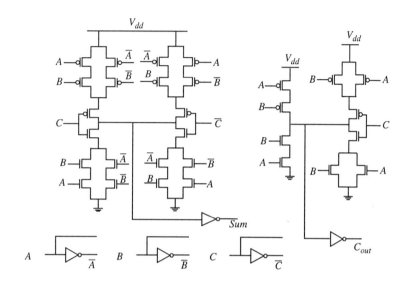

Transistor count (conventional CMOS) : 40

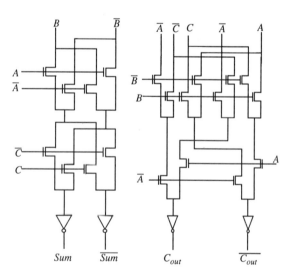

Transistor count (CPL) : 28

Figure 7.21: CMOS vs. Pass Gate Logic: adder example [Yano90].

However, a CPL implementation as shown in Figure 7.21 has two basic problems. First, the threshold drop across the single channel pass transistors results in reduced current drive and hence slower operation at reduced supply voltages; this is important for low-power design since it is desirable to operate at the lowest possible voltages levels. Second, since the "high" input voltage levels at the regenerative inverters is not V_{dd}, the PMOS device in the inverter is not fully turned off, and hence direct-path static power dissipation could be significant. To solve these problems, reduction of the threshold voltage has proven effective, although if taken too far will incur a cost in dissipation due to subthreshold leakage and reduced noise margins. The power dissipation for a pass-gate family adder with zero-threshold pass transistors at a supply voltage of 4V was reported to be 30% lower than a conventional static design due to the reduced node capacitance, with the difference being even more significant at lower supply voltages due to the reduction in voltage swing [Yano90].

7.4.3 Synchronous vs. Asynchronous

In synchronous designs, the logic between registers is continuously computing every clock cycle based on its new inputs. To reduce the power in synchronous designs, it is important to minimize switching activity by powering down execution units when they are not performing "useful" operations. This is an important concern since logic modules can be switching and consuming power even when they are not being actively utilized.

While the design of synchronous circuits requires special design effort and power-down circuitry to detect and shut down unused units, self-timed logic has inherent power-down of unused modules, since transitions occur only when requested. However, since self-timed implementations require the generation of a completion signal indicating the outputs of the logic module are valid, there is additional overhead circuitry. There are several circuit approaches to generate the requisite completion signal. One method is to use dual-rail coding, which is implicit in certain logic families such as the DCVSL[Jacobs90][Chu87]. The completion signal in a combinational macrocell made up of cascading DCVSL gates consists of simply ORing the outputs of only the last gate in the chain, leading to small overhead requirements. However, for each computation, dual-rail coding guarantees a switching event will occur since at least one of the outputs must evaluate to zero. We found that the dual rail DCVSL family consumes at least two times more in energy per input transition than a conventional static family. Hence, self-timed implementations can prove to be expensive in terms of energy for datapaths that are continuously computing.

7.5 Physical Design

7.5.1 Layout Optimization

Optimizing the layout to minimize the parasitics is important to minimizing the physical capacitance. Figure 7.22 three different approaches to laying out a large device that might be used for large buffers and clock drivers[Shoji88].

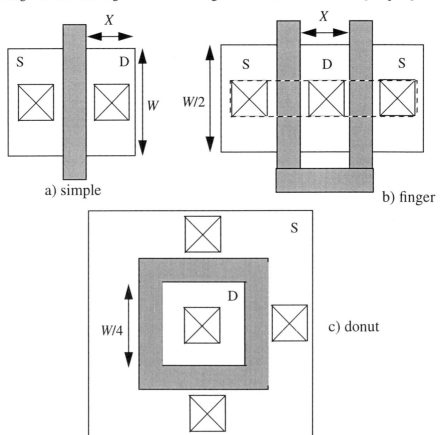

Figure 7.22: Three layout approaches for a large W/L transistor.

These structures have the same gate capacitance but different drain diffusion capacitances. The first case is a straight forward layout and the capacitance is given from Equation 58,

$$C_{drain} = C_{area} \, WX + C_{jsw} \, (W+2X) \tag{241}$$

where C_{area} is the area capacitance per unit area and C_{jsw} is the side-wall capacitance per unit length.

The second layout structure is a finger type structure that is commonly used to reduce the diffusion capacitance. Here, the side wall capacitance and the drain area are reduced compared to the simple layout. The capacitance is given by:

$$C_{drain} = C_{area} \bullet W/2 \bullet X + C_{jsw} \, (2X) \tag{242}$$

This example layout shows only two fingers, but for large sized devices several fingers could be used.

Finally, the third structure is a ring type structure which eliminates any side-wall capacitance in the drain of the device. The drain capacitance is given by

$$C_{drain} = C_{area} \bullet W^2/16 \tag{243}$$

Table 7.5 shows the results of drain capacitance for a device with $W/L = 20$ in a 1.2μm technology (i.e $W = 24$μm). For this technology $C_{area} = 0.3$fF/μm^2 and $C_{jsw} = 0.8$fF/μm for an NMOS device. Also the minimum contact size is 4λ. As seen from the table, significant reduction in physical capacitance is possible through layout optimization.

Table 7.5 The influence of layout optimization on physical capacitance.

Layout Style	Area μm^2	Perimeter contributing to sidewall in μm	C_{drain}
Simple	5λx24 = 72	30	45.6fF
Finger	6λx12= 43.2	7.2	18.72fF
Ring	6x6=36	0	10.8fF

Another optimization that can performed at the layout level is for regular tilable datapath structures, where the cells are of parameterized bit-widths. Here, all the data-signals are buses interconnecting the various cells in the datapath. To minimize the physical capacitance, it is important to minimize both the cell size, and the routing channels needed for the buses. The approach used in the low-power cell-library [Burd94] (used in the designs described in Chapter 9) was to route all buses in metal2, and to not use any metal2 for routing within a cell. Unfortunately, this forces some of the modules, such as the adder, to use poly for intracell routing, which has two to three times the capacitance per unit area. However, the inter-block bus routing becomes much more efficient and the capacitance increase from using poly is small compared to the overall decrease attributable to the more compact layout and smaller global buses. Each cell is 64λ high, which allows seven metal2 feedthroughs over the cell. These are used both as I/O ports to the cell and for the global data-path routing over cells. Utilizing this strategy reduced the area of several sample datapaths, on average, by 35% over the previously used data-path library not optimized for low-power. The area decrease of the datapath modules translates into a reduction of global routing wire length and their associated parasitic capacitances.

7.5.2 Place and Route

At the layout level, the place and route should optimized such that signals that have high switching activity (such as clocks) should be assigned short wires and signals with lower switching activities can be allowed progressively longer wires. Current design tools typically minimize the overall area or wire lengths given a timing constraint, which does not necessarily reduce the overall capacitance switched. Recently CAD efforts are starting to address the problem of physical design for low-power in which signal transition activities are used to drive place and route [Hirendu93][Chao94]. An example of the benefits of this optimization is presented in Chapter 9.

7.6 Summary

In the previous chapter, an aggressive voltage scaling strategy was used to minimize the energy to perform a given function. Another approach to low power design is to reduce the switching activity to the minimal level required to perform the computation. In CMOS circuits, since energy is only consumed when capacitance is being switched, power can be reduced by minimizing this capacitance at all levels of the design including algorithms, architectures, logic design, circuit design, and physical design.

At the algorithmic level, the basic computational complexity must be optimized. The switched capacitance can be minimized through reducing the number of operations needed to perform a given function (e.g. using hardwired shift-add operations instead of real multiplications or using algebraic transformations to minimize complexity), optimized data representation (binary vs. gray-coding, two's complement vs. sign-magnitude, etc.), and minimizing bitwidths (as will be discussed in the next chapter). Algorithmic optimization typically has the greatest impact on minimizing the switched capacitance.

At the architecture level, there are various degrees of freedom in minimizing the switched capacitance. This includes using sign-magnitude for multiply-accumulate architectures to minimize the number of transitions, ordering of operations to change dynamic range, balancing signal paths at the module level to minimize glitching transitions, and optimizing resource assignment by keeping correlated data on the same hardware and uncorrelated data on different hardware units.

At the logic level, logic minimization and logic-level power down are used to minimize the switched capacitance. At the circuit level, pass transistor logic can be used to minimize the physical capacitance of the circuit to implement function like adders, multiplier, etc. Also, at the circuit level, dynamic vs. static logic or synchronous vs. self-timed styles can be explored. At the physical design level, the place and route can be optimized such that signals that have high switching activity can be assigned short wires while signals that have low switching activity can be allowed to have long wires.

REFERENCES:

[Alidina94] M. Alidina, J. Monteiro, A. Ghosh, and M. Papaefthymiou, "Precomputation-Based Sequential Logic Optimization for Low Power," *1994 International Workshop on Low-power Design*, pp. 57-62, April 1994.

[Burd94] T. Burd, Low-power Cell Library, M.S. Thesis, U.C. Berkeley, June 1994.

[Callaway92] T. Callaway and E. Swartzlander, Jr., "Optimizing Arithmetic Elements for Signal Processing," *VLSI Signal Processing V*, pp. 91-100, IEEE Special Publications, 1992.

[Chao94] K. Chao and D. Wong, "Low-power Considerations in Floorplan Design," *1994 International Workshop on Low-power Design*, pp.45-50, April 1994.

[Chu87] K. Chu and D. Pulfrey, "A Comparison of CMOS Circuit Techniques: Differential Cascode Voltage Switch Logic Versus Conventional Logic," *IEEE Journal of Solid-State Circuits*, pp. 528-532, August 1987.

[Fang92] W.C. Fang, C.Y. Chang, and B.J. Sheu,"A Systolic Tree-Searched Vector Quantizer for Real-Time Image Compression," *VLSI signal processing IV*, New York: IEEE Press, 1992.

[Gersho92] A. Gersho and R. Gray, *Vector Quantization and Signal Compression*, Kluwer Academic Publishers, 1992.

[Hodges88] D. Hodges and H. Jackson, *Analysis and Design of Digital Integrated Circuits*, McGraw-Hill, Inc., 1988.

[Hirendu93] V. Hirendu and M. Pedram, "PCUBE: A Performance Driven Placement Algorithm for Lower Power Designs," Euro-DAC '93, pp. 72-77, 1993.

[Jacobs90] G.Jacobs and R.W. Brodersen, "A Fully Asynchronous Digital Signal Processor Using Self-Timed Circuits," *IEEE Journal of Solid-State Circuits*, pp. 1526-1537, December 1990.

[Monteiro93] J. Monteiro, S. Devadas, and A. Ghosh, "Retiming Sequential Circuits for Low Power," *International Conference on Computer-Aided Design*, pp. 398-402, 1993.

[Shen92] A. Shen, A. Ghosh, S. Devadas, and K. Keutzer, "On Average Power Dissipation and Random Pattern Testability of CMOS Combinational Logic Networks," *International Conference on Computer-Aided Design*, pp. 402-407, 1992.

[Shoji88] M. Shoji, *CMOS Digital Circuit Technology*, Prentice-Hall,1988.

[Stan94] M. Stan and W. Burleson, "Limited-weight Codes for Low-power I/O," *1994 International Workshop on Low-power Design*, pp. 209-214, April 1994.

[Su94] C. Su, C Tsui, and A. Despain, "Low-power Architecture Design and Compilation Techniques for High-Performance Processors," pp. 489-498, *Compcon 1994*.

[Tiwari93] V. Tiwari, P. Ashar, S. Malik, "Technology Mapping for Low Power," *Design Automation conference*, pp. 74-79, 1993.

[Tsui93] C. Tsui, M. Pedram, and A. Despain, "Technology Decomposition and Mapping Targeting Low Power Dissipation," *Design Automation conference*, pp. 68-73, 1993.

[Veendrick84] H.J.M. Veendrick, "Short-Circuit Dissipation of Static CMOS Circuitry and Its Impact on the Design of Buffer Circuits," *IEEE Journal of Solid-State Circuits*, Vol. SC-19, pp. 468-473, August 1984.

[Yano90] K. Yano *et al.*, "A 3.8ns CMOS 16x16 Multiplier Using Complementary Pass Transistor Logic," *IEEE Journal of Solid-State Circuits*, pp. 388-395, April 1990.

[Yuan89] J. Yuan and C. Svensson, "High-Speed CMOS Circuit Technique," *IEEE Journal of Solid-state Circuits*, pp. 62-70, February 1989.

8

Computer Aided Design Tools

In the previous chapters, approaches were presented to minimize the power consumption though supply voltage scaling and the reduction of switched capacitance. The focus of this chapter is on automatically finding computational structures that result in the lowest power consumption for DSP applications that have a specific throughput constraint given a high-level algorithmic specification. The basic approach is to scan the design space utilizing various algorithmic flowgraph transformations, high-level power estimation, and efficient heuristic/probabilistic search mechanisms. While algorithmic transformations have been successfully applied in high-level synthesis with the goal of optimizing speed and/or area, they have not addressed the problem of minimizing power.

There are two approaches taken to explore the algorithmic design space to minimize power consumption [Chandrakasan95]. First, is the exploitation of concurrency which enables circuits to operate at the lowest possible voltage without loss in functional throughput. Second, computational structures are used that minimize the effective capacitance that is switched at a fixed voltage: through reductions in the number of operations, the interconnect capacitance, and internal bit widths and using operations that require less energy per computation.

8.1 Previous Work

8.1.1 Power Estimation

Most of the previous work done in power estimation falls under four categories: circuit-level estimation, switch-level estimation, gate-level probabilistic/statistical estimation, and architectural power estimation. The primary trade-off between these approaches is the computational complexity vs. accuracy. Circuit-level approaches result in the most accurate estimates, while being the most computationally intensive.

8.1.1.1 Continuous Time Circuit-Level Power Estimation

At the lowest level of design, power can be estimated using a circuit simulator such as SPICE. The design is described at a very low-level and the simulator accounts for short-circuit and leakage components of power. The major problem with this approach is the long simulation time which limits the number of patterns that can be applied. Therefore, this approach is only viable for analyzing small circuits (e.g. when characterizing library cells).

Figure 8.1 shows a circuit that can be used to measure the energy consumed by a circuit using the SPICE simulator. The current drawn by the circuit under test from the power supply is monitored by the current controlled current source and integrated on the capacitor C. The resistance R is provided for DC-convergence reasons and should be chosen as high as possible to minimize leakage. Through proper choice of parameters for the capacitor, C and the gain for the current controller current source, k, the voltage on the capacitor can be made equal to the energy drawn from the supply.

The voltage across the capacitor is given below (under the assumption that the initial voltage on the capacitor C is zero):

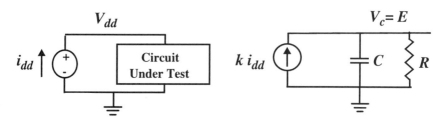

Figure 8.1: Circuit for measuring the power-delay product.

$$C\frac{dV_c}{dt} = ki_{dd}(t) \qquad (244)$$

$$V_c = \frac{k}{C}\int_0^T i_{dd}(t)\,dt \qquad (245)$$

The energy consumed by the circuit under test is:

$$E = V_{dd}\int_0^T i_{dd}(t)\,dt \qquad (246)$$

Therefore, from Equations 245 and 246, by setting $V_{dd} = k/C$, the voltage across the capacitor will equal the energy consumed in the interval $[0,T]$.

Figure 8.2 shows an example of estimating the energy consumed for a 1-bit full adder circuit. The integrating capacitor value is chosen to be 1pF and k is chosen to be V_{dd} (=1.5V for this example). Therefore, the voltage across the capacitor represents the cumulative energy (in pJ) for the 1-bit full-adder circuit. The graph in Figure 8.2 shows the voltage across the capacitor for a sequence of input patterns ($IN_0\ IN_1\ CIN$). The jumps in the voltage curve represent the amount of energy drawn for that particular input transition. For example, for the input transition 111 -> 000, the change in voltage across the capacitor is 0.84V (the energy drawn for this transition is 0.84pJ). From this, the capacitance switched for this transition can be determined as $0.84pJ/(1.5)^2 = 0.37pF$. The dependence of power on data sequencing is obvious from this figure as different transitions draw different amounts of energy; for example, 100 ->010 does not draw very much energy since neither output switches, while 111 -> 000 draws a lot of energy since both outputs switch. The average energy drawn per transition is 3.64pJ/8 (total energy/number of input transitions) = 0.44pJ.

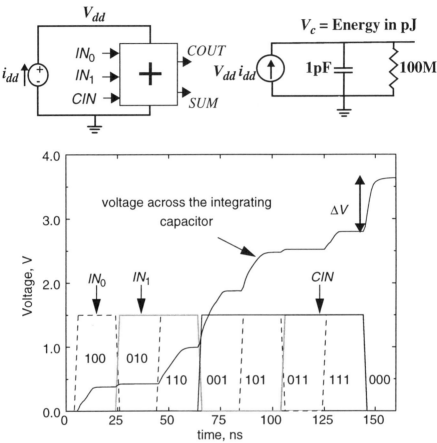

Figure 8.2: Measuring capacitance switched for a 1-bit full-adder example.

8.1.1.2 Switch-level Power Estimation

An approach for estimating the power consumption in CMOS circuits using a switch-level simulator is presented in [Kimura91]. The basic idea is to monitor the number of times each node in the circuit transitions during the simulation period. C_{avg} is given by $\Sigma (N_i /N) C_i$, where N_i is the total number of power con-suming transitions for node i, N is the number of simulation cycles, and C_i is the physical capacitance of node i. This approach using the IRSIM simulator [Salz89] was used to estimate power at the layout level for the examples presented later. This analysis ignores short-circuit and leakage power components. The results from a few fabricated chips, indicate that the predicted power (with calibrated models from test chips) from IRSIM is within 30% of the measured power. A commercial version, called PowerMill from Epic Design, also exists that takes into account the extra power due to finite rise/fall times [Deng94].

An important parameter for power estimation through simulation is the number of simulation cycles that are needed for accurate estimation of the switched capacitance. In the case of SPICE simulation, only a small number of patterns can be applied due to the time consuming nature of the simulator. Also, only simulation of small circuits is possible, restricting the application to the characterization of cell-libraries. Switch-level simulation is significantly faster and large circuits (whole chips) can be simulated. Figure 8.3 shows the average capacitance switched per cycle as a function of the number of sample periods simulated for an FIR filter. For this example which has no feedback and for the input data pattern used, a good indication of the average power consumption can be obtained by simulating for 50-100 cycles. In general, determining the number of cycles is a non-trivial task, strongly dependent on the topology (signal correlations) and data patterns. A methodology and tool to determine the minimum number of simulation cycles required has been developed [VanOostende93].

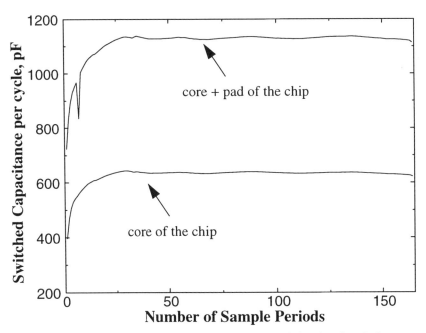

Figure 8.3: Switched capacitance vs. number of simulated periods.

8.1.1.3 Gate-level Probabilistic Power Estimation

Gate-level probabilistic approaches estimate the internal node activities of a network given the distribution of the input signals [Cirit87][Najm91] [Ghosh92]. Once the signal probability for each node in the network is determined, the total average capacitance switched is then estimated as

$$C_{total} = \sum_{i=1}^{numnodes} \left(p_{i0} \right)\left(1 - p_{i0} \right) \cdot C_i \tag{247}$$

where *numnodes* is the total number of nodes in the network, p_{i0} is the probability that node i will be in the *ZERO* state, and C_i is the physical capacitance associated with node i. The total power is then estimated as $C_{total} * V_{dd}^2 * f_{clk}$. Most gate-level simulators typically ignore the interconnect component of loading capacitance and model the capacitance as the sum of the output capacitance of the driver plus the input gate capacitance of all the gates being driven. Two other major problems with these simulators are in modeling reconvergent fanout (which introduces correlations between signals) and glitching activity (which arrises due to signal skew). Significant efforts to address these problems have been made [Ghosh92].

8.1.1.4 Architectural Power Estimation

Using concepts abstracted from the gate level, an architectural estimation technique has been developed based on high level statistics such as mean, variance, and autocorrelation [Landman93]. While, the above mentioned gate level tools focus on the power consumed by boolean logic gates (NOT, AND, OR, etc.) as a function of their input bit probabilities, the architectural tool considers module (adder, multiplier, register, etc.) power consumption as determined by input word statistics.

As described in Chapter 3, there exists a direct relationship between bit level probabilities and word level statistics. The transition probability as a function of the bit number can be represented as a simple piecewise linear model with breakpoints BP_0 and BP_1 (see Figure 3.22). The important features of this model - the values of the breakpoints and the signal and transition probabilities - can all be extracted from three statistical parameters: the mean, μ; the variance, σ^2; and the lag one correlation coefficient, $\rho_1 = \text{cov}(X_t, X_{t+1})/\sigma^2$. Similar to gate level techniques, given the statistics of the module inputs, the statistics of the outputs are calculated by statistical propagating; this way the model parameters (μ, σ^2, and ρ_1) for each bus in the architecture can be derived. The top half of Figure 8.4 presents the appropriate propagation equations for the case of an addition and a constant multiplication (the two key operations of linear, time-invariant systems). Once again, the issue of reconvergent fan-out tends to complicate matters by introducing correlations between module inputs which is not handled by these equations; however, this is overcome by abstracting heuristic techniques similar to those applied at the gate level. For example, the revised equations at the bottom of Figure 8.4 account for signal correlations.

$$\mu_z = C\mu_x$$
$$\sigma_z^2 = C^2\sigma_x^2$$
$$\rho_1^z = \rho_1^x$$

$$\left.\begin{array}{l}\mu_z = \mu_x + \mu_y \\ \sigma_z^2 = \sigma_x^2 + \sigma_y^2 \\ \rho_1^z = (\rho_1^x\sigma_x^2 + \rho_1^y\sigma_y^2)/\sigma_z^2\end{array}\right\} \text{ if } X \perp Y$$

Revised Adder Equations for Correlated Inputs:

$$\mu_z = \mu_x + \mu_y$$
$$\sigma_z^2 = \sigma_x^2 + 2\rho^{xy}\sigma_x\sigma_y + \sigma_y^2$$
$$\rho_1^z = (\rho_1^x\sigma_x^2 + [\rho_1^{xy}+\rho_1^{yx}]\sigma_x\sigma_y + \rho_1^y\sigma_y^2)/\sigma_z^2$$

Figure 8.4: Statistical parameter propagation [Landman93].

To use these approaches mentioned above in a high-level synthesis framework, the high-level representation of the algorithm has to be mapped to a low-level description (gate or transistor level), which is very time consuming. Even going to the architecture level (which involves allocation/assignment/scheduling) is too time consuming if the goal is to explore 100's or 1000's of possible structures in a time efficient manner. The estimation time for each new topology is time-consuming itself. Hence, power must be estimated efficiently from an even higher level of abstraction. In this chapter, techniques will be described to estimate power from an algorithmic level so that the design space can be quickly and efficiently explored.

8.1.2 High-level Transformations

Over the last few years, several high-level synthesis systems have incorporated comprehensive sets of transformations, coupled with powerful optimization strategies. Example systems with elaborate applications of transformations are Flamel [Trickey87], SAW [Walker89], SPAID [Haroun89], and HYPER [Rabaey91a].

Among the set of transformations used by the Flamel design system are loop transformations, height reduction and constant propagation. SAW uses among other transformations in-line expansion, dead code elimination, four types of transformations for conditional statements and pipelining as supporting steps during behavioral and structural partitioning. The SPAID system set of transformations includes retiming and pipelining, interleaving, substitution of multiplications with constants by addition and shifts and algebraic transformations. HYPER uses more than 20 different transformations whose application is supported by several probabilistic optimization algorithms. All systems provide interactive frameworks where the designer explores the influence of the transformation mechanism or the optimization algorithms for a specific transformation.

The systems described above use transformations to optimize design parameters such as area and throughput. The system presented in this chapter will address the problem of how high-level flowgraph transformations can be used to reduce the power consumption of a VLSI implementation.

8.2 Application of Transformations to Minimize Power

Transformations are changes to the computational structure in a manner that the input/output behavior is preserved. The number and type of computational modules, their interconnection and the sequencing of operations are optimized. The use of transformations makes it possible to explore a number of alternative architectures and to choose those which result in the lowest power. The control-data flow graph format will be used to represent computation [Rabaey91a], in which nodes represent operations, and edges represent data and control dependencies. Transformations have primarily been used until now to optimize either the implementation area or the system throughput. The goal here is to optimize a different function, namely the power dissipation of the final circuit while meeting the functional throughput of the system. Two key approaches are used to reduce power for a fixed throughput: reducing the supply voltage by

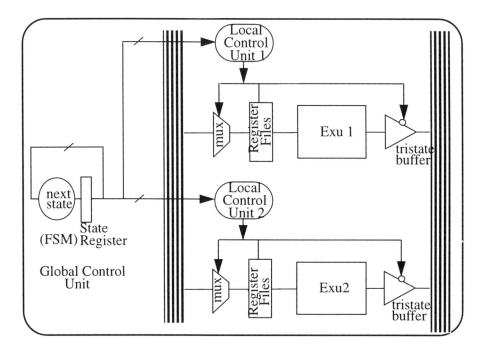

Figure 8.5: Hardware model used in HYPER.

utilizing speed-up transformations (Section 8.2.1) and reducing the effective capacitance switched (Sections 8.2.2 - 8.2.5).

Figure 8.5 shows the hardware model used in HYPER. A time-multiplexed hardware model is used in which multiple operations on a data-control flowgraph can be mapped to the same hardware unit. A datapath module contains execution units, register files, multiplexors, and buffers. Data stored in the register files is read out to the execution unit whose result is written back into register files in the same datapath or in other datapaths. The register file may be preceded by multiplexors depending on whether the execution unit receives data from different sources at different control cycles in the sample period. The execution units on different datapaths communicate through global busses and the outputs of the execution units are tri-stated if two different units share a common global bus. A distributed control approach (which contains a global controller and several local controllers) is used to sequence through the operations in the control cycles. The critical path will be stated in the number of control steps, which is the number of cycles required to implement the given function.

8.2.1 Speed-up transformations

This is probably the single most important type of transformation for power reduction. It is not only the most common type of transformation, but often has the strongest impact on power. The basic idea is to reduce the number of control steps, so that slower control clock cycles can be used for a fixed throughput, allowing for a reduction in supply voltage. In Chapter 4, for the simple adder-comparator example presented, parallel and pipelined architectures were used to reduce the supply voltage and power. This simple example did not have any feedback and the computation was therefore easily parallelizable. However, there are a wide class of applications inherently recursive in nature, ranging from simple ones, such as infinite impulse response (IIR) and adaptive filters, to more complex cases such as systems solving non-linear equations and adaptive compression algorithms. There is an algorithmic bound on the level to which pipelining and parallelism can be exploited in these feedback systems since the delay in the feedback loop is set by the algorithm and additional delay cannot be added without algorithmic transformations.

In general, exploiting concurrency in an algorithm is not a trivial task and high-level computation transformations are required to exploit concurrency. Many transformations profoundly affect the amount of concurrency in the computation. This includes retiming/pipelining, algebraic transformations and loop transformations. For linear systems, a sequence of computational transformations applied in a well defined order can be used to exploit arbitrary parallelism and operate circuits at very low supply voltages.

To illustrate the application of speed-up transformations to lower power, consider a first order IIR filter, as shown in Figure 8.6a. A time-multiplexed model is assumed in which each operation is assumed to take once control cycle and therefore this structure has a critical path of 2 (i.e., for this example, $T_{sample}/T_{clock} = 2$). The goal is to reduce the number of cycles to process one sample since this will allow a reduction of the power supply. Due to the recursive bottleneck [Messerschmitt88] [Parhi89] imposed by the filter structure, it is impossible to reduce the critical path using retiming or pipelining. Also, the simple structure does not provide opportunities for the application of algebraic transformations and applying a single transformation is not enough to reduce power in this example.

Figure 8.6b shows the effect of applying loop unrolling (where two output samples are computed in parallel based on two input samples) on the initial flow-graph.

$$Y_{N-1} = X_{N-1} + A * Y_{N-2} \tag{248}$$

$$Y_N = X_N + A * Y_{N-1} = X_N + A * (X_{N-1} + A * Y_{N-2}) \tag{249}$$

The critical path is doubled, but two samples are processed in parallel; therefore, the effective critical path is unchanged and therefore the supply voltage cannot be altered. The effective capacitance switched does not change either since the number of operations is doubled for processing two input samples in parallel. Since neither the capacitance switched nor the voltage is altered, the power of this implementation remains unchanged. Therefore, loop unrolling by itself does not reduce the power consumption.

However, loop unrolling enables several other transformations (distributivity, constant propagation, and pipelining) which result in a significant reduction in power dissipation. After applying loop unrolling, distributivity and constant propagation in a systematic way, the output samples can be represented as:

$$Y_{N-1} = X_{N-1} + A * Y_{N-2} \tag{250}$$

$$Y_N = X_N + A * X_{N-1} + A^2 * Y_{N-2} \tag{251}$$

The transformed solution has a critical path of 3 (Figure 8.6c) for processing two samples. Therefore, the effective critical path is reduced and therefore the voltage can be dropped. This results in a small reduction of the power consumption (by 20%).

However, pipelining can now be applied to this structure, reducing the critical path further to 2 cycles (Figure 8.6d). Since the final transformed block is working at half the original sample rate (since we are processing 2 samples in parallel), and the critical path is same as the original datapath (2 control cycles), the supply voltage can be dropped to 2.9V (the voltage at which the delays increase by a factor of 2, see Figure 3.24). However, note that the effective capacitance increases since the transformed graph requires 3 multiplications and 3 additions for processing 2 samples while the initial graph requires only one multiplication and one addition to process one sample, or effectively a 50% increase in capacitance. The reduction in supply voltage, however, more than compensates for the increase in capacitance resulting in an overall reduction of the power by a factor of 2 (due to the quadratic effect of voltage on power).

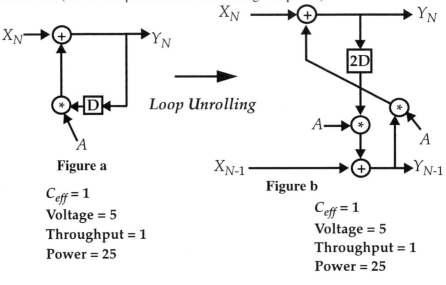

Figure a

C_{eff} = 1
Voltage = 5
Throughput = 1
Power = 25

Figure b

C_{eff} = 1
Voltage = 5
Throughput = 1
Power = 25

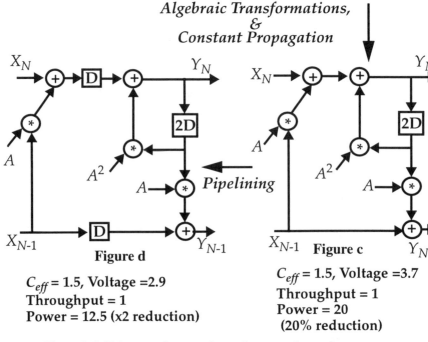

Figure d

C_{eff} = 1.5, Voltage =2.9
Throughput = 1
Power = 12.5 (x2 reduction)

Figure c

C_{eff} = 1.5, Voltage =3.7
Throughput = 1
Power = 20
(20% reduction)

Figure 8.6: Using speedup transformations to reduce power.

   This simple example can used to illustrate that optimizing for throughput
will result in ***different*** solution than optimizing for power. For this example, arbitrary speedup can be achieved by continuing to apply loop unrolling combined
with other transformations (algebraic, constant propagation, and pipelining). The
speedup grows linearly with the unrolling factor, as shown in Figure 8.7. If the
goal is to minimize power consumption while keeping the throughput fixed, the
speedup can be used to drop the supply voltage. Unfortunately, the capacitance
grows linearly with unrolling factor (since the number of operations per input
sample increases) and soon limits the gains from reducing the supply voltage.
This results in an "optimum" unrolling factor for power of 3, beyond which the
power consumption starts to increase again.

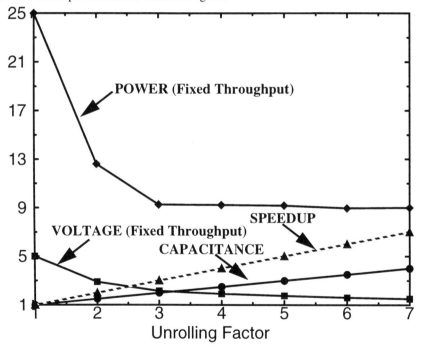

Figure 8.7: Speed optimization is *different* than power optimization.

   This example brings out two very important points: First, the application of
a particular transformation can have conflicting effects on the different components of power consumption. For example, a transformation can reduce the voltage component of power (through a reduction in the critical path) while
simultaneously increasing the capacitance component of power. Therefore, while

speed-up transformations can be used to reduce power by allowing for reduced supply voltages, the "fastest" solution is often NOT the lowest power solution. Second, the application of transformations in a combined fashion almost always results in lower power consumption than the isolated application of a single transformation. In fact, it is often necessary to apply a transformation which may temporary increase the power budget, in order to enable the application of transformations which will result in a more dramatic power reduction.

8.2.2 Operation Reduction

The most obvious approach to reduce the switching capacitance, is to reduce the number of operations (and hence the number of switching events) in the data control flow graph. While reducing the operation count typically has the effect of reducing the effective capacitance, the effect on its critical path is case dependent. To illustrate this trade-off, consider evaluating second and third order polynomials. Computation of polynomials is very common in digital signal processing, and Horner's scheme (the final structure in our examples) is often suggested in filter design and FFT calculations when very few frequency components are needed [Goertzel68].

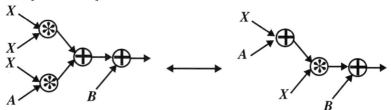

Figure 8.8: Reducing capacitance while maintaining throughput.

First we will analyze the second order polynomial $X^2 + AX + B$. The left side of Figure 8.8 shows the straightforward implementation which requires two multiplications and two additions and has a critical path of 3 (assuming that each operation takes one control cycle). On the right side of Figure 8.8, a transformed version having a different computational structure (obtained using algebraic transformations) is shown. The transformed graph has the same critical path as the initial solution and therefore the two solutions will have the same throughput at any given supply voltage. However, the transformed flowgraph has one less multiplication, and therefore has a lower capacitance and power.

Figure 8.9 illustrates a situation where a significant reduction in the number of operations is achieved at the expense of a longer critical path. This example once again involves the computation of a polynomial, this time of the form X^3+AX^2+BX+C. Again, by applying algebraic transformations we can transform the computation to the Horner's scheme. The number of multiplications reduces by two, resulting in a reduction of the effective capacitance. However, the critical path is increased from 4 to 5, dictating a higher supply voltage than in the initial flowgraph for the same computational throughput. Once again, this example shows that a transformation can have different effects on capacitance and voltage, making the associated power minimization task a difficult optimization problem.

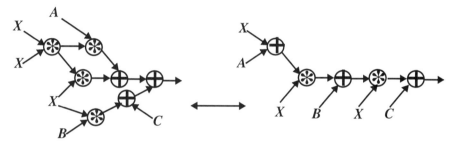

Figure 8.9: Reducing capacitance at the expense of a higher supply voltage.

Transformations which directly reduce the number of operations in a data control flow graph include: common subexpression elimination, manifest expression elimination, and distributivity.

8.2.3 Operation Substitution

Certain operations inherently require less energy per computation than other operations. A prime example of a transformation which explores this trade-off is strength reduction, often used in software compilers, in which multiplications are substituted by additions [Aho77]. Although this situation is not as common as the ones presented in the previous section, sometimes it is possible to achieve significant savings using this type of trade-off.

Unfortunately, this type of transformation often comes at the expense of an increase in the critical path length. This point is illustrated in Figure 8.10, which shows the frequently used application of redundancy manipulation, distributivity and common subexpression in complex number multiplication. A_i and A_r are constants. While the second implementation has a lower effective capacitance, it has a longer critical path. Fixing the available time to be a constant, we see that the second implementations requires at least three control cycles (assuming each operation take one control cycle), while the first one requires only two. This implies that the voltage in the first implementation can be dropped lower than the voltage in the second implementation, since the same computational throughput can be met with a 50% slower clock rate.

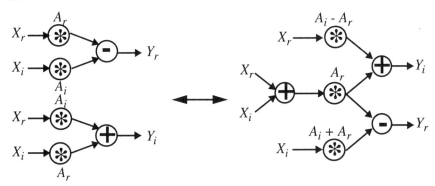

Figure 8.10: Replacing multiplication with an addition.

Another powerful transformation in this category is conversion of multiplications with constants into shift-add operations. Table 8.1 and Table 8.2 shows the power breakdown for a 11 tap FIR filter before and after this transformation. The power consumed by the execution units is reduced to one-eighth of the original power and the power consumed in the registers is also reduced. A small penalty is paid in the controller but there is a gain in the interconnect power due to a reduction in the length of the bus lines (since the final implementation has a smaller area), resulting in a total power savings of 62%.

Table 8.1 Breakdown of power consumed for a 11 tap FIR filter before constant multiplications.

Component	Switched Capacitance, pF	Energy @ 2.5V, nJ	%
Exu	739.65	4.62	64.80
Registers & Clock	179.57	1.12	15.73
Control	65.45	0.41	5.73
Bus	156.69	0.98	13.74

Table 8.1 Breakdown of power consumed for a 11 tap FIR filter before constant multiplications.

Component	Switched Capacitance, pF	Energy @ 2.5V, nJ	%
Total	1141.36	7.13	100.00

Table 8.2 Breakdown of power consumed for a 11 tap FIR filter after constant multiplications.

Component	Switched Capacitance, pF	Energy @ 2.5V, nJ	%
Exu	93.07	0.58	21.63
Registers & Clock	161.4	1.00	37.50
Control	83.79	0.52	19.47
Bus	92.10	0.58	21.40
Total	430.36	2.69	100.00

A shift-add multiplication can be implemented using a direct-mapped full parallel approach - in which there is a one to one correspondence between the operations on the flowgraph and the operators on the hardware implementation - or a time-multiplexed approach - in which there is one adder and one shifter and a controller which controls the shifter. In the parallel implementation, there is no control required since the shift operation is hardwired in which the shift operation degenerates to routing. In addition to control overhead, the time-multiplexed approach also has the overhead of one register access for each shift-add operation. Therefore, if there are many 1's in the coefficient, the time-multiplexed shift-add implementation can consume more power compared with a full array multiplier. Several algorithms and tools have been developed to minimize the number of shift-add operations for constant coefficients [Jain85].

Another power reduction technique is to scale the coefficients so as to minimize the number of shift-add operations. When multiplication of multiple coefficients with a single input variable is involved, common-subexpression elimination can be combined with scaling to minimize the number of shift-add operations. The basic idea is illustrated in the example shown in Figure 8.11, which involves multiplying an input with two coefficients. The left side shows a brute-force implementation in which the two outputs are computed in parallel. This requires 5 shifts and 3 additions. Another approach is shown on the right, which exploits the common terms in the constants to share some of the shift-add operations; here, A is computed first and is then used in the computation of B. The overall number of shift-add operations that have to be performed reduced.

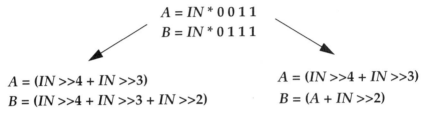

$$A = IN * 0\ 0\ 1\ 1$$
$$B = IN * 0\ 1\ 1\ 1$$

$$A = (IN >> 4 + IN >> 3)$$
$$B = (IN >> 4 + IN >> 3 + IN >> 2)$$

$$A = (IN >> 4 + IN >> 3)$$
$$B = (A + IN >> 2)$$

Figure 8.11: Example of common sub-expression elimination.

To illustrate the application of shift-add conversion with scaling on a larger example, consider the color space conversion matrix from $RGB \rightarrow YIQ$ which is used in most compression algorithms. Figure 8.12 shows the basic operation being performed. On the encoder, a digital RGB vector is converted to digital YIQ and sub-sampled. This conversion involves a matrix multiplication operation with constant coefficients. On the decoder, the YIQ vector is converted back to RGB. For this operation, there are two main optimizations that can be performed on the multiplications. First, the coefficients on the encoding matrix can be scaled; for example, C_{21}, C_{22}, and C_{23} can all be scaled by $1/\alpha_i$, as long as the coefficients on the decoder matrix D_{12}, D_{22}, and D_{32} are scaled by α_i. This gives some opportunity for exploring the design space to minimize the number of shift-add operations. Of course, the scaling operation should not come at the expense of "visual image degradation". The second optimization is to exploit multiplication of inputs with multiple coefficients; for example, D_{12}, D_{22}, and D_{32} are all multiplied with I'. This allows the application of the common sub-expression elimination to share some of the shift-add terms.

Figure 8.13 shows the plot of the number of shift-add operations that have to be performed for the three coefficients D_{12}, D_{22}, and D_{32} which are all multiplied with I'. The top curve indicates the number of operations with just scaling of coefficients and does not include the effect of exploiting common terms in the three coefficients. The curve below indicates the number of operations that has to be performed using a strategy that combines scaling with common sub-expression elimination. It is clear from this plot that slight variations in the coefficients can result in significant reduction of the computational complexity.

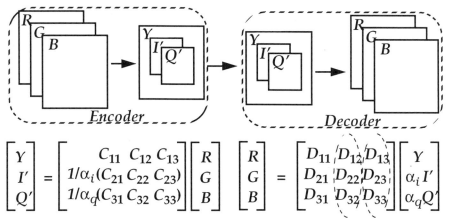

$$\begin{bmatrix} Y \\ I' \\ Q' \end{bmatrix} = \begin{bmatrix} C_{11} \; C_{12} \; C_{13} \\ 1/\alpha_i(C_{21} \; C_{22} \; C_{23}) \\ 1/\alpha_q(C_{31} \; C_{32} \; C_{33}) \end{bmatrix} \begin{bmatrix} R \\ G \\ B \end{bmatrix} \qquad \begin{bmatrix} R \\ G \\ B \end{bmatrix} = \begin{bmatrix} D_{11} \; D_{12} \; D_{13} \\ D_{21} \; D_{22} \; D_{23} \\ D_{31} \; D_{32} \; D_{33} \end{bmatrix} \begin{bmatrix} Y \\ \alpha_i I' \\ \alpha_q Q' \end{bmatrix}$$

Figure 8.12: Color conversion matrix for video applications.

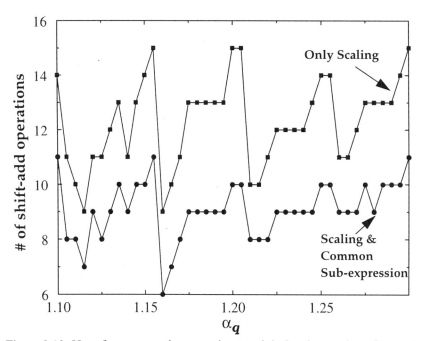

Figure 8.13: Use of common sub-expression to minimize the number of operations.

8.2.4 Resource Utilization

It is often possible to reduce the required amount of hardware, while pre-serving the number of control steps [Potkonjak91]. This is possible because after certain transformations, operations are more uniformly distributed over the available time, resulting in a denser schedule. These transformations include retiming for resource utilization, associativity, distributivity and commutativity. Figure 8.14 shows the result of applying retiming on a second order IIR filter. Both the

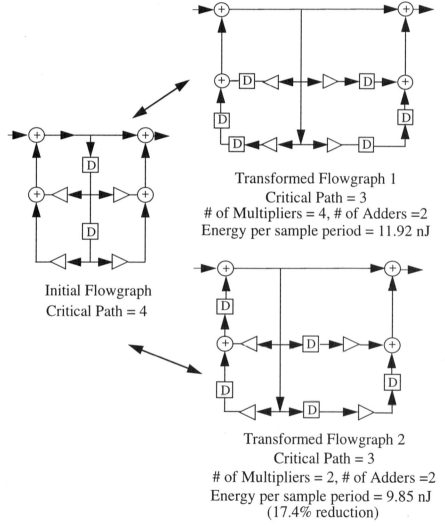

Transformed Flowgraph 1
Critical Path = 3
# of Multipliers = 4, # of Adders =2
Energy per sample period = 11.92 nJ

Initial Flowgraph
Critical Path = 4

Transformed Flowgraph 2
Critical Path = 3
# of Multipliers = 2, # of Adders =2
Energy per sample period = 9.85 nJ
(17.4% reduction)

Figure 8.14: Retiming for improving resource utilization: 2nd order IIR filter.

transformed graphs, 1 and 2, are obtained from retiming and have a critical path of 3; however, the transformed graph 2 can be scheduled with only 2 multipliers while the first needs 4 multipliers (since all the four multiplications can be performed only in the 3rd control step).

Given that there is a degree of freedom in choosing the amount of resources used for a fixed throughput (i.e at a fixed supply voltage), the question then becomes is the solution with the minimal amount of resources, as chosen for the minimal area implementation, the "best" for low-power. On one hand, reducing the amount of resources can reduce the wiring capacitance since there are fewer interconnects and/or fewer functional elements and registers, which are obstacles during floorplanning and routing. However, the amount of multiplexors and control logic circuitry will typically increase with more time-sharing of resources. Therefore the optimization strategy (and hence the power estimation model) must take into account the trade-off between interconnect capacitance and control circuitry (which determines the effective capacitance being switched).

8.2.5 Wordlength Reduction

The number of bits used strongly affects all key parameters of a design, including speed, area and power. It is desirable to minimize the number of bits during power optimization for at least three reasons:

- Fewer bits result in fewer switching events and therefore lower capacitance.

- Fewer bits imply that the functional operations can be done faster, and therefore the voltage can be reduced while keeping the throughput constant.

- Fewer bits not only reduce the number of transfer lines, but also reduce the average interconnect length and capacitance.

The influence of various transformation on numerical stability (and therefore the required wordlength) varies a lot. While some transformations, for example retiming, pipelining and commutativity, do not affect wordlength, associativity and distributivity often have a dramatic influence [Golberg91]. In some cases, it is possible to reduce both the number of power expensive operations and the required wordlength. In other cases, however, a reduction in wordlength comes at the expense of an increased number of operations.

To illustrate the importance of wordlength optimization, consider the direct form and parallel form implementations of an 8th order Avenhaus bandpass filter

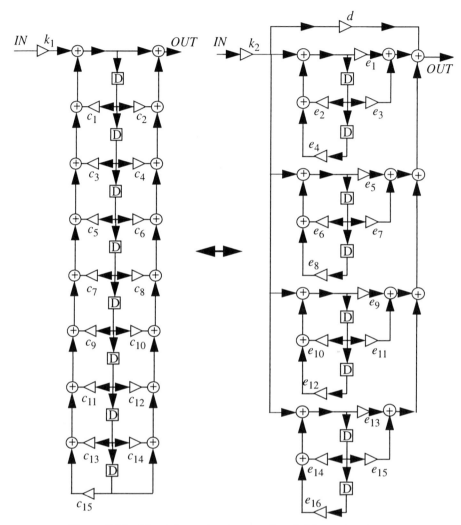

Figure 8.15: Direct form structure vs. Parallel form structure: 8th order Avenhous filter.

(as shown in Figure 8.15). The critical path of the direct form, after multiplication to shift and addition substitution is 20 clock cycles while the critical path of the parallel form is 28 cycles, assuming that each operation, in both cases, takes one cycle. However, the numerical stability of the parallel form is significantly higher, resulting in wordlength requirements of only 11 bits, while the direct form solution requires 23 bits. The effective critical path for the direct form solution is 980 ns, while parallel form has a critical path of only 610 ns. Therefore, taking wordlength requirements into consideration, changes the situation from one where the critical path in the parallel form is almost 50% longer, to one where the critical path in the direct form is more than 50% longer. Similarly, a detailed analysis of these two alternatives shows that the final implementation of the parallel filter is reduced by a factor of four for a given throughput.

The results from section 8.2 can be summarized in the following requirements for the application of transformations for power reduction:

- Efficient implementation of known and new transformations so that power is an explicit part of the cost function and the development of the cost function itself.
- Development of a transformation framework and search mechanism which will determine the order and extent to which the transformations are applied so that final result is globally optimal.

8.3 Cost Function

The goal is to develop an objective function that is highly correlated to the final (and unknown) power dissipation of the circuit. The objective function should be very easy to compute since it has to be evaluated many times during the optimization process. A detailed estimation, while being accurate, will require hardware mapping and compilation steps to convert a flowgraph to layout, making it impractical during the optimization process. Hence a model correlated to the power must be developed strictly from the flowgraph level. This involves a statistical study of the effects of various high level parameters on interconnect capacitance, control capacitance, etc.

The power consumption of a circuit (represented in a control data flowgraph) is given by:

$$P_{total} = C_{total} * V^2 * f_{sampling} \qquad (252)$$

The goal of power optimization in this work is to keep the throughput constant (i.e., $f_{sampling}$ is fixed) by allowing the supply voltage to vary. For a fixed sample period, power optimization is equivalent to minimizing the total energy switched, $C_{total} V^2$, where V is appropriate voltage required to meet the throughput rate.

8.3.1 Capacitance estimate

The total capacitance switched depends on four components:

$$C_{total} = C_{exu} + C_{registers} + C_{interconnect} + C_{control} \qquad (253)$$

The capacitance estimation is built on top of an existing estimation routine in HYPER that determines bounds and activity of various execution, register, and interconnect components as well as the implementation area [Rabaey91b, Schultz92]. The details of the capacitance estimation routines are described below.

8.3.1.1 Execution Units

The capacitance switched by the various execution units is estimated by multiplying (over all types of units utilized) the number of times the operation performed by the unit occurs per sample period with the average capacitance per access of the unit type. The total capacitance is hence given by:

$$C_{exu} = \sum_{i=1}^{numtypes} N_i \cdot C_i \qquad (254)$$

where *numtypes* is the total number of operation types, N_i the number the times the operation of type i is performed per sample period (or the activity), and C_i is average capacitance per execution of operation type i. The average capacitance per execution has been characterized for the various modules (through SPICE and IRSIM simulations, using models that were calibrated with results from experimental measurements) for a uniformly distributed set of inputs. In general the probabilities are not uniform, however, this assumption is made to simplify the cost evaluation. The capacitance values are parameterized as a function of bit-width and are accessed when computing the power contributed due to the execution units. Table 8.3 shows the high-level capacitance models for various modules in the hardware library.

Table 8.3 Capacitance models for some of the library hardware units for a 1.2μm CMOS technology.

Component	Capacitance Model (in fF)	Parameters
ripple adder	-46 + 151 N	N: bitwidth
carry select	-158 + 214 N	N: bitwidth
comparator	181 N	N: bitwidth

Table 8.3 Capacitance models for some of the library hardware units for a 1.2μm CMOS technology.

Component	Capacitance Model (in fF)	Parameters
multiplier	$253 * N_1 * N_2$	N_1, N_2: input bitwidths
shifter	$N* 28.7 + \log(M+1)*$ $(43.8 + 12.2N + .06\,N^2 + 0.24MN -$ $0.18\,SN)$	M: Maximum shift allowed N: bitwidth S: Shift
register file	$87 + 51\,R + 35\,N + 8\,R\,N$	N: bitwidth R: number of registers

The capacitance per access for a ripple carry adder and a carry-select adder as a function of the bitwidth are shown in Figure 8.16.

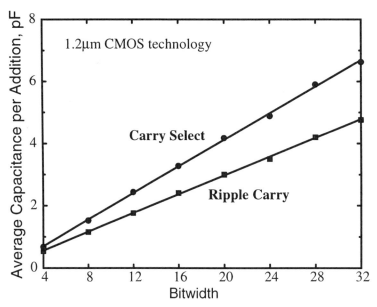

Figure 8.16: Average capacitance per addition for two different modules.

The model presented above assumes the capacitance contribution due the execution units is relatively independent of allocation; however, as described in

Chapter 7, this not always a good assumption since time-multiplexing of hardware can increase the capacitance switched.

8.3.1.2 Registers

Registers are treated the same way as the execution units. The existing estimation program gives information about the total number of register accesses (read/write) within a given sampling period. This is essentially the "activity" of the registers. For the purposes of calculating the register energy, it is enough to know this "activity" and the actual number of physical registers is not required. The number of register accesses is multiplied with the average capacitance per register access to yield a register contribution given by:

$$C_{register} = N_{registers} \cdot C_{registers} \tag{255}$$

While the total number of registers is not important in calculating the register switching capacitance, it will affect floorplanning and chip area and therefore the interconnect capacitance. Gated clocks are used and the clock capacitance is taken into account during the characterization of the registers. Each register has a control slice that locally buffers the incoming clock.

8.3.1.3 Interconnect

While estimating the power consumed by the execution units and registers is quite simple and accurate, estimating the interconnect component is a very difficult and challenging task. Driven by yield, floorplanning and synthesis considerations for throughput and area optimization, several elaborate prediction models for total chip and interconnect area have been built and successfully used [Kurdahi89][Heller77]. However, high-level synthesis adds additional requirements on the prediction tools next to accuracy; during the optimization process in high level synthesis, it is necessary to estimate the final cost frequently and therefore computationally intensive models are prohibited, regardless of their precision.

Estimating the interconnect component of the final implementation should not only take into account the effects of a wide spectrum of high level synthesis tools, such as assignment, allocation, scheduling and partitioning into macro blocks, but also the effects of many low-level CAD tools, such as placement, floorplanning and global and detailed routing. Of course, an accurate model for such a complex system can be built only when a particular set of design tools is targeted. As mentioned earlier, we targeted the HYPER high-level synthesis tools and the Lager IV silicon assembler [Shung91]. We used the scalable CMOS design rules provided by Mosis and targeted feature sizes of 1.2μm and 2μm.

The selection of this particular suit of design tools, enabled us to somewhat simplify the estimation process. We concentrated our attention only on the inter-block (between macro blocks, e.g. different datapaths) routing capacitance. The effect of intra-block (between logic modules inside a datapath) routing capacitance is already taken into account during the calculation of execution units and register contribution, by incorporating an average loading capacitance (determined for the datapath compiler used in the LagerIV silicon assembler).

Building a model which will take into account the effects of the various tools mentioned above is a formidable task. An extensive experimental study, followed by in-depth statistical analysis and verification is the only viable solution which will satisfy the contradictory requirements of modeling a complex system, with high accuracy in a computationally efficient manner.

The model for interconnect capacitance was built using fifty examples which were mapped from their Silage descriptions to layout using the HYPER synthesis system and the LagerIV silicon assembler. The selected examples cover a wide variety of DSP applications including linear and nonlinear filters (including FIR, direct form, cascade, parallel, continuous function, ladder, wave digital, Rao-Kailath IIR, polynomial and homomorphic filters), fast transformations algorithms (FFT and DCT), several video and image processing algorithms and audio examples. Selecting the set of examples for building the model was guided by the goal of including as diverse and as typical examples as possible. While the smallest example had only 12 operations, the largest one had more than 400 operations. One half of the examples were pipelined to various extents, and a subset of the rest were transformed using several transformation in different orders. The examples cover a wide variety in the ratio of critical path to available time, amount of parallelism, types of used operations, level of multiplexing, size of the final implementations and other parameters.

After assembling the results for half the examples (for 25 examples), we started building the statistical model. It immediately became apparent that the best correlation is one between the total interconnect capacitance and the implementation area predicted by HYPER. It is widely recognized that the quality of the prediction model is inversely proportional to number of parameters used during the prediction model building [Breiman84]. The number of parameters is equal to the sum of the number of input variables in the model, and amount of data needed to describe the model. We built the interconnect capacitance model using only one variable as the predictor (estimated area of the chip) and the complexity of the curve which fits data was minimized as much as possible. Although it appeared that both the line and quadratic polynomial, and in particular the third order polynomial fit the data well, none of these models passed the strict statistical test for goodness of fit and resubstitution validation procedure. However, a piecewise linear least square fit showed both excellent accuracy and robustness.

The three segment piece-wise linear fit used to model the interconnect is characterized by 4 points, which are (2.95, 16.5), (10, 77.6), (23.5, 217) and (80, 1257). The model is valid from a predicted area of 2 mm^2 (which is equal to the size of a chip with one execution unit, with a few registers and interconnects) to a predicted area of 90 mm^2 (which is equivalent to a chip can that can accommodate more than 30 execution units, with more than hundred registers and multiplexers).

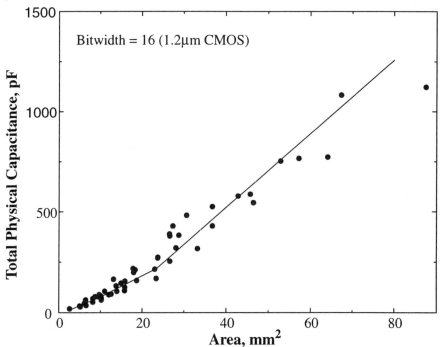

Figure 8.17: Statistical estimation of interconnect capacitance.

The measurement on all 50 examples, showed that the average error of the piecewise linear model is less than 18% percent, and the maximum error is smaller than 33%. Robustness of the model is illustrated by the fact that during 10 random resubstitution of 25 different values for prediction, the largest average error was below 20%, and the maximum error did not exceed 40%.

During the development of prediction tools, often consistency (for two different randomly selected instances, the one with lower predicted value indeed has lower measured value) is more important than accuracy. Consistency of our single prediction variable (predicted area) was, despite the very simple and robust model, higher than 96%. Figure 8.17 shows both the measured (for all 50 examples) and predicted values of the interconnect capacitance as a function of the estimated chip area.

Once the physical interconnect capacitance is accurately estimated, it is easy to establish a good interconnect power consumption model. The interconnect capacitance component is then given by:

$$C_{interconnect} = \alpha * C_{total} / N \qquad (256)$$

where α is the average activity (the total number of interconnect accesses multiplied by an average signal transition probability), C_{total} is the total estimated interconnect capacitance of the chip and N is an estimate of the number of physical interconnects (after bus-merging). The HYPER system provides accurate estimates of the number of interconnects and activity.

The interconnect model was built using the automatic place and route features of the LagerIV placement and layout tool (Flint).

8.3.1.4 Control Logic

The high level estimate for power consumed by control, like that for interconnect, is complicated since the control is not defined until after scheduling and hardware mapping. Neither the number of control blocks nor the their function / size is known at the estimation stage, making it impossible to estimate any properties of the control theoretically. After scheduling, the control is defined and optimized by the *hardware mapper* and further during the *logic synthesis process* before mapping to layout. Like interconnect, therefore, the control needs to be estimated statistically.

The distributed control model used by HYPER is specially suited for low power. The control model incorporates a central finite state machine that generates the state information which is distributed to the local controllers. Control signals (e.g. LOAD for the register file) for the datapath are generated in the local controllers. Bus capacitance on the control lines is reduced by placing the local controllers close to the datapaths. Other than the lines carrying global state information, there are no global control lines.

Statistical models were built to estimate the power consumed by the global and local controllers. The models were generated using several DSP algorithms. Both initial and transformed versions were used to get a complete description of the sample space. Fully pipelined versions were not considered since the critical path and hence the number of states is one, eliminating the need for a controller. Each of the benchmark examples were mapped to SDL (structural description) using HYPER and then into layout using the LAGER IV silicon compiler. IRSIM was used to extract the switching capacitances required to build the statistical model. This process has been automated for characterizing new libraries or the effects of new tools.

Global Control Model:

The amount of capacitance switched in the global controller was found by simulating each benchmark circuit for a whole sample period after the initial conditions were set up. The capacitance switched per sample period in the global controller was found to be directly related to the number of states, N_{states}, and is given by:

$$C_{FSM} = \alpha_1 N_{states} + \alpha_2 \qquad (257)$$

For a 1.2μm technology, α_1 is 4.9fF and α_2 is 22.1fF. Figure 8.18 shows the total capacitance switched by the global controller as a function of the number of states. Note that the number of states not only determines the number of cycles the FSM goes through, but also the number of output bits it drives. Therefore, the total number of transitions and hence the capacitance switched, is strongly dependent on the number of states.

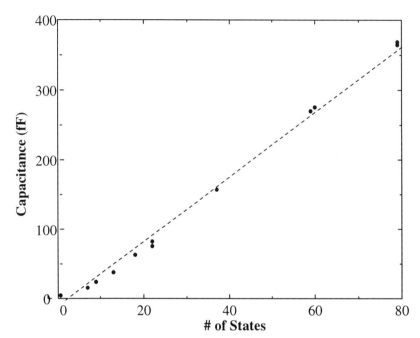

Figure 8.18: Total capacitance switched vs. # of states for the global controller.

Local Control Model:

Since there are several local controllers in each design, the local controllers account for a larger percentage of the total capacitance than the global controller. A set of fifty examples were used to build the capacitance model for the local controller. Table 8.4 shows the data obtained for four different examples implemented using a 1.2μm CMOS technology.

Table 8.4 Results from various local controller units of the selected sample set.

Name	Name (exu-unit)	# of outputs	Total Switched Cap (pF)	# of trans	length, width (in λ)	Area in μ^2
Wave-let (xlarge) 3 States 2 Inputs	add160	5	0.928	14	195, 94	6598.80
	add1610	5	0.943	14	219, 86	6780.24
	add1611	5	0.721	14	76, 46	1258.56
	shr160	11	1.787	34	227,166	13565.5
	shr1610	9	1.509	30	211,174	13217.0
	shr1611	5	0.943	14	211, 86	6532.56
	sub160	5	0.691	14	203, 94	6869.52
	sub1610	7	1.129	22	203,118	8623.44
	sub1611	7	0.980	18	219,118	9303.12
	transfer160	5	0.845	18	203, 78	5700.24
	transfer161	5	0.930	14	203,126	9208.08
	transfer162	5	0.808	18	195, 70	4914.00
IIR 22 States 5 Inputs	adder120	53	82.738	544	451,1504	244189.44
	adder121	47	76.981	528	427, 1232	189383.04
	shr120	30	42.553	296	403, 1048	152043.84
	shr121	25	55.556	302	363, 976	127543.68
	shr122	24	43.253	312	403, 1096	159007.68
	sub120	59	49.855	364	411, 1232	182286.72
	sub121	45	55.556	380	427, 1088	167247.36
	transfer	19	14.706	64	307, 504	55702.08
Dis-crete Cosine trans-form; 36 states; 6inputs	adder80GL	51	128.808	644	539, 1726	334913.04
	shr80GL	35	101.891	428	475, 1566	267786.00
	subtractor 80GL	53	145.349	704	547, 1806	355637.52

Table 8.4 Results from various local controller units of the selected sample set.

Name	Name (exu-unit)	# of outputs	Total Switched Cap (pF)	# of trans	length, width (in λ)	Area in μ²
Aven-haus Ladder Filter 79 States 7inputs	adder 140GL	7	27.68	174	251,344	31083.84
	adder 1410GL	7	24.916	170	243,256	22394.88
	adder 1411GL	7	13.410	174	235,232	19627.20
	adder 1412GL	7	27.635	174	251,280	25300.80
	shr140GL	12	97.841	420	395,1128	160401.60
	shr141GL	13	119.757	560	355,1008	128822.40
	shr142GL	12	102.044	426	403,1000	145080.00
	shr143GL	12	143.883	542	395,1120	159264.00
	subtractor 140GL	25	97.104	430	87,1000	139320.00
	subtractor 141GL	17	78.879	406	371,952	127149.12
	subtractor 142GL	17	81.948	426	355,896	114508.80
	subtractor 143GL	11	71.729	374	339,744	90797.76

The capacitance switched in the local controllers were determined by applying the state input sequence to each of the local controllers and simulating using IRSIM over a whole sample period. Correlations between the capacitance switched and several parameters like the number of states in the control, the number of outputs of the controller, and the total number of transitions on the output control signals were measured. The capacitance of the control was found to be highly correlated to the number of transitions on the output control signals. Note that, in this case, unlike the global controllers, the number of states gives no information about the number of transitions on the output nodes which depends on the glue-logic to be implemented. The transitions must therefore be separately accounted for. Using statistical tools to fit the data to a polynomial function, it was found that the total capacitance switched, in one sample period, for any local

controller is a linear function of the number of transitions, the number of states
and the bus factor and is given by:

$$C_{lc} = \beta_0 + \beta_1 N_{trans} + \beta_2 N_{states} + \beta_3 B_f \qquad (258)$$

where N_{trans} is the number of transitions, N_{states} is the number of states, B_f is the
bus factor (explained below), and C_{lc} is the capacitance switched in any local
controller in one sample period. B_f represents the activity factor on busses and is
defined as the ratio of the number of bus accesses to the number of busses. It is a
measure of the average number of times busses are accessed. It represents a measure on the number of multiplexors and each multiplexor requires control signals
from the controller. For a 1.2 μm technology, β_0, β_1, β_2 and β_3 are 72, 0.15, 8.3
and 0.55 respectively. Figure 8.19 shows the correlation between the actual and
predicted capacitance switched per sample period for several different examples.

This model, though accurate to within 20% (as indicated by resubstitution
validation procedure) is not sufficient since it assumes that the number of transitions is known. The number of transitions depends on assignment, scheduling,
optimizations performed by the hardware mapper and the logic optimization

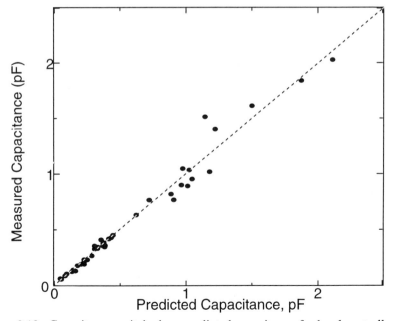

Figure 8.19: Capacitance switched vs. predicted capacitance for local controllers.

tool(misII), the standard cell library used, the amount of glitching, and the statistics of the inputs. It is impossible to determine the combined effects of all these elements apriori. A statistical model relating the number of transitions to high level parameters was therefore derived. The important high level parameters that affected the number of transitions are:

- Size/complexity of the graph, i.e. the number of nodes and the number of edges. This is because the number of reads from and write to registers is dependant on the number of edges and nodes in the graph respectively. Control signals need to be generated for every read and write process.

- The number of execution units times the number of states. This is because each execution unit receives the clock and clock inverse every control cycle. In the hardware model used, the clock inverse is generated in the local controllers and fed to the execution units.

The HYPER high level synthesis system provides a lower bound on the number of execution units within as accuracy of 10% [Rabaey91b]. The lower bound on the busses tracks the actual number of busses closely and the maximum number of busses tracks the total number of bus accesses. The numbers predicted by HYPER were therefore used in building the correlation model. The fit function obtained for the total number of output transitions on local controllers based on the three high level parameters mentioned above is given by:

$$N_{trans} = \gamma_1 + \gamma_2 \left(N_{nodes} + N_{edges} \right) + \gamma_3 \left(S \times N_{Exu} \right) \qquad (259)$$

where N_{trans} is the number of transitions on the outputs of the local controllers, S is the number of control cycles per sample period, N_{edges} and N_{nodes} are the number of edges and nodes respectively in the CDFG and N_{exu} is an estimate for the total number of execution units. For a 1.2μm technology γ_1, γ_2 and γ_3 are 178.7, 7.2 and 2.0 respectively. Figure 8.20 shows the total number of transitions in one sample period for the different examples vs. the fit function for the transitions based on high level parameters.

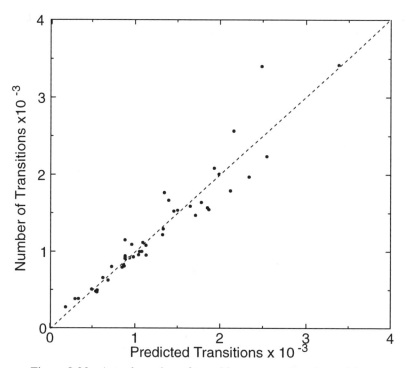

Figure 8.20: Actual number of transitions vs. predicted transitions
for local controllers.

This statistical estimate takes into account the effects of logic synthesis and optimization. Effect of undetermined factors such as glitching are also included. Each datapath element in the cell library has a built in control-slice to locally buffer the incoming control signals. The internal gate and routing capacitance is taken into account when characterizing the leafcells. For example, the multiplexor select signal for a 16-bit datapath is buffered in the control slice. Therefore, the controller only drives a single (typically minimum sized) gate. This information was used to determine the loading capacitance during simulations.

8.3.2 Supply Voltage Estimation

The power supply voltage at which the flowgraph implementation will meet the timing constraints is estimated. The initial flowgraph which meets the timing constraints is typically assumed to be operating at a supply voltage of 5V with a critical path of $T_{initial}$ (the initial voltage will be lower if $T_{initial} < T_{sam}$-

$_{pling}$). After each move, the critical path is re-estimated, and the new supply voltage at which the transformed flowgraph still meets the time constraint, $T_{sampling}$, is determined. For example, if the initial solution requires 10 control steps (and let's assume that this is the same as the sampling period) running at a supply voltage of 5V, then a transformed solution that requires only 5 control steps can run at a supply voltage of 2.9V (where the delay increases by a factor of 2, Figure 3.24) while meeting the same constraints as the initial graph.

A model for delay as a function of V_{dd} is derived from the curve shown in Figure 8.21. Let the speedup of a transformed solution be defined as:

$$Speedup = T_{sampling} / T_{criticalpath} \qquad (260)$$

where $T_{criticalpath}$ is the critical path of the transformed solution and $T_{sampling}$ is the throughput constraint.

Since, the major power reduction during CDFG optimization using transformations is attributed to a reduction of the supply voltage and since the supply voltage has to be constantly evaluated (every time the objective function is called), it is important to model delay-V_{dd} relationship using an accurate and computationally efficient procedure. This relationship (of delay-V_{dd}) was modeled using Neville's algorithm for rational function interpolation and extrapolation [Press88]. Neville's algorithm provides an indirect way for constructing a polynomial of degree N-1 so that all of the used points are exactly matched. Figure 8.21 also shows the accuracy of the interpolated data (plotted for speedup values ranging from 0.86 to 26 with increments of 0.1) when compared to the experimental data.

Figure 8.22 shows an overview of the power estimation routine. It takes a control dataflow graph as input and computes the power using the existing estimation routines (critical path, bounds on resources and activity) along with the newly developed capacitance and voltage estimation routines (as discussed in this section).

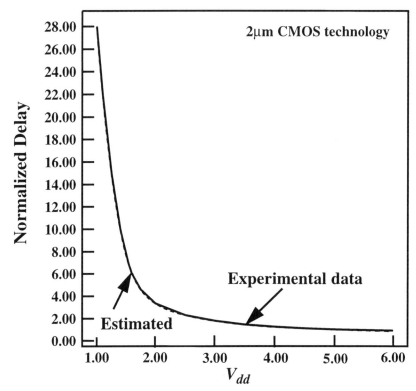

Figure 8.21: Accuracy of V_{dd} estimation.

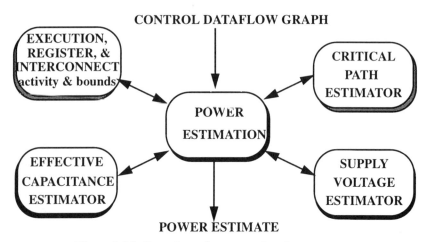

Figure 8.22: Overview of power estimation.

8.4 Optimization Algorithm

The algorithm development for power minimization using transformations was greatly driven by several considerations and constraints. The basic strategy involved:

- Efficient tailoring of existing library transformations for power optimization.

- Fully utilizing the information about the design solution space obtained through extensive experimentation and theoretical insights (so that the algorithm can quickly come out of deep local minima).

- Efficient use of computer resources (CPU time and memory) so that large real-life examples can be addressed.

- Developing modular software so that future enhancements or changes can easily be incorporated.

The computational complexity analysis of the power minimization problem showed that even highly simplified versions of the power optimization tasks using transformations are NP-complete. For example, we have shown that retiming for power minimization is an NP-complete problem. Computational complexity of retiming for power optimization is particular interesting and somewhat surprising, since retiming for critical path reduction has several algorithms of polynomial-time complexity [Leiserson91].

The computational complexity of the power minimization problem implies that it is very unlikely, even when the set of applied transformations is restricted, that a polynomial time optimal algorithm can be designed. Two widely used alternatives for design of high quality suboptimal optimization algorithms are probabilistic and heuristic algorithms. Both heuristic and probabilistic algorithms have several distinctive advantages over each other. While the most important advantage of heuristic algorithms is a shorter run time, probabilistic algorithms are more robust and have stronger mechanisms for escaping local minima's.

An extremely wide spectrum of probabilistic algorithms are commonly used in various CAD and optimization problems. Among them, the most popular and successful are simulated annealing, simulated evolution, genetic algorithms, and various neural network algorithms (e.g. Boltzmann machine). Although recently there has been a considerable effort in analyzing these algorithms and the type of problems they are best suited for (for example a deep relationship between simulated annealing and solution space with fractal topology have been verified both experimentally and theoretically [Sorkin91]), algorithm selection for the task at hand is still mainly an experimental and intuitive art.

In order to satisfy all major considerations for the power minimization problem, we decided to use a combination of heuristic and probabilistic algorithms so that we can leverage the advantages of both approaches. Before we get into the algorithmic details, we will briefly outline the set of transformations used.

The transformation mechanism is based on two types of moves, global and local. While global moves optimize the whole DCFG simultaneously, local moves involve applying a transformation on one or very few nodes in the DCFG. The most important advantage of global moves is, of course, a higher optimization effect; the advantages of local moves is their simplicity and small computational cost. We used the following global transformations (i) retiming and pipelining for critical path reduction (ii) associativity (iii) constant elimination and (iv) loop unrolling. In the library of local moves we have implemented three algebraic transformations: associativity (generalized to include properties of inverse elements as introduced in [Potkonjak91]), and commutativity as well as local retiming. All global moves minimize the critical path using polynomial optimal algorithms. Their use is motivated by the need for shorter run-times. For example, the global pipelining move for a 7th order IIR filter with 33 nodes took 1.8 CPU seconds on a SPARC2 workstation. Extensive experimentation showed that although the best solutions in semi-exhaustive searches is rarely one with the minimum critical path, it is very often topologically very close (a small number of moves is needed to reach it starting from a solution with the shortest critical path).

The Leiserson-Saxe algorithm for critical path minimization using retiming is used. The algorithm is modified in such a way that it automatically introduces the optimal number of pipeline stages needed to minimize the cost function for power. Pipelining is accomplished by allowing the simultaneous addition of the equal number of delay elements on all inputs or all outputs but not both. The Leiserson-Saxe algorithm is strictly based on minimizing the critical path and inherently does not optimize the capacitance being switched for a given level of pipelining. The modified algorithm determines the level of pipelining where the power is minimized and this is used as a "good" starting point for optimization using local moves (as will be discussed below). The global move for associativity involves the minimization of the critical path using the dynamic programming algorithm. Loop unrolling does not involve any optimization, instead it enables transformations which a reveal large amount of concurrency for other transformations. For each local move, we defined the inverse local move which undoes the effect of the initial move.

As previously mentioned, the algorithm for power minimization using transformations has both heuristic and probabilistic components. While the heuristic part uses global transformations, the probabilistic component uses local moves. The heuristic part applies global transformations one at the time in order to provide good starting points for the application of the probabilistic algorithm. The probabilistic algorithm conducts a search in a broad vicinity of the solution provided by the heuristic part with the goal of minimizing the effective capacitance (trading-off control, interconnect and activity). The underlying search mechanism of the probabilistic part is simulated annealing with the goal that local minima's can be escaped by using uphill moves and the effects of various transformations can be efficiently combined so as to minimize the capacitance component of power.

Initially constant propagation is applied to reduce both the number of operations and the critical path, followed by the modified Leiserson-Saxe algorithm and the dynamic programming algorithm for the minimization of the critical path using associativity. This step is followed by simulated annealing which tries to improve the initial solution by applying local moves probabilistically. For simulated annealing we used the most popular set of parameters [Algoritmica91]. For example, we used a geometric cooling schedule with a stopping criteria which terminates the probabilistic search when no improvement was observed on three consecutive temperatures. After each move, a list of all possible moves is generated for each local transformation. The transformation and particular move to be applied is then selected randomly. In addition to the above algorithm, loop-unrolling can be used as preprocessing step.

HYPER-LP also allows optimization using a user specified sub-set of transformations (for example, optimizing with only pipelining). Figure 8.23 shows an overview of the HYPER-LP system.

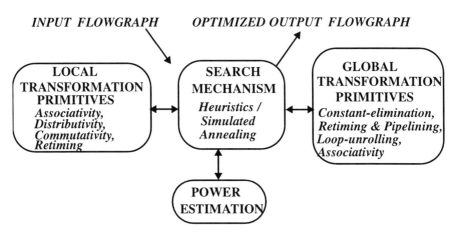

Figure 8.23: Overview of the HYPER-LP system.

8.5 Examples and Results

The techniques described in the previous sections will now be applied to three DSP examples to demonstrate that a significant improvement in power can be achieved. Three examples of distinctly different computational structures are presented in this section and are optimized for power using the HYPER-LP system. Multiplications with constants were converted to shift-add operations. The improvement from this transformation was not taken into account since this it is a standard transformation performed for area or throughput optimization.

8.5.1 Example #1: Wavelet Filter

The first example is a wavelet filter, used in speech and video applications. The implementation contains high-pass and low-pass filters that are realized as 14th order FIR modules using a wordlength of 16-bits (determined by high-level simulation). This example is representative of a wide class of important signal processing applications such as FFT, DCT (Discrete Cosine Transform), matrix multiplication, cyclic convolution and correlation that have no feedback loops in the signal flowgraph. For this class of applications, retiming and pipelining provides an efficient and straight forward way to reduce critical paths (and thus voltage) while preserving throughput and the number of operations.

The initial reference design is a fully time-multiplexed area-optimized realization of the filter, with minimal amount of resources (1 adder, 1 subtractor, 1 shifter, and 4 global busses). The number of global busses were minimized using global bus-merging (i.e., busses that are never accessed simultaneously can be collapsed) with the goal of minimizing the area. This solution has a critical path of 22 clock cycles and can be clocked at a maximum frequency of 1.5Mhz (since the critical path for this 16-bit datapath is around 30ns). Using HYPER-LP, it was found that retiming by itself was sufficient to reduce the critical path of this design to 3 control cycles. Since the throughput requirement is 1.5Mhz, the clock period can be made approximately 7 times as long (= 22 / 3) while meeting the timing constraint and therefore the supply voltage can be reduced to 1.5V.

Since the amount of resources increases significantly (since the allocation is done with available time = critical path = 3 cycles), we expect the chip area and hence the interconnect component of power to increase. This is seen from Figure 8.24, which shows a plot of the average capacitance per bit obtained from layout extraction as a function of the bus number for the initial 22 cycle implementation (which contained 4 busses) and for the final 3 cycle implementation (which con-

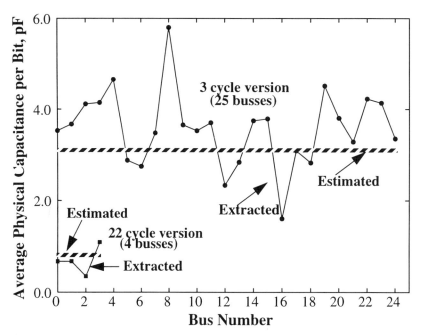

Figure 8.24: Bus capacitance for the wavelet filter (estimated and extracted).

tained 25 busses). The figure also shows the capacitance obtained from the estimation of interconnect using the model presented in section 8.3.1.

The global bus capacitance switched increased by a factor of 3 from 400pF to 1200pF. However, the total capacitance switched was actually *smaller* than the minimal area version! There are several factors contributing to this decrease in capacitance:

- The amount of multiplexor, tri-state and control switches (which contributed to 45% of the total power of the minimal area design) were reduced significantly.

- The initial design uses bus-merging to significantly reduce the interconnect area. However, this does not come for free as the loading for the tri-state buffers driving the bus increases. Also all of the multiplexors whose inputs are connected to the shared bus will switch every time the bus value transitions, resulting in extra power. The problem is not as severe in the final design since there is little room for bus-sharing since there are fewer control cycles.

- The average transition activity (which is not modeled by the estimator presented in Section 8.3) and hence the capacitance per access for the logic modules (adder, subtractor, and shifter) was significantly lower in the final version. The reason is that in the initial version, since the modules were fully time-multiplexed, the inputs to the modules were coming from different sources (and hence uncorrelated) and had a higher probability of transitioning. In the final version (in which many units were allocated), there are longer periods of inactivity during which the inputs to the modules do not change and hence the units do not switch, resulting in lower activity.

The total power hence reduced by a factor of 18 (these results are obtained from the IRSIM switch level simulator using a linear RC model using real input data). Table 8.5 shows the statistics of the initial and final designs. Note that area has been traded for lower power.

Table 8.5 Results for the Wavelet Chip (1.2μm CMOS technology)

Version	f_s MHz	# of Control Cycles / Clock Period	Supply Voltage	Total Average Capacitance	Power, mW	Area mm^2
Initial	1.5	22 / 30ns	5V	2870pF	107	8.5

Table 8.5 Results for the Wavelet Chip (1.2μm CMOS technology)

Version	f_s MHz	# of Control Cycles / Clock Period	Supply Voltage	Total Average Capacitance	Power, mW	Area mm^2
Low Power Design	1.5	3 / 220ns	1.5V	1735pF	5.8	62.9

8.5.2 Example #2: IIR Filter

The second example is a 16-bit 7th order IIR filter with a 4th order equalizer designed using the filter design program Filsyn [Comsat82]. This example is representative of the largest class of examples where feedback is an inherent but not particularly limiting part of the computation structure.

Unlike the previous example, when the transformation set of the HYPER-LP system was limited only to retiming, no reduction in the power consumption was observed. However, when the transformation set was enhanced to include both retiming and pipelining, the tool was very effective and achieved a reduction in the number of control steps used from 65 to 6 cycles. The implementation area increased as the number of control cycles decreased. Figure 8.25a shows a plot of the number of execution units vs. the number of control cycles. The number of registers increased from an initial number of 48 (for 65 control cycles) to 102 (for 6 control cycles). This is a typical example where area is traded for lower power. It is immediately evident that the price paid for reducing critical path is the increased registers and routing overhead. Figure 8.25b shows a plot of the normalized clock cycle period (for a fixed throughput) as a function of number of control cycles. We see that reducing the number of control cycles through transformations allows for an increase in clock cycle period or equivalently a reduction in supply voltage. However, at very low voltages, the overhead due to routing increases at a very fast rate and the power starts to increase with further reduction in supply voltage. The power reduced approximately by a factor of 8 as a result of dropping the voltage from 5V to approximately 1.5V. Figure 8.26 shows a plot of Power vs. V_{dd} for a fixed throughput. Note that the low-power solution is somewhere between the minimal area solution and the maximum throughput solution.

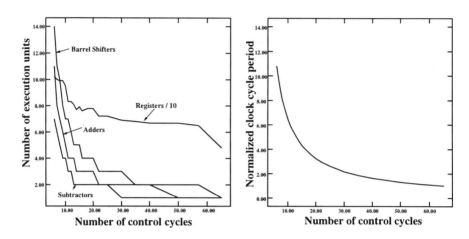

Figure 8.25: Plot of # of units (a) and clock cycle period (b) vs. # of control cycles
for the IIR example

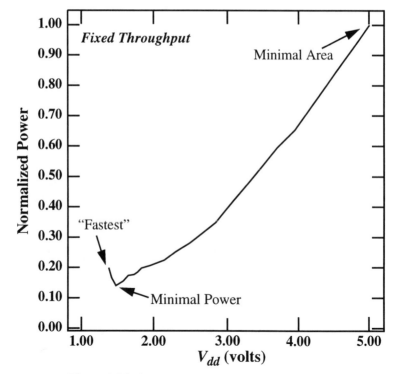

Figure 8.26: Power vs. V_{dd} for the IIR example.

8.5.3 Example #3: Volterra Filter

The third example is a second order Volterra filter. This is a particularly challenging example since pipelining cannot be applied due to a recursive bottleneck imposed by long feedback loops (optimizing with pipelining as the only transformation results in power reduction only by a factor of 1.5). For this case, loop unrolling, retiming and associativity transformations proved to be important in alleviating the recursive bottleneck, allowing for a reduction in critical path and voltage.

Figure 8.27 shows the final layouts of optimizing the Volterra filter for power as well as for area while keeping the throughput fixed. This final implementation has approximately nine times better power characteristic and the same area as the initial implementation.

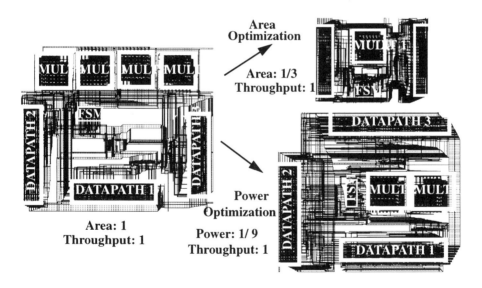

Figure 8.27: Optimizing the volterra filter.

Table 8.6 shows a summary of power improvement after applying transformations relative to initial solutions that met the required throughput constraint at 5V for the representative examples described in this section. The results indicate that a large reduction in power consumption is possible (at the expense of area) compared to present-day methodologies. Also interesting was the fact that the

optimal final supply voltage for all the examples was much lower than existing
standards and was around 1.5V.

Table 8.6 Summary of results.

Example	Power Reduction	Area Increase
Wavelet Filter	18	7.4
IIR Filter	8	6.4
Volterra Filter	9	1

8.6 Summary

The focus of this chapter is on automatically finding computational structures that result in the lowest power consumption for DSP applications that have a
specific throughput constraint given a high-level algorithmic specification. The
synthesis approach consisted of applying transformation primitives in a well
defined manner in conjunction with efficient high-level estimation of power consumption. The basic approach is to scan the design space utilizing various algorithmic flowgraph transformations, high-level power estimation, and efficient
heuristic/probabilistic search mechanisms. The estimation of power consumption
involved estimating the total capacitance switched and the operating voltage. The
estimation of the total capacitance switched includes four components: execution
units, registers, interconnect and control. Both analytical and statistical
approaches were used in the capacitance modeling. The synthesis system was
used to optimize power for several design examples and the results indicate that
more than an order of magnitude reduction in power is possible over current-day
design methodologies (that tend to minimize area for a given throughput) while
maintaining the system throughput. This chapter has addressed some key problems in the automated design of low-power systems, and provides a good starting
point for addressing other research problems like detailed power estimation, module selection, partitioning, and scheduling for power optimization.

REFERENCES

[Aho77] A. V. Aho and J.D. Ullman, *Principles of compiler Design*, Reading, Addison-Wesley,
    1977.

[Algoritmica91] *Algoritmica*, Special Issue on Simulated Annealing, A. Sangiovanni-Vincentelli, ed. Vol. 6, No. 3, pp. 295-478, 1991.

[Breiman84] L. Breiman, J.H. Friedman, R.A. Olshen, and C.J. Stone: "Classification and Regression Trees", Wadsworth & Brooks, CA 1984.

[Chandrakasan95] A. P. Chandrakasan, R. Mehra, M. Potkonjak, J. Rabaey, and R. W. Brodersen, "Optimizing Power Using Transformations," *IEEE Transaction on CAD*, pp. 13-32, January 1995.

[Cirit87] M. Cirit, "Estimating Dynamic Power Consumption of CMOS Circuits," *International Conference on Computer-Aided Design*, pp. 534-537, 1987.

[Comsat82] Comsat General Integrated Systems, Super-Filsyn Users Manual, March 1982.

[Deng94] A. Deng, "Power Analysis for CMOS/BiCMOS Circuits," *1994 International Workshop on Low-power Design*, pp. 3-8, April 1994.

[Ghosh92] A. Ghosh, S. Devedas, K. Keutzer, and J. White, "Estimation of Average Switching Activity," *Design Automation Conference*, 1992.

[Goertzel68] G. Goertzel, "An algorithm for the Evaluation of Finite Trigonometric Series," *Amer. Math. Monthly*, Vol. 65, No. 1, pp. 34-35, 1968.

[Goldberg91] D. Goldberg, "What every computer scientist should know about floating-point arithmetic," *ACM Computing Surveys*, Vol. 23, No. 1, pp. 5-48, 1991.

[Haroun89] B.S. Haroun and M.I. Elmasry, "Architectural Synthesis for DSP Silicon Compilers," *IEEE Transaction on CAD*, Vol. 8, No. 4, pp. 431-447, 1989.

[Heller77] W.R. Heller *et al.*, "Prediction of wiring space requirements for LSI," *Design Automation Conference*, pp. 20-22, 1977.

[Jain85] R. Jain, J. Vandewalle, and H. J. DeMan, "Efficient and accurate Multiparameter Analysis of Linear Digital Filters Using a Multivariable Feedback Representation," *IEEE Transaction on CAS*, CAS-32:225-235, March 1985.

[Kimura91] N. Kimura and J. Tsujimoto, "Calculation of Total Dynamic Current of VLSI Using a Switch Level Timing Simulator (RSIM-FX)," *CICC*, 1991.

[Kurdahi89] F.J. Kurdahi and A.C. Parker: "Techniques and Area Estimation of VLSI Layouts," *IEEE Transaction on CAD*, Vol. 8, No. 1, pp. 81-92, Jan 1989.

[Landman93] P. E. Landman and J. M. Rabaey, "Power Estimation for High Level Synthesis," *EDAC '93*, pp. 361-366, Feb. 1993.

[Leiserson91] C.E. Leiserson and J.B. Saxe, "Retiming Synchronous Circuitry," *Algorithmica*, Vol. 6., No. 1, pp. 5-35, 1991.

[Messerschmitt88] D. Messerschmitt, "Breaking the Recursive Bottleneck," 1988, in *Performance Limits in Communication Theory and Practice*, J.K. Skwirzynski, pp. 3-19, 1988.

[Najm91] F. Najm, "Transition Density, A Stochastic Measure of Activities in Digital Circuits," *Design Automation Conference*, pp. 644-649, 1991.

[Parhi89] K.K. Parhi, "Algorithm Transformation Techniques for Concurrent Processors," *Proceedings of the IEEE*, Vol. 77., No. 12, pp. 1879-1895.

[Potkonjak91] M. Potkonjak and J. Rabaey, "Optimizing the Resource Utilization Using Transformations," *International Conference on Computer-Aided Design*, pp. 88-91, November 1991.

[Potkonjak92] M. Potkonjak and J. Rabaey, "Maximally Fast and Arbitrarily Fast Implementation of Linear Computations," *International Conference on Computer-Aided Design*, pp. 304-308, 1992.

[Press88] W.H. Press, B.P. Flannery, S.A. Teukolsky, and W.T. Vetterling, *Numerical Recipes in C*, Cambridge University Press, 1988.

[Salz89] A. Salz and M. Horowitz, "IRSIM: An Incremental MOS Switch-level Simulator," *Design Automation Conference*, pp. 173-178, June 1989.

[Schultz92] D. Schultz, "The Influence of Hardware Mapping on High-Level Synthesis," U.C. Berkeley M.S. report, ERL M92/54.

[Shung91] C. S. Shung *et al.*, "An Integrated CAD System for Algorithm-Specific IC Design," *IEEE Transactions on CAD*, April 1991.

[Sorkin91] G.B. Sorkin, "Efficient Simulated Annealing on Fractal Energy Landscapes," *Algoritmica*, Vol. 6, No. 3, pp. 367-418, 1991.

[Rabaey91a] J. Rabaey, C. Chu, P. Hoang, and M. Potkonjak, "Fast Prototyping of Data Path Intensive Architecture," *IEEE Design and Test*, Vol. 8, No. 2, pp. 40-51, 1991.

[Rabaey91b] J. Rabaey and M. Potkonjak, "Estimating Implementation Bounds for Real Time Application Specific Circuits," *European Solid-State Circuits Conference*, Milano, Italy, pp. 201-204, September 11-13, 1991.

[Trickey87] H. Trickey, "Flamel: A high-Level Hardware Compiler," *IEEE Transaction on CAD*, Vol. 6, No. 2, pp. 259-269, 1987.

[VanOostende93] P. Van Oostende, P. Six, J. Vandewalle, and H. De Man, "Estimation of Typical Power of Synchronous CMOS Circuits Using a Hierarchy of Simulators," *IEEE Journal of Solid-State Circuits*, vol. 28, no. 1, Jan. 1993.

[Walker89] R.A. Walker and D.E. Thomas, "Behavioral Transformation for Algorithmic Level IC Design," *IEEE Transaction on CAD*, Vol 8. No.10, pp. 1115-1127, 1989.

[Walker91] R.A. Walker, R. Camposano, *A Survey of High-Level synthesis Systems*, Kluwer, MA, 1991.

9

A Portable Multimedia Terminal

The near future will bring the fusion of three rapidly emerging technologies: personal communications, portable computing, and high bandwidth communications. Over the past several years, the number of personal communications services and technologies has grown explosively. For example, voiceband communications systems such as mobile analog cellular telephony, radio pagers, and cordless telephones have become commonplace, despite their limited nature and sometimes poor quality of transmission. In portable computing, "notebook" computers more powerful than the desktop systems of a few years ago are commonplace. Wide area high bandwidth networks that are in the planning stages will provide the backbone to interconnect multi-media information servers. Wireless communications through radio and IR links are being used as LAN replacement as well as to provide ubiquitous audio communications. However, there has yet to be an integration of these diverse services, in part because of the difficulty in providing a portable terminal that can process the high speed multimedia data provided by the network servers. The circuitry to support the wireless link for such a terminal is being investigated and a low-power solution to this problem is felt to be feasible [Sheng92]. The focus of this chapter is on the processing between this wireless modem and the terminal I/O devices with emphasis on how this processing can be realized with minimal possible power consumption.

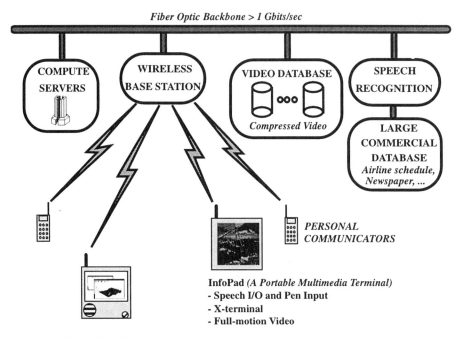

Figure 9.1: Overview of a future personal communications system.

In Figure 9.1, a schematic view of a wireless multimedia access system developed at U.C. Berkeley called the InfoPad is shown. In addition to providing portable voice communication to the user, this system will provide portable access to the wide variety of information and computation services that will be offered on the future wide band network. This will include access to large commercial databases (which will contain news, financial information, educational information, etc.), video databases (which will enable video on demand), and compute servers (which can be used to run applications such as spreadsheets and word processors). By coupling to this high-bandwidth network through a low-latency interface, the InfoPad terminal gives the user equally good service no matter where the person and the service provider is located in a building. Simplified entry mechanisms such as voice-recognition and handwriting-recognition offer novel access to the above functions. The design of an effective user interface to access such a vast information storehouse is a critical issue. By using speech recognition and pen-based input, supported by large, speaker-independent recognition servers placed on the network, such interfacing and information access can be tremendously simplified. Placing the recognition units on the network conserves power in the portable, and enables much larger and much more complex recognition algorithms to be employed.

Figure 9.2 summarizes the functionality that must be provided by the portable terminal to support the access of various multimedia information servers. This includes the interface to a high speed wireless link, text/graphics output, simplified user interfaces like pen input and speech I/O, and finally support for one-way full-motion video. The key design consideration for such a portable multimedia terminal is the minimization of power consumption. The portable multimedia form factor must be slim and light weight, while being durable and tolerant of the users environment. The focus of this chapter is on the low-power approaches used in the design of a chipset that performs the various I/O functions required by the InfoPad terminal. Minimizing power consumption of the terminal electronics requires a design methodology that supersedes the technology and circuit levels and addresses architecture, algorithm and system partitioning as well.

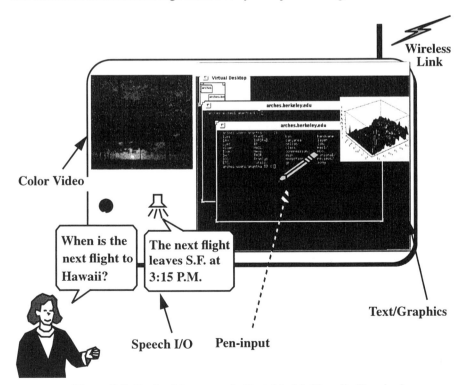

Figure 9.2: Desired features of a Portable Multimedia Terminal.

9.1 Technology Trends for Low Power Portable Systems

9.1.1 Battery Technology

The energy capacity of batteries has been improving over the last two decades, but at a very slow pace. The most popular rechargeable battery in use is nickel-cadmium, which has two times more capacity today than it did two decades ago. There have been significant efforts to come up with rechargeable high capacity batteries which are environmentally safe. The key competition for nickel-cadmium batteries has been nickel-metal hydride (NiMH) which provides improved energy density with the promise of reducing or eliminating regulatory concerns regarding the disposal of toxic heavy metals [Eager91]. Nickel-metal hydride batteries posses many operating similarities to traditional nickel-cadmium, but have intrinsically different cell chemistry. Figure 9.3 shows the energy density for three different battery technologies over the last two decades. As shown on this figure, improvements in battery technology only provides a marginal solution to portability requirements, and therefore we cannot rely solely on battery improvements.

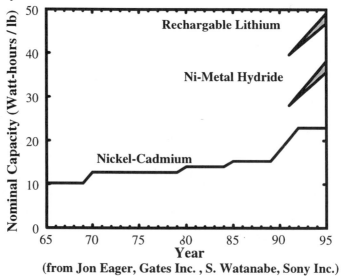

(from Jon Eager, Gates Inc. , S. Watanabe, Sony Inc.)

Figure 9.3: Trends in battery technology for the last 20 years.

9.1.2 Display Technology

Improvements in display technology will be a key element to the implementation of future low-power portable systems. Currently, two display technologies are used for laptop computers: passive displays and active-matrix displays. Passive displays consume significantly lower power (2-3W) while having a large response time (150mS). For applications that require a fast response time, like full-motion video, active matrix displays are required. In color active matrix displays, the majority of power goes into the backlight - required due to the poor transmission characteristics of the color filters - which requires more than 10W.

Fortunately, improvements are being made in color display technology that uses reflective technology. SHARP recently introduced a prototype reflective color active matrix display. They have used a technology that forms high-reflectivity contacts using an organic dielectric film deposited on top of the TFT pixels. Typically, reflective LCDs place the reflector on the outside of the LCD panel glass, but in their new device, the pixel electrodes double as reflectors, making it possible to view the screen even from oblique angles while delivering a bright, parallax-free display without generating double images. Since they don't require a backlight, the power consumption is only 50mW. Figure 9.4 shows a 5" (320 x 240 pixels) prototype display which currently has 4 colors, with 512 color versions in development.

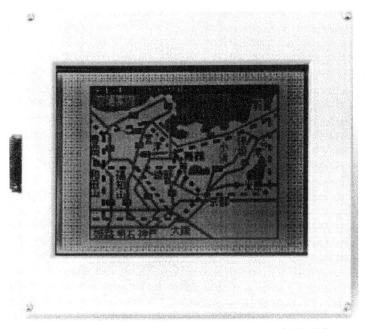

Figure 9.4: Low-power color TFT display (courtesy of SHARP).

9.1.3 Packaging Technology

Implementing systems using present day technology - single chip packaging using printed circuit boards for interconnection - results in a third or more of the total power being consumed at the chip input/output (I/O), since the capacitances at the chip boundaries are typically much larger than the capacitances internal to the chip. Typical values range from a few 10's of femtofarads at the chip internal nodes, to 10's of picofarads at the chip interface attributed to pad capacitance (approximately 5-10 pF/pin) and PCB traces (3-4 pF/in)[Benson90].

Advances in packaging technology will tremendously improve system performance by lowering power consumption, reducing system delays, and increasing packing density. The emerging multichip module (MCM) technology is one important example in that it integrates many die onto a single high-density interconnect structure, hence reducing the size of inter-chip capacitances to the same order-of-magnitude as on-chip capacitances [Benson90]. This reduces the requirements on the CMOS output drivers, and thus can reduce total power by a factor of 2 to 3. Since this style of packaging also allows chips to be placed closer together, system delays are reduced and packing density increases. Thus, with MCM's, the majority of the power is consumed within the functional core of the chip itself, as opposed to the interface.

9.2 System Partitioning for Low-power

Future systems will allow users to be connected in a wireless fashion to fixed multimedia servers. This provides a major degree of freedom for power minimization at the system level which involves the partitioning of computation between the portable terminal and the computer servers on the backbone network. One approach to implement the various general purpose applications (X-server, word processing, etc.) for the portable terminal is to use a general purpose processor. This approach can result in power consumption levels of 10's of Watts; for example, the DEC Alpha chip consumes 30W running at 100MHz. The problem of implementing low power, general purpose processing is primarily being addressed by power down strategies of subsystems, with a limited amount of voltage scaling. This is effective to some extent, but because of the difficulty in determining when an application is really idle, there also needs to be modifications to the software and operating system to fully utilize processor inactivity. Unfortunately, a primary figure-of-merit for computers, that use one of the industry standard microprocessors, has become the clock rate, which results in architectures which are antagonistic to low power operation. On the other hand, as discussed in Chapter 4, parallel architectures operating at reduced clock rates, can have the *same* effective performance in terms of instructions/second at substantially reduced power levels.

Another strategy to reducing the power required for general purpose computation is to effectively **remove** it from the portable. This is accomplished by providing communications capability to computational resources, either at a fixed base station at the other end of a wireless link or back on the network at available compute servers. This is simply the use of the well known client-server architecture, in which the portable unit degenerates to an I/O server (e.g. a wireless X-terminal). Client-server computing environments, such as the MIT X-Window system, have made it clear that computation need not be done on the local machine (the display server) that a user is operating; instead, the computation is done by programs executing on many remote machines (clients), that simply issue graphics commands to the server to display their results. Many inexpensive "X-terminals" already exist. Unlike TTY terminals that can only communicate with a single host machine, such X-terminals possess all of the necessary networking capability to communicate with as many remote servers as needed. It is upon this model that the Infopad multimedia terminal is based; remote computation servers will be used to run applications such as spreadsheets, word processors, and so forth, with the results being transmitted to the terminal.

Figure 9.5, illustrates this trade-off for partitioning the X-server computation. On the left side is a conventional implementation in which the X-server runs on a general purpose processor and it communicates with client applications using X-protocol. This approach as mentioned above is very power inefficient. Another approach, as shown on the right, is to use an X-server split between the terminal and the backbone; the basic strategy is to transmit raw pen data over the uplink, perform the processing (X-server as well as client applications) on the base station or on a remote compute server sitting on the wired backbone and transmit incremental bit-maps (the updates to the text/graphics display) from the X-server over the down-link. Hence the processing for supporting text/graphics in the terminal degenerates to frame-buffer interface and display interface.

The approach of transmitting bitmaps for text/graphics versus X-protocol will indeed tend to slightly increase the overall bandwidth for the downlink traffic. However, the savings in power more than compensates for the slight increase in bandwidth. Also, since some client applications like Frame-maker use a

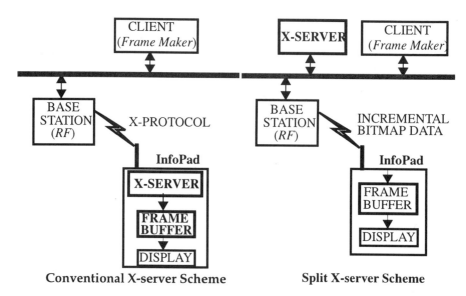

Figure 9.5: Partitioning of computation: X-server example.

bit-map approach anyway, the overall bandwidth of the bit-mapped approach compares favorably with the X-protocol approach. Table 9.1 shows a breakdown of bandwidth for various data types. The link is very asymmetrical as seen from the data rates of the uplink and the downlink; no high-rate signals are intended for transmission from the mobile to the base station.

Table 9.1 Bandwidth requirement for various data types.

Data Type	Downlink	Uplink
Compressed Color Video	690Kbs	-
Speech	64Kbs	64Kbs
Pen	-	4Kbs
Text/Graphics	Variable < 100Kbs	-

To realize a 1Mbs bandwidth per user, cellular networking techniques are used to achieve spatial reuse [Sheng92]. Unlike conventional sized cells, which have a radius on the order of kilometers, we are using very small picocells (with a radius on the order of 5m). Using picocells implies that we will need about

100Mhz of RF spectrum to support a high density of users (~8-10 users in a 10m radius) while using conventional sized cells would require more than 1 Ghz of RF spectrum.

The use of picocells is another example of significant power reduction that can be achieved by optimizing at the system level. The transmit power is scaled down as the cells move closer together to reduce interference and the power consumed in the portable's transmitter to drive the antenna is correspondingly reduced. Whereas existing cellular systems utilize 1 watt transmit power for voiceband RF links in 5 mile cells, a picocellular system with 10 meter cells will require less than one milliwatt to maintain the link.

Similarly to the X-server computation, certain I/O processing tasks such as speech and handwriting recognition are moved to the backbone. PCM sampled speech is transmitted back and forth over the wireless link. Since video is accessed over a wireless network with limited available spectrum (< 1Mbits/sec allocated per user), video is transmitted in a compressed format and therefore decompression has to be performed on the terminal. For I/O processing tasks that have to be performed on the terminal such as video decompression, a system level low-power approach is used to minimize power consumption.

9.3 Architecture of the Portable Multimedia Terminal

The portable terminal is designed to transmit audio and pen input from the user to the network on a wireless uplink and to receive audio, text/graphics and compressed video from the backbone on the downlink. Since the general purpose applications are performed by compute servers on the backbone network, the terminal electronics only have to provide the interface to I/O devices, as shown in Figure 9.6 [Chandrakasan94].

Six chips (a protocol chip, a bank of six 64 Kbit SRAMs for text/graphics frame-buffering, and four custom chips for the video decompression) provide the interface between a high speed digital radio modem and a commercial speech codec, pen input circuitry, and LCD panels for text/graphics and full-motion video display. The chips provide protocol conversion, synchronization, error correction, packetization, buffering, video decompression, and digital to analog conversion. Through extensive optimization for power reduction, the total power consumption is less than 5mW.

The protocol chip communicates between the various I/O devices in the system. On the uplink, 4 Kbps digitized pen data and 64 Kbps, μ-law encoded speech data are buffered using FIFO's, arbitrated and multiplexed, packetized, and transmitted in a serial format to the radio modem. On the down link, serial data from the radio at a 1Mbs rate is depacketized and demultiplexed, the header information (containing critical information such as data type and length) is error corrected and transferred through FIFO's to one of the three output processing modules: speech, text/graphics, and video decompression.

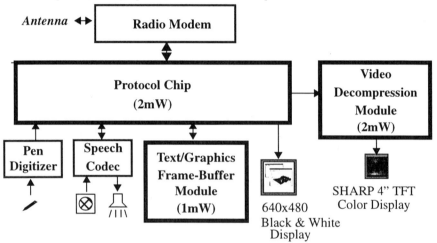

Figure 9.6: Block Diagram of the InfoPad Terminal.

The speech module inside the protocol chip transmits serial data from the FIFO to a codec at a rate of 64Kbs. The text/graphics module also inside the protocol chip handles the protocol used to transmit the bitmaps for text/graphics. The text/graphics module corrects error sensitive information using a Hamming (7,4,1) code, and generates the control, timing and buffering for conversion of the 32 bit wide data from the text/graphics frame-buffer (implemented using six low-power 64Kbit SRAM chips) to the 4-bits at 3MHz required by a 640x480, passive LCD display. The final output module is the video decompression, which is realized using four chips. The decompression algorithm is vector quantization, which involves memory look-up operations from a codebook of 256 4x4 pixel patterns. Compressed *YIQ* video is buffered using a ping-pong scheme (one pair of memories for *Y* and one pair for *IQ*), providing an asynchronous interface to the radio modem and providing immunity against bursty errors. The amount of

RAM required is reduced over conventional frame-buffer schemes by a factor of 24 by storing the video in compressed format. The *YIQ* decompressed data is sent to another chip that converts this data to digital *RGB* and then to analog form using a triple DAC that can directly drive a 4 inch active matrix color LCD display. A controller chip performs the video control functions, including the synchronization of the various chips and the color LCD display, control of the ping-pong memories, and loading of the code-books. It uses an addressing scheme that eliminates the need for an output line buffer.

9.4 Low Power Implementation of the I/O Processing Modules

This chipset exploited signal statistics to reduce the total number of transitions. The physical capacitance is reduced by utilizing a low-power cell library that features minimum sized transistors and optimized layout. In addition to minimizing the switched capacitance, an architecture driven voltage scaling strategy is used to scale the supply voltage to the 1-1.5V range while meeting the computational throughput requirements. Since the circuits operate at a supply voltage that is less than the sum of the NMOS and PMOS threshold voltages ($V_{tn} = 0.7$, $V_{tp} = -0.9$), the devices can never conduct simultaneously for any possible input voltage, *eliminating* short-circuit power. Also, due to the high values of the threshold voltage, the leakage power is negligible.

9.4.1 Protocol Chip

The protocol chip, shown in Figure 9.7, communicates between the various I/O devices in the system. It provides the interface between a commercial radio modem and the pen digitizer, the speech codec, and the 640x480 text/graphics LCD display. It also controls the custom frame-buffer of the text/graphics display and provides the necessary interface between the frame-buffer and the LCD display. The radio receiver and transmitter modules are isolated from the I/O processing modules using 32-bit FIFOs. The text/graphics FIFO has one extra bit to delimit packets.

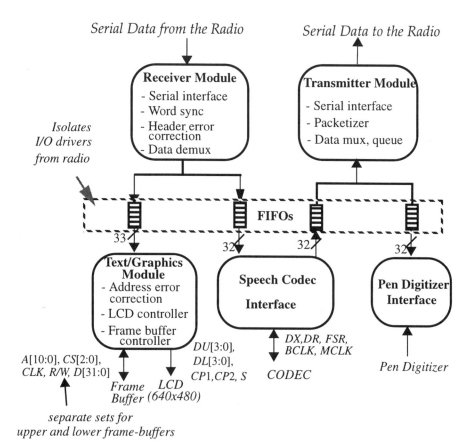

Figure 9.7: Block diagram of the protocol chip.

9.4.1.1 Parallelism Enables Operation in the 1-1.5V Range

Reducing the power supply voltage has the greatest impact on the power consumption. Unfortunately, this simple solution to low power design comes at the cost of increased gate delays and therefore lower functional throughput. Figure 9.8 shows the normalized access time for a 64Kbit SRAM as a function of supply voltage V_{dd}.

As presented in Chapter 4, an architecture driven voltage scaling strategy using parallelism and pipelining to compensate for the increased gate delays at reduced supply voltages can maintain functional throughput. Parallelism and

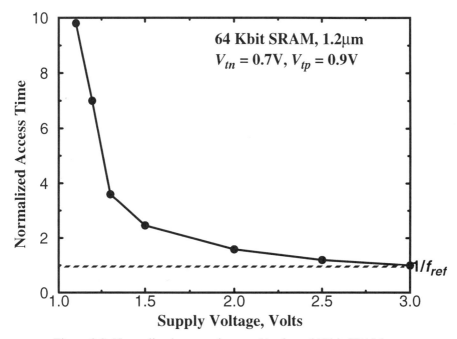

Figure 9.8: Normalized access time vs. V_{dd} for a 64Kbit SRAM.

pipelining were used extensively in this chipset for both arithmetic computation and memory access, allowing the supply voltages to scale down to the 1-1.5V range.

Parallelizing arithmetic computation was presented in previous chapters. Here an application of parallelism to memory access is presented, which involves reading black and white pixel data from the text/graphics frame-buffer memory to the 640x480 LCD display. The LCD display is a split-panel display in which the top half (640x240) and bottom half are addressed in parallel. Each half requires 4-bits (or 4-pixel values) at a rate of 3MHz.

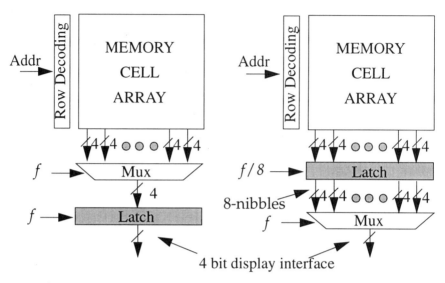

Figure 9.9: Parallel memory access enables low-voltage operation.

Figure 9.9 shows two alternate schemes for reading the required 4-bits of data from memory at the throughput rate, f. On the left side is the serial access scheme in which the 4-bits of data are read out in a serial format and the memory is clocked at the throughput rate f. For this single-ported SRAM frame-buffer implementation (in which the read/write data and address busses are shared) utilizing a serial access scheme requires a supply voltage of 3V. Another approach (shown on the right side) is to exploit the sequential data access pattern for the frame-buffer and access several 4-bit nibbles in parallel, allowing the memory to be clocked at a lower rate. For example, reading eight nibbles in parallel from the memory allows the memory to be clocked at $f/8$ without loss in throughput. The latched data (8 nibbles) in the protocol chip is then multiplexed out at the throughput rate. This implies that the time available for the memory read operation for the parallel implementation is 8 times longer than the serial scheme and therefore the supply voltage can be dropped for a fixed throughput. The parallel version can run at a supply voltage of 1.1V (which corresponds to the voltage at which the gate delays increase by a factor of 8 relative to the serial access scheme running at 3V from Figure 9.8) while meeting throughput requirements. It is interesting that architectural techniques can be used to drive the supply voltage to such low levels even with a process that has $V_{tn} = 0.7$V and $V_{tp} = -0.9$V. If the process could be modified to reduce the threshold voltage, the power supply voltage and therefore the power consumption can be further reduced.

In general, for signal processing applications that do not have feedback in the computation, arbitrary amount of parallelism can be exploited to reduce the power supply voltage. However, exploiting parallelism often comes at the expense of increased silicon area (for example, there is an increase in the number of I/O pins for the SRAM chip going from a serial memory architecture to a parallel memory access architecture) and capacitance.

9.4.1.2 Reduced Swing FIFO

Reducing the power supply voltage is clearly a very effective way to reduce the energy per operation since it has a quadratic impact on the power consumption. At a given supply voltage, the output of CMOS logic gates make rail to rail transitions. An approach to reducing the power consumption further is to reduce the swing on the output node. For example, using an NMOS device to pull up the output will limit the swing to V_{dd}-V_t, rather than rising all the way to the supply voltage. The power consumed for a $0 \rightarrow V_{dd}$-V_t transition will be $C_L V_{dd}$ (V_{dd}-V_t) (which was derived in Chapter 3), and therefore the power consumption reduction over a rail to rail scheme will be $\propto V_{dd} / (V_{dd}$-$V_t)$. This scheme of using an NMOS device to reduce the swing has two important negative consequences: first, the noise margin for output high (NM_H) is reduced by the amount V_t, which can reduce the margin to 0V if the supply voltage is set near the sum of the thresholds. Second, since the output does not rise to the upper rail, a static gate connected to the output can consume static power for a high output voltage (since the PMOS of the next stage will be "on"), increasing the effective energy per transition.

Therefore, to utilize the voltage swing reduction, special gates are needed to restore the noise margin to the signal, and eliminate short-circuit currents. These gates require additional devices that will contribute extra parasitic capacitances. Figure 9.10 shows a simplified schematic of such a gate, used in the FIFO memory cells of the low-power cell library used in this chipset [Burd94]. This circuit uses a precharged scheme that clips the voltage of the bit-line (which has several transistors similar to M5 connected to it) to V_{dd}-V_t, where $V_t > V_{t0}$ due to the body effect. The devices M1 and M4 precharge the internal node (the input of the inverter) to V_{dd}, and the bit-line to V_{dd}-V_t. During evaluation (φ = "1"), if V_{in} is high, the bit-line will begin to drop, as shown in the SPICE output next to the schematic. Because the capacitance ratio of the bit-line to the internal node is very large, once the bit-line has dropped roughly 200mV to sufficiently turn on M3, the internal node quickly drops to the potential of the bit-line, providing signal amplification. Thus, this circuit conserves energy greatly by reducing the voltage swing on the high-capacitance bit-line, and reduces delay by providing signal amplification.

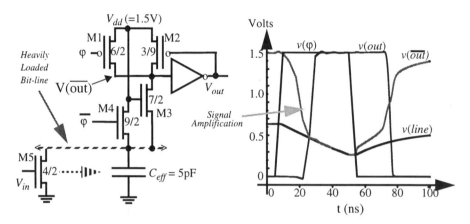

Figure 9.10: Signal swing reduction for memory circuits: FIFO example.

9.4.1.3 Level-conversion Circuitry to Interface with I/O Devices

The protocol chip operates at a supply voltage of 1.1V and communicates with I/O devices running at higher voltages - for example the text/graphics module in the chip core has to communicate with the 640x480 LCD display running at 5V or the speech I/O module of the core has to communicate with the speech codec running at 5V. Level-conversion I/O pads (a tri-state level-conversion I/O pad is shown in Figure 9.11) are used to convert the low voltage-swing signals from the core of the low-power chips (running at 1.1V) to the high-voltage signal swings required by I/O devices (e.g. 5V required by displays) or vice-versa. For communication between the protocol chip and the frame-buffer chips, normal low-voltage low-power pads were used. With signal swings of only 1.1V, the I/O power was significantly reduced as well.

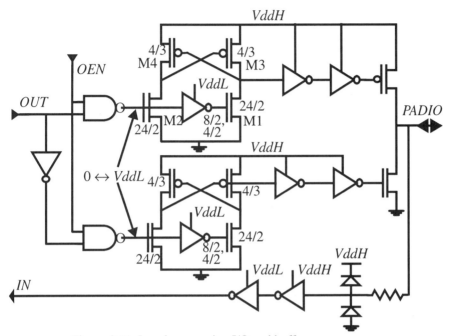

Figure 9.11: Level-conversion I/O pad buffer.

This tristate output buffer uses two supply voltages, the low-voltage supply, *VddL*, that is tied to the pad-ring and the high-voltage supply, *VddH*, coming in through another unbuffered pad. The low-to-high conversion circuit is a PMOS cross-coupled pair (M3, M4) connected to the high supply voltage, driven differentially (via M1, M2) by the low-voltage signal from the core. The N-device pull-downs, M1 and M2, are DC ratioed against the cross-coupled P-device pullups, M3 and M4, so that a low-swing input (*VddL*=1V) guarantees a correct output transition (*VddH*=5V). That is, the PMOS widths are sized so that the drive capability of the NMOS can overpower the drive of the PMOS, and reverse the state of the latch. The ratio is larger than just the ratio of mobilities, because the PMOS devices are operating with $V_{GS}=V_{ddH}$, and the NMOS is operating with $V_{GS}=V_{ddL}$. This level-converting pad consumes power only during transitions and consumes no DC power. The remaining buffer stages and output driver are supplied by *VddH*. This level-conversion pad will work only with an NWELL process since it requires isolated wells for the PMOS devices that are connected to the high voltage.

9.4.1.4 Use of Gated Clocks to Reduce the Switching Activity

At the logic level, gated clocks are used extensively to power down unused modules. For example, consider the error correction in the text/graphics module. The basic protocol for the text/graphics module that is sent over the radio link and through the text/graphics FIFO is a base address for the frame-buffer followed by bit-mapped data - the length information is decoded in the depacketizer module and a control field in the FIFO (a 33rd bit indicating End of Packet - *EOP*) is used to delimit the packets. Figure 9.12 shows the protocol of the text/graphics packet.

D32	D31 D30 ... D0
0	Base Address
0	32 Bit TG DATA
	o o o
0	32 Bit TG DATA
1	32 Bit TG DATA
0	Base Address
0	32 Bit TG DATA

FIFO DATA

- **End of Packet Flag (*EOP*):**
  Bit 32: 1 = Last Data Entry in Packet

- **Base Address (in unencoded format):**
  Bit 13: Bank #, 0 = Lower & 1 = Upper
  Bit [12:0]: Address

Figure 9.12: Text/graphics protocol.

The base address is the starting location in the frame-buffer for the bit-mapped data. The first entry is a start address which indicates the location where the pixel data that follows is written into. The pixel data is written into consecutive locations starting from the start address in the frame-buffer. While address information is sensitive to channel errors (since errors can cause the data to be written in the wrong area of the screen), bit-mapped data is not very sensitive since errors in data appear as dots on the screen. Therefore, to enable a bandwidth efficient wireless protocol, only the address information is error corrected and the bit-mapped data is left unprotected.

The capacitance switched in the error correction module is minimized by using gated clocks. Figure 9.13 shows the block diagram of a power efficient implementation of the error correction function. A register is introduced at the output of the FIFO and a gated clock (coming from the controller) is used to enable the error correction module to process only the address information; the ECC is shut down during the rest of the time. The gated clock is generated by a controller which uses the state information of the FIFO (which is based on the $\overline{EMPTY}$ signal and the *End of Packet* signal from the FIFO). In this manner, the inputs to the ECC are not switching when the data portion of the protocol is accessed from the FIFO. Since typically, the address is only a small portion of the text/graphics packet, significant power savings is possible. Gated clocks were used in many circuits to tune the frequency and hence clock load for each module. Gated clocks are not usually used because of the timing skew introduced; this is not a problem in our approach since we use a low clock rate methodology.

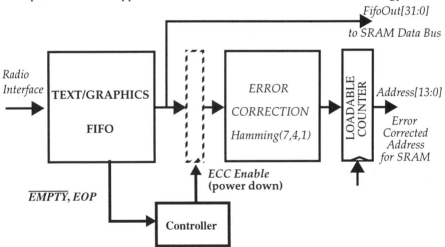

Figure 9.13: Gated clocks are used to shut down modules when not used.

9.4.1.5 Optimizing Placement and Routing for Low-power

At the layout level, the place and route is optimized such that signals that have high switching activity (such as clocks) are assigned short wires and signals that have low switching activity are allowed to have long wires. Figure 9.14 shows an example that involves the routing of large data and control busses from the text/graphics module of the chip core to the pads on the south side of the protocol chip, shown in Figure 9.16. The text/graphics module on the protocol chip communicates to both the text/graphics frame-buffer and to the text/graphics 640x480 LCD display. The split-panel display requires 8-bits (4 for the top half and 4 for the bottom half) at a rate of 3Mhz, while each SRAM module uses 32-bits clocked at 375KHz (using the parallel access scheme described in Section 9.4.1.1). An activity factor (for both 0->1 and 1->0 transitions) of 1/2 is assumed for the data. The address bits are also clocked at 375KHz but they have a much lower switching activity since the accesses are mostly sequential, coming from the output of a counter. The address bus is time-multiplexed between the read and write addresses for the SRAM, but since the write into the text/graphics frame-buffer is relatively infrequent, the address bus usually carries the read address and is therefore very temporally correlated (the number of transition for a counter output per cycle is 1 + 1/2 + 1/4 +... ≈ 2 since the lsb switches every cycle, the next lsb switches every other cycle, etc.). As seen from the plot of physical capacitance as a function of distance from core to pads, the display data and display clock, which have high switching activity, are assigned the shortest wire lengths, while the SRAM address, which has an activity factor 16 times lower, is allowed to have the longest wires.

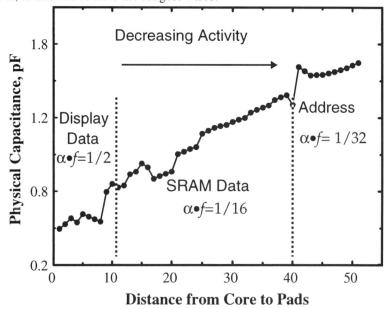

Figure 9.14: Optimizing placement for low-power: routing large busses.

9.4.1.6 Low-power Cell Library

The entire chipset was designed using an extensive cell-library that was optimized for low-power operation [Burd94]. The library contains parameterized datapath cells (adders, registers, shifters, counters, register files, buffers, etc.), multiple strength and multiple input standard cells for random logic (NAND, NOR, registers, buffers, etc.), low-power memories (FIFO, SRAM), low-voltage pads (clock pads, low-voltage I/O pads, level-converting pads, etc.) and low-voltage high efficiency switching regulators (Chapter 5). Some of the key design approaches for the cell library are outlined below:

- Device sizing: Minimum sized devices were used in most datapath and standard cells. As analyzed in Chapter 4, minimum sized devices should be used to minimize the physical capacitance for circuitry inside the datapath blocks where self-load, rather than interconnect capacitance dominates. When driving large interconnect capacitances, for example when communicating between datapaths, large sized devices are used.

- Reduced swing circuitry: as described earlier, reduced swing circuitry is used limit the voltage swing in memory circuits (FIFO and SRAM) and reduce the power consumption in a linear fashion.

- Self-timed Circuits: The memory circuits use self-timing to eliminate spurious transitions on high capacitance data busses. This guarantees the minimum number of transitions per memory access.

- Efficient Layout: For tilable datapath cells, all buses are routed in metal 2, and metal 2 is not used for routing within a cell. Unfortunately, this forces some of the modules, such as the adder, to use poly for intracell routing, which has two to three times the capacitance per unit area. However, this scheme reduces the amount of cell stretching inside each datapath (for routing channels) which reduces the overall datapath area. This in turn results in smaller global bus lengths and therefore lower global interconnect capacitance. That is, the inter-block bus routing becomes much more efficient, and the capacitance increase from using poly is small compared to the overall decrease attributable due to smaller global buses.

- Integrated control slice: the bit-sliced datapaths have integrated control slices which perform various functions. This reduces or sometimes eliminates the amount of standard-cell control that is used. This reduces power since devices in the control slice can be optimized for the load that has to be driven. For standard cell control, the load is the load for the cells in the bit-slice plus the extra interconnect capacitance.

9.4.1.7 Clocking strategy

Highly pipelined systems have a significant portion of the total power consumed by the clock. Therefore, it is important to minimize the number of global clock nets (one approach is to use gated clocks), as well as all the gate capacitance switching at the clock rate. To accomplish this, the TSPC (True Single Phase Clocking) methodology was used in the design of all register elements, with the basic register shown in Figure 9.15. The two main advantages of this style are that only one global clock net is needed and there is only a total of four minimum sized transistors attached to the clock net per register. This provides almost a 50% reduction in clock power over non-overlapping two-phase latches. The library also contains a dual-edge triggered register which allows the clock frequency to be halved for the same throughput as normal register. This once again reduces the power consumed in the clock routing.

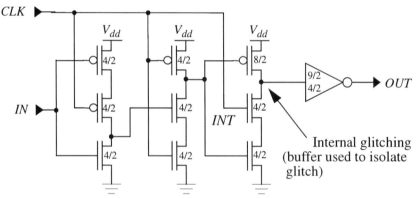

Figure 9.15: True single phase clock pipeline register.

There are four key design considerations when using the TSPC register: First, the rise-time of the clock signal is critical to proper operation and therefore a control slice attached to each datapath register provides local clock buffering. Second, it is a dynamic register, such that the clock has a minimum operating frequency. Measurements have shown the minimum to be on the order of 500 Hz, which satisfies the constraints of most applications. A modified version of the register exists with two extra transistors that provide static feedback while the clock is low. Thus, the register can be used for gated-clock configurations, without signal leakage. Third, there is partial internal glitching due to a momentary race condition. This is was not found to be a problem since the glitch is small and the output does not switch. Finally, the internal node *INT* will have transitions if the input is a constant LOW since *INT* will discharged during evaluation and pre-charged when the clock is low. This is typically not significant for data which is switching around frequently. If the TSPC is used as a register for control signals that are LOW most of the time, then the complement of the signal should be sent through the register and a static inverter should be used at the output to achieve correct polarity.

9.4.1.8 Chip Layout and Statistics

The protocol chip was designed using the LagerIV silicon compiler. Figure 9.16 shows the layout of the protocol chip. Table 9.2 shows the statistics of the

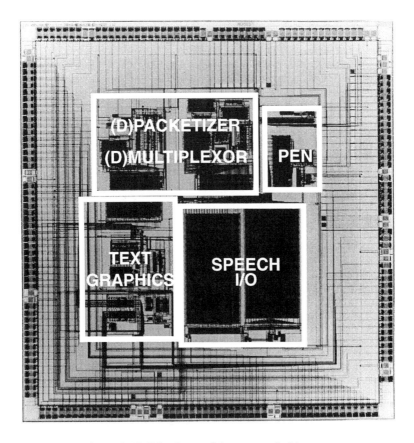

Figure 9.16: Die photo of the protocol chip.

protocol chip. The chip operates all the way down to a supply voltage of 1.1V and consumes only 1.9mW at 1.5V.

Table 9.2 Statistics for the protocol chip.

Parameter	Value
Technology	1.2μm CMOS
Size	9.4mm x 9.1mm
Number of Transistors	136000
Minimum Operating Voltage	1.1V
Power Consumption at 1.5V	1.9mW
Throughput	1Mbs serial for radio 3MHz for 4-bit text/graphics 4Kbs for pen input 64Kbs for speech I/O

Table 9.3 shows a summary of the various power reduction techniques that were applied to the protocol chip and the amount of power reduction that is achieved using each technique.

Table 9.3 Summary of power reduction techniques applied to the protocol chip.

Approach	Power Reduction	Comments
Supply Voltage Scaling (Parallelism and Pipelining)	x21	1.1V operation vs. 5V operation
Optimized Cell Library	x3-4	Transistor sizing, Clocking, Reduced swing FIFO, Self-timed Glitch free FIFO, Layout optimization, Low-power Pads
Gated Clocks	x2-3	Gated clocks reduce both the clock load and power in logic

Table 9.3 Summary of power reduction techniques applied to the protocol chip.

Approach	Power Reduction	Comments
Activity Driven Place and Route	x1.5-2	This reduction is just for the global busses. Standard placement tools were used for the logic generation

9.4.2 Text/Graphics Frame-buffer Module [Burstein93]

The frame-buffer for the split panel text/graphics display contains six 64Kbit SRAM chips (Figure 9.17), which were synthesized from the low power cell library's tileable SRAM module. This module meets several constraints to make it a useful component of a wide variety of low power systems.

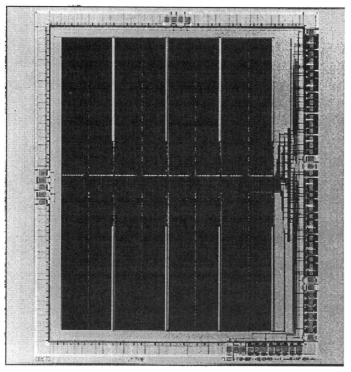

Figure 9.17: Die photo of the text/graphics frame-buffer.

First, the SRAM module builds memories over a wide variety of sizes. At one extreme, it can create entire memory chips that are only limited by maximum die size, such as the frame-buffer memories used for the text/graphics display. At the other extreme, the same module synthesizes smaller (hundreds or thousands of bit), area and power-efficient memories that are placed on the same chip as datapath and control systems (as in the video decompression chips described in the next section). The designer can specify both the number of words and the bit width of the words. The designer specifies the number of words, which may be any size over two, and the number of bits per word, which may be from two to sixty four. The designer can also control the aspect ratio by specifying the number of blocks at the architecture level, as detailed below.

Next, the SRAM must match the timing and electrical constraints of the rest of the library. Since our low power strategy stresses running highly parallel and pipelined systems operating at low clock rates and lower voltages, the SRAM only needed a cycle time of 100ns - 200ns at 1.5 volts to be useful in our low power systems; there is no advantage to designing faster components that consumed more power, since our systems have fixed data throughputs. Another major principle of low power operation is to use a low supply voltage: 1.5 volts is the target supply for the low power library. Although it is optimized to meet its speed requirement at 1.5 volts, the SRAM functions over a wide range of supply voltages. The text/graphics frame buffer chips have been used with supplies as low as 1.1 volts by exploiting parallelism. The SRAM also can operate with supplies up to 5V, with increases in both speed and power consumption.

The final requirement is to make the timing and control of the SRAM as simple and general purpose as possible, so that system designers will not have to be concerned with clocks internal to the SRAM. It triggers off of the rising edge of a single clock; all other timing signals are generated internal to the SRAM, using self-timing circuitry that scales with the size of the SRAM. Also, the SRAM can use either a single, bi-directional bus or can use separate input and output buses.

9.4.2.1 Memory Architecture

The architecture of the SRAM is designed both to minimize power consumption and to enhance scalability. At the highest level of the architecture, the SRAM is organized into a parameterizable number of independent, self-contained blocks, each of which reads and writes the full bit width of the entire memory. For example, the 64 Kbit (2k word by 32 bit) frame buffer chips contain eight 8 Kbit blocks, each organized as 256 words by 32 bits. Since one of the fundamental power saving techniques is to minimize the effective capacitance switching per clock cycle, the SRAM only activates circuits in one of these blocks per clock cycle (Figure 9.18). The power savings from this architecture are twofold. First, there is less overhead capacitance, since only one set of control signals and decoders switch at one time. Second, and more fundamentally, by having wide data widths into each block, the memory has minimal column decoding, and hence less column capacitance to switch per bit. From an ideal power consumption perspective, it would be best to have no column decoding at all; as a practical concession to pitch matching, this design uses 2 to 1 column decoding.

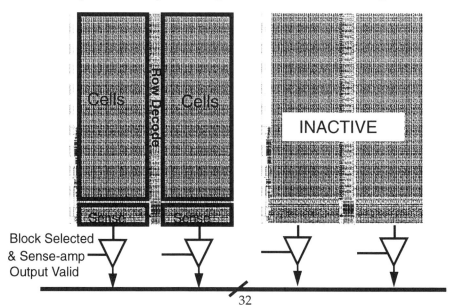

Only one block in the SRAM is enabled per access

Figure 9.18: SRAM block organization.

The trade-off for this savings in power is a loss in speed. Since only one block is activated per clock cycle, each block has a decoder (composed of static logic) that must activate its block before that block can begin any operations. Thus, the delay through these block level decoders is added to the total SRAM delay. However, this loss of speed causes no penalty as long the SRAM meets the overall system throughput needs.

Using self-contained blocks also makes the architecture flexible and expandable. Since the number of words per block and the total number of blocks are parameterized, the designer can control the aspect ratio by trading off between the number and size of the blocks. Adding blocks to the memory increases its number of words, with a minimal effects on power consumption and circuit design. Since circuits in only one block switch at a time (except for the block level decoders, which consume minimal power) the only increase in power consumption is from the increased wiring capacitance between blocks. The only circuits that need to change are the block level decoders. All of the other control, timing, decoding, and sensing circuitry in the blocks are insensitive to the number of blocks in the memory.

9.4.2.2 Circuit Level Optimization

At the circuit level, the SRAM's most important power saving technique is to reduce the voltage swing on the bitlines. As shown in (Figure 9.19), the bitlines are precharged through NMOS devices, so the maximum bitline voltage is the supply voltage minus the NMOS threshold voltage (with body effect). Compared to full swing bitlines, this reduces power consumption by only 20% for V_{dd}=5V, but as much as 50% for V_{dd}=1.5V and 75% for V_{dd}=1.2V. (It was not practical to limit the minimum voltage on the bitlines by timing the wordline signals because the timing would have to work for many sized memories, over many voltages, and for different fabrication technologies.) This precharge strategy also allows the column select transistors to double as cascode amplifiers, creating voltage gain between the bitlines and the input to senseamp.

Another important power saving technique is to eliminate glitching on the data bus during read operations. The output driver of the senseamp shown in (Figure 9.19) is tristated at the beginning of each clock cycle. Even after the output enable signal is true, the output remains tristated until the senseamp's cross coupled latch has resolved the data value, at which time either the NMOS or the PMOS output device drives the data bus.

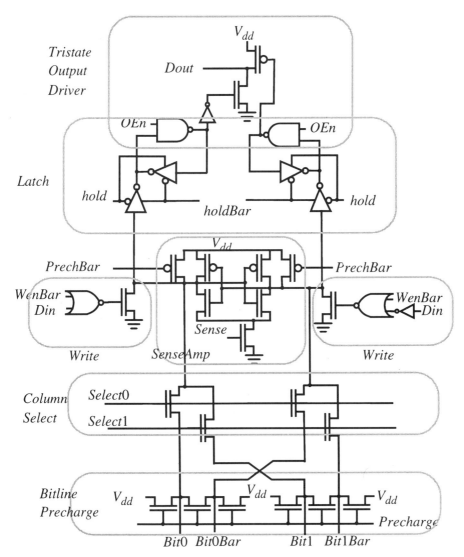

Figure 9.19: Precharge, sense-amp and "Glitch" free tri-state buffer.

One of the major handicaps of using a high threshold process (V_{tp}=-0.9V, V_{tn}=0.7V) at low supply voltages is that the low gate overdrive voltage for the PMOS (V_{gs} - V_t) exacerbates their lower carrier mobility, creating a large imbalance between the current drive of the PMOS compared to the NMOS devices. A partial remedy is to make sure that large capacitances are never driven by more than one PMOS device in series. The senseamp output driver meets this requirement, in addition to eliminating glitches. Another partial remedy is to use a large $Wp{:}Wn$ ratio -- approximately 6 -- in static inverters that are in the critical path. Ideally, it would be better to have NMOS and PMOS threshold equal in value, and as low as possible. Ultimately, the minimum threshold voltage would be set by the maximum acceptable power consumption from leakage current. However, since most of the leakage occurs in the memory cells, it would be advantageous to have a dual threshold process, using the high V_t transistors in the memory cells and the low V_t transistors elsewhere.

9.4.2.3 Chip Statistics

Table 9.4 shows a summary of the frame-buffer chip used to store text/graphics pixel data.

Table 9.4 Statistics for the text/graphics frame-buffer chip.

Parameter	Value
Technology	1.2µm CMOS
Size	7.8mm x 6.5mm
Number of Transistors	428,800
Memory Size	64Kbits
Word Width	32bits
Minimum Operaung Voltage	1.1V
Power Consumption at 1.5V (including loading of two other chips)	500µW
Throughput	375kHz read (write and read cycles are interleaved)

Table 9.5 shows a summary of the various power reduction techniques that were applied to the frame-buffer chip and the amount of power reduction that is achieved using each technique.

Table 9.5 Summary of power reduction techniques applied to the frame-buffer chip.

Design Approach	Power Reduction	Comments
Supply Voltage Scaling	x21	1.1V operation vs. 5V operation
Self-timed Outputs Drivers	x2	Assuming that the words are temporally uncorrelated and the output capacitance dominates the global bus capacitance - reasonable since it drives off chip interconnect.
Bit Swing Reduction	x3.7	This reduction is only for the memory array. The global bus has rail to rail swing. Comparison is for 1.1V swing vs. 300mV swing
Block Decoding	x8	This reduction is just for the memory array. This is the power reduction due to enabling only one block vs. enabling all blocks in the SRAM - as done in high speed applications

9.4.3 Video Decompression Chipset

The final output module is the video decompression module which includes all of the circuitry required to take a compressed video stream from the radio and convert it to the analog data format required by a 4" (128x240 pixels) color active matrix display. This section presents the design of a set of 4 chips to perform video decompression for this low-resolution display.

9.4.3.1 Algorithms Selection for Low-power Video Decompression

The choice of algorithm is the most highly leveraged decision in meeting the power constraints. The ability for an algorithm to be parallelized is critical and the basic complexity of the computation must be highly optimized. Minimizing the number of operations to perform a given function is critical to minimizing the overall switching activity and therefore the power consumption. The task of selecting the algorithm for the portable terminal depends not only on the traditional criteria of achievable compression ratio and the quality of reconstructed images, but also on computational complexity which directly impacts the power.

Most compression standards (for example, JPEG and MPEG) are based upon the Discrete Cosine Transform (DCT). The basic idea in intra-frame schemes such as JPEG is to apply a two-dimensional DCT on a blocked image (typically 8x8) followed by quantization to remove correlations within a given frame. In the transform domain, most of the image energy is packed into only a few of the coefficients, and compression is achieved by transmitting only a carefully chosen subset of the coefficients. One main characteristic of the DCT is the symmetric nature of the computation; i.e., the coder and decoder have equal computational complexity.

Although the computational complexity of the DCT can be optimized by restructuring algorithms, it still requires several arithmetic and memory operations per pixel. Table 9.6 shows a comparison of a few algorithms that can be used to implement the DCT [Clarke85][Feig90][Rao90]. Minimizing the operation count is important in minimizing the switching events and hence the power consumption.

Table 9.6 Computational complexity of the Discrete Cosine Transform.

DCT Algorithm	Multiplies (8x8)	Additions (8x8)
Brute Force	4096	4096
Row-Col. DCT	1024	1024
Chen's Algorithm	256	416
Lee's Algorithm	192	464
Feig's Algorithm	54	462

An alternative compression scheme is vector quantization (VQ) coding, which is asymmetrical in nature and has been unpopular due to its complex coder requirements. Figure 9.20 shows the basic idea behind intra-frame vector quantization compression. On the encoder, a group of pixels is blocked into a vector and compared (using some metric such as Mean Square Error) against a set of predetermined reproduction vectors (a set of possible pixel patterns) and the index of

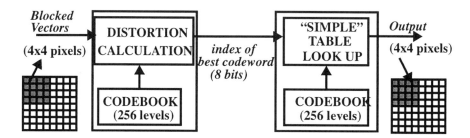

Figure 9.20: Video compression/decompression using vector quantization.

the best match is output from the encoder. The decoder has a copy of all possible reproduction vectors (codebook) and the index of the best codeword is used to reconstruct the image using a simple lookup table operation.

Vector quantization has been unpopular due to its complex coder requirements. However, since VQ algorithms require a very simple memory-lookup decoder they are well suited for single-encoder, multiple-decoder systems. If one-way video communication is desired, the VQ solution provides a means of implementing real-time decompression using very little computation and power; while a DCT based decompression system based may require many operations per pixel, a VQ based decoder will only require a single memory lookup operation [Gersho92]. Table 9.7 shows a comparison of a typical DCT and a VQ algorithm. In addition to minimizing the number of operations in the decoder, the power consumption can be further reduced by optimizing the memory structure and organization for reduced voltage operation and minimal switched capacitance.

Table 9.7 Asymmetrical algorithm required for low-power implementation.

Method	Coder	Decoder
Transform (DCT) (Row - Col. fast algo) [Rao90]	4 multiply 8 additions 6 mem. access	4 multiply 8 additions 6 mem. access
Vector Quantization (Tree search - differential CB) [Fang90]	8 multiply 8 additions 8 mem. access	1 mem. access

Figure 9.21 shows a block diagram of the video decompression module implementing the VQ algorithm. One chip performs all of the control functions for the video decompression and interface to the radio, two chips perform the frame-buffering and the video decompression based on the vector quantization table look-up algorithm, and one chip performs the color space conversion and analog interface to the display.

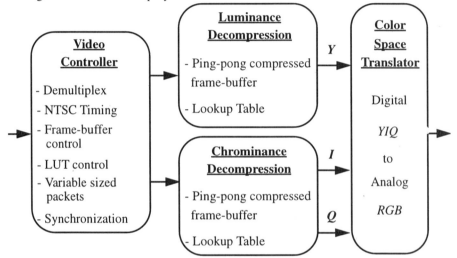

Figure 9.21: Block diagram of the video decompression.

9.4.3.2 Luminance Video Decompression

For this implementation, the image is segmented into 4x4 blocks (i.e., the vector size is 16) and there are 256 entries in the codebook. The original image on the encoder side is represented in the *RGB* domain using 6-bits for each color plane - using 6-bits to represent video data instead of 8-bits results in very little visual distortion on this low-resolution display. Color information is transformed to the *YIQ* domain and each plane is individually coded with separate codebooks. The *I* and *Q* color components (called the chrominance components) are sub-sampled in both the horizontal and vertical dimensions. Therefore, the *YIQ* representation (*Y*:1, *I*:1/4, *Q*:1/4) gives 2:1 compression over the *RGB* representation (*R*:1, *G*:1, *B*:1). On each plane, VQ results in 12:1 compression since only 8-bits are transmitted for each 4x4 block (choosing 1 out of 256 codes) instead of 16 x 6 bits for the true data. A total of 24:1 compression is achieved (32:1 if the quantization from 8bit to 6bit representation is taken into account) and therefore to support full-motion color video at 30 frames/sec, this system consumes a bandwidth of 690 Kbits/sec on the wireless link.

Figure 9.22 shows a block diagram of the luminance decompression chip. The incoming compressed video data (VQ codewords for the image) is stored using a ping-pong addressing scheme. For example, when the compressed video coming from the RF link is being stored in BANK0, the compressed data stored in BANK1 is read out to be decompressed using the lookup table (LUT). After a full frame of compressed video is assembled in BANK0, the $R/\overline{W}$ signal (signal *WBANK*) of the two frame-buffers toggles, resulting in data being read from BANK0 to the lookup table while new compressed video data is written into BANK1. This ping-pong addressing scheme provides an asynchronous interface between the radio clock and the video system and provides immunity against bursty channel errors; if a frame of compressed video is dropped or if higher priority data is sent over the link (such as text/graphics data or speech which require data to be sent with minimal latency), the complete compressed frame that is already stored in the terminal is decompressed and displayed until a new frame of compressed video is assembled in the other frame-buffer. Also, since the refresh rate of the color display is 60Hz while the image is updated only at 30Hz, the ping-pong frame-buffer is actually required unless the bandwidth on the link is doubled.

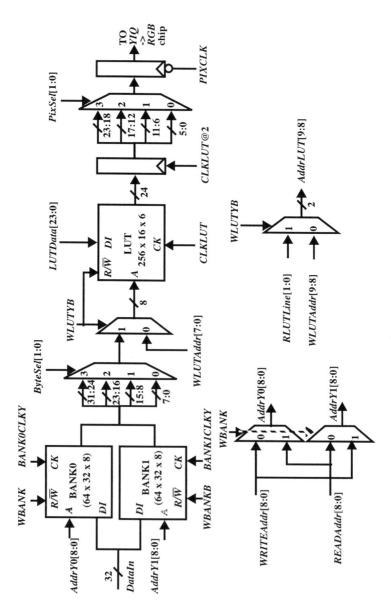

Figure 9.22: Block diagram of the luminance decompression chip.

The video is stored in a compressed rather than uncompressed format to reduce the amount of memory in the system by a factor of 24. Rather than decompressing and storing the data once and displaying the decompressed image twice (since as mentioned above, the refresh rate is 60Hz while the image is updated at 30Hz), the image is stored in a compressed format and is read out and decompressed twice; that is, there is no decompressed frame-buffer. The compressed frame-buffer is read out in a parallel fashion, where four 8-bit VQ codewords are read in parallel (once again the access pattern of data is know and is exploited) though only one is used at a given time; this once again enabled operation at supply voltage as low as 1.1V. In this implementation, the video frame-buffer was clocked at 156kHz while meeting the throughput rate of 2.5MHz. The output of the frame-buffer is multiplexed at 4:1, and is used to index the lookup table which generates the decompressed data.

The chip uses an addressing scheme that eliminates the need for an output line buffer which is typically used to convert a block data format to the raster format required by the display. Figure 9.23 shows the numbering of codewords in the frame-buffer and the ordering of pixels inside each block. Each time a codeword is accessed from the frame-buffer, only 4 pixels are read out from the lookup table, creating a raster output which can be sent directly to the display. That is, pixels P0-P3 corresponding to each codeword between CW0 and CW31 are read from the lookup table and then P4-P7 are read once again for codewords CW0 through CW31, and then P8-P11, and finally P12-P15. Thus, each codeword is accessed four times per image. This approach increases the frequency of codeword access relative to a scheme which reads each codeword only once (and stores P0-P15 in a line-buffer), but since the codewords are accessed at a much lower frequently relative to the lookup table data, and since the line-buffer access has been eliminated, the overall power due to memory access is reduced by approximately a factor of 1.5. This is an example of architectural restructuring to reduce the number of operations.

The control of the luminance decompression chip (address generation, clock generation, multiplexor selection, etc.) is performed by the video controller chip. The multiplexors for read/write address lines are integrated in the decompression chip. *RLUTLine*[1:0] controls the row of pixels that are read for a given codeword. The lookup table is programmable over the radio link through the video controller chip.

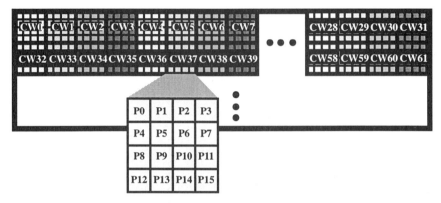

Figure 9.23: Addressing of the frame-buffers and lookup table.

Figure 9.24 shows the die photograph of the luminance decompression chip. The chip consumes only 115µW running at 1.5V to support a video through-put rate of 2.5MHz for the 4" color display.

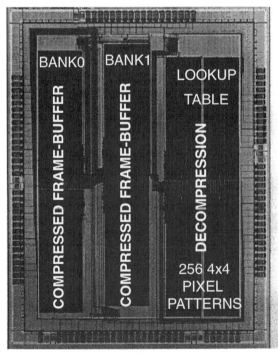

Figure 9.24: Die photo of the luminance decompression chip.

Table 9.8 shows a summary of the power reduction techniques used in the luminance decompression chip.

Table 9.8 Summary of power reduction techniques applied to the luminance decompression chip.

Design Approach	Power Reduction	Comments
Algorithm Selection	x5-10	Due to operation reduction: Comparison of VQ decoding to typical DCT algorithms
Supply Voltage Scaling (through parallelism)	x21	1.1V operation vs. 5V operation
Memory Restructuring	x1.5	Elimination of Output Line Buffer Memory
Self-timed Outputs Drivers	x2	Elimination of glitching on the output of data busses. This number is for just the memory output busses
Bit Swing Reduction	x3.7	This reduction is only for the memory array. The global bus has rail to rail swing. Comparison is for 1.1V swing vs. 300mV swing

9.4.3.3 Chrominance Decompression Chip

The color video decompression chip is implemented using a very similar architecture to the luminance architecture. Since the chrominance components, I and Q, are each sub-sampled by a factor 4, the amount of frame-buffer memory required is reduced by a factor of 4 for each component. Also, since they use the same addressing scheme, the I and Q data are stored in the same physical frame-buffer and the data is interleaved; that is, the 32-bit frame-buffer word is organized as: CWI0 CWI1 CWQ0 CWQ1. The 4:1 mux to select the codeword in the Y decompression chip is replaced by two 2:1 muxes, one for I (to select between CWI0 and CWI1) and one for Q (to select between CWQ0 and CWQ1). The chrominance (IQ) decompression chip also has two separate (256x16x6) lookup table memories. The IQ data is stored in the memory in a sign-magnitude format to reduce switching activity when accessed. The chrominance chip was implemented using two of the luminance chips to minimize the number of chips that had to be fabricated.

9.4.3.4 Video Controller

The video controller performs all the control functions for the video decompression module. It generates all of the timing for the NTSC display, interfaces to the radio, controls the frame-buffers and lookup tables and performs synchronization for the system. A summary of the functions performed is outlined below:

NTSC sync generation for the 4" display: The LQ4RA01 4" color display is responsive to a standard composite sync signal with negative polarity of the same amplitude level as that of the video composite signal. The standard sync found in NTSC format has extra timing information such as an extra half line in one field (to distinguish between even and odd fields), and pre and post equalization half-line pulses during vertical sync that are not a necessity for proper operation of the LCD. A significant simplification of this protocol which involves block sync (with no equalization pulses) and the elimination of the half-line can be used to obtain a sync that still provides adequate synchronization information. The implemented sync signal is simply a scaled digital combination of *VSYNC* (vertical sync) and *HSYNC* (horizontal sync). The original and modified sync signals are shown in Figure 9.25.

Decodes and demultiplexes the radio data: similar to the function performed in the protocol chip, the video controller chip decodes packets from the radio and demultiplexes between a frame-buffer FIFO and LUT FIFO. Contained inside each frame-buffer packet (that is sent over the frame-buffer FIFO) is the encoding of type information (TYPE = 0 => data is for the *Y* frame-buffer, and TYPE= 1 => data is for the *IQ* frame-buffer). Similarly, inside the LUT packet is encoding information about the LUT data type (*Y, I* or *Q*).

Controls reading and writing of the frame-buffer memories for both *Y* and *IQ* decompression chips: it generates the read and write addresses for the ping-pong frame-buffers. It also generates the multiplexor control signals that selects the output of the frame-buffers (*ByteSel* in Figure 9.22). It also performs the $R/\overline{W}$ control (i.e., controls when the frame-buffers switch between reading to writing) and generates the clocks for the SRAMs.

Controls loading and reading of the lookup table memories for *Y, I* and *Q* data: it generates the write address (*WLUTAddr* in Figure 9.22) and part of the read address (*RLUTLine* in Figure 9.22) for the lookup table memories. It also generates the multiplexor control signals that selects the output of the LUT (*Pix-Sel* in Figure 9.22).

Support for variable sized packets: a decompressed frame of video can be broken into multiple packets and the size of the packets is variable, providing a flexible platform to test the effects of packet sizes. Also, this allows a form of inter-frame coding in which only the differences between the current frame and the frame corresponding to two image ago is sent. This is effective only if the BER is fairly low for the reasons explained earlier.

Matches pipeline delays in the system: since the system is pipelined, the output sync signals for the display must be delayed to avoid offsets of pixel data on the display. The timing signals needed by the color space converter are also generated in this chip.

Figure 9.26 shows the die photograph of the video controller chip.

Field One (Odd) Composite Timing

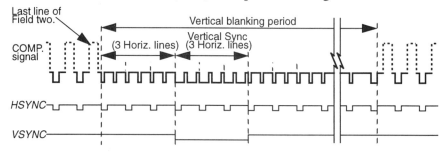

Field Two (Even) Composite Timing

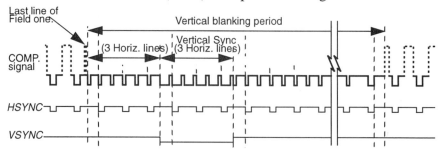

(Both Fields) Modified (Simplified) Timing

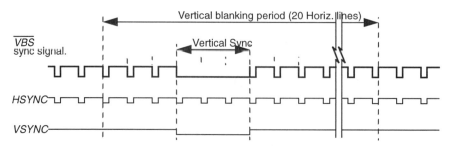

Figure 9.25: Simplification of the NTSC system timing.

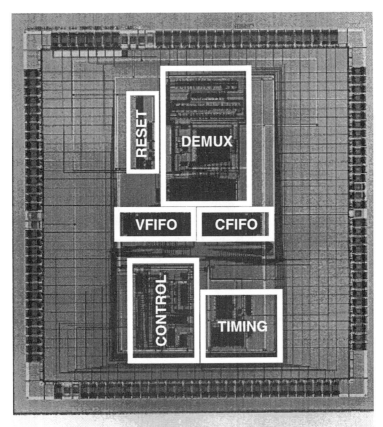

Figure 9.26: Die photograph of the video controller chip.

9.4.3.5 Color Space Converter and Digital to Analog Converter

The digital *YIQ* information from the video decompression chips are sent to a color space converter which converts it to analog *RGB* to drive the 4" color LCD display from SHARP. The digital *YIQ* is first converted to digital *RGB* and then a triple digital to analog converter directly drives the display.

Digital YIQ to Digital RGB Conversion

In the *YIQ* to *RGB* translation, which involves multiplication with constant coefficients, the switching events are minimized at the algorithmic level by substituting multiplications with hardwired shift-add operations (in which the shift operations degenerate to wiring) and by optimally scaling coefficients. As described in Chapter 8, the multiplication of multiple coefficients with the same input was exploited to minimize number of shift-add operations. In this way, the 3x3 matrix conversion operation degenerated to 8 additions. The implementation was fully parallel and therefore there was no controller. For I/O communication (between the decompression chips and the color space chip) and in the matrix computation, sign-magnitude representation is chosen over two's complement to reduce the toggle activity in the sign bits.

At the architecture level, time-multiplexing was avoided as it can destroy signal correlations, increasing the activity. Figure 9.27 shows two alternate schemes for transmitting the I and Q data from the decompression chips to the color space converter chip. On the left is a fully parallel version in which I and Q have separate data busses. Also shown is the data for I and Q for a short segment in time. As shown, the data is slowly varying and therefore has low switching activity in the higher order bits. On the right is a time-multiplexed version in which there is a single time-shared bus in which the I and Q samples are inter-leaved. As seen from the signal value on the data bus, there is high switching activity resulting in higher power. Figure 9.28 transition activity for time-multi-plexed activity vs. fully parallel architecture.

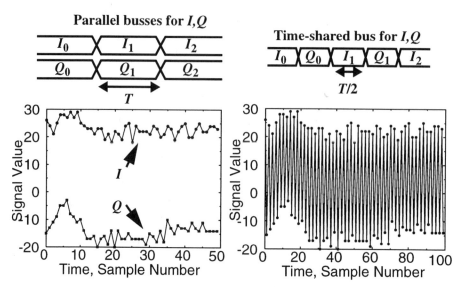

Figure 9.27: Time-multiplexing can destroy signal correlation increasing activity.

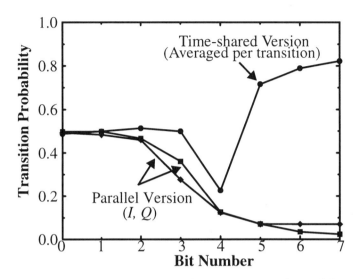

Figure 9.28: Transition activity for time-multiplexing activity.

The digital *YIQ* -> digital *RGB* consumes only 100μW at 1.5V. This is more than three orders of magnitude lower power compared to existing commercial solutions. Table 9.9 shows the power reduction achieved from various techniques compared to a commercial color space converter.

Table 9.9 Summary of power reduction techniques applied to the *YIQ* -> *RGB* conversion

Design Approach	Power Reduction	Comments
Frequency Reduction (14MHz -> 2.5MHz)	x5.6	The display resolution used was small and therefore the throughput requirement was much lower.
Supply Voltage Scaling (Parallelism)	x11	1.5V operation vs. 5V operation
Optimized Cell Library	x2-3	Minimum sized devices, Single Phase Clocking.

Table 9.9 Summary of power reduction techniques applied to the *YIQ -> RGB* conversion

Design Approach	Power Reduction	Comments
Hardwired Shift-add Coefficient Optimization	x7	The commercial implementation allowed programmability of coefficients (NTSC, PAL, etc.). Fixed application allowed optimization of the coefficients for minimal computational complexity - no real multipliers.
Fully Parallel Implementation	x1.5-2	Hardware assignment that keeps uncorrelated data on different units.
Sign-Magnitude over Two's Complement Representation	x1.2	For this application data was correlated, but there was still some statistical improvement.
Integrate Processing and DAC's	x1.4	The DAC (as described in the next sub-section) was integrated onto the same chip and eliminated I/O power.
Bitwidth Reduction	x1.3	Exploiting the low resolution of the display only 6bits per plane was used instead of the typical 8bits.

Low-voltage Digital to Analog Converter

A low-voltage, low-throughput 6-bit DAC has been developed to drive the 4" SHARP LCD. The LCD display takes pixel data in the analog *R,G,B* format which has voltage levels compatible with the NTSC format (i.e., V_{pp} = 0 to 0.7V). The DAC, shown in Figure 9.29, has an architecture based on a conventional non-weighted current switched array. Based on a decoded 6-bit digital word, an appropriate number of current sources are turned ON and summed. The output voltage is obtained by passing current through an external resistor. Since the settling time requirement is quite low and since the LCD display has a high impedance capacitive load, the external resistor was chosen to be approximately 1KΩ rather than 75Ω. This reduces the power consumption by more than an order of magnitude since the average current drawn from the supply is reduced by more than order of magnitude. The row decoding and column decoding logic is identical to the one presented in [Miki86]. The decoding logic was implemented using minimum sized low-power standard cells.

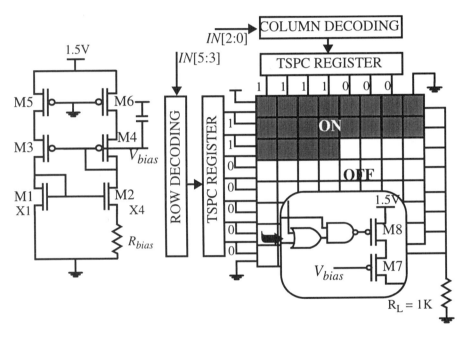

Figure 9.29: Low-voltage 6-bit Digital to Analog converter.

A current source that operates at a reduced supply voltage of 2.7V has been developed [Miki92]. The current source used in this DAC is similar. However, due to the modest DAC throughput requirements for this 4" display, and to statistically reduce the power consumption by a factor of 2 (on average only half of the current sources will be ON) a single ended architecture is used instead of the differential scheme. The current cell consists of stacked PMOS devices (M7 and M8) with the top transistor M8 being digitally switched, thus operating in the linear region. Therefore, effectively, the output resistance of the current source is the output resistance of a single transistor (M7) degenerated with a source resistor. To increase the output resistance of M7, the length of the device was made non-minimum. In order to operate the DAC down to a supply voltage of 1.5V and to meet the 0.7Vpp requirement for the output, the W/L of M8 was made large to keep the voltage dropped across M8 to less than 100mV. Also, the bias voltage was set up so that the V_{gs}-V_t for M7 was approximately 200mV, allowing supply voltages to be as low as 1.2-1.3V. The power consumption of the DAC is dominated by the analog power and for a 1.5V supply: $P_{avg} = V_{dd} \bullet I_{avg} = 1.5 \bullet 1/2 \bullet 0.7/1.2K = 440\mu W$. The power measured in the actual system was lower since the DAC was shut down (the digital input was forced to 0) during the horizontal and vertical blanking periods.

The digital decoding inside each current cell is implemented using a single complex CMOS gate $\overline{(A + B) \bullet C}$ to eliminate glitching on the gate of M8 as compared to path balancing approaches [Fournier91]. The device sizes on the complex gate are minimum to minimize the dynamic power consumption.

Figure 9.29 also shows the schematic of the bias circuitry which is a bootstrapped current reference [Gray84]. The top current mirror, M4 thru M6, forces the bias currents in M1 and M2 to be equal. Transistors M5 and M6 operate in the linear region and emulate the voltage drop across the switch transistor in the current cell array (M8). For the bottom current source:

$$V_{gs1} = V_{gs2} + I_{out} R_{bias} \tag{261}$$

$$V_{tn} + \sqrt{\frac{2 \cdot I_1}{K_n \cdot \dfrac{W}{L}1}} = V_{tn} + \sqrt{\frac{2 \cdot I_2}{K_n \cdot \dfrac{W}{L}2}} + I_2 \cdot R_{bias} \tag{262}$$

Assuming $V_{tn} = V_{tp}$, and $I_1 = I_2$ (forced by the top current mirror M3-M6),

$$I_{ref} = \frac{2}{K_n R_{bias}^2} \left(\frac{1}{\sqrt{\dfrac{W}{L}1}} - \frac{1}{\sqrt{\dfrac{W}{L}2}} \right)^2 \tag{263}$$

Note that the reference current ($I_{ref} = I_1 = I_2$) is to first order independent of supply variations (this is due to the bootstapped techniques used). The bias current is set by sizing M1 and M2 and by the choosing R_{bias}. The current source can operate at low supply voltages even with a process that has a standard threshold voltage. The minimum operating voltage for this current reference is given by:

$$V_{ddmin} = V_{ds6} + |V_{tp4}| + V_{dsat4} + (V_{gs2} - V_{tn}) + I_{ref} R_{bias} \tag{264}$$

In the above equation, V_{ds6} is very small (< 100mV since M6 is in the linear region and by device sizing), V_{tp} for the MOSIS 1.2μm CMOS process is typically around 0.9V, and the last two terms can be made small (100mV - 200mV) by device sizing and choice of bias current. Therefore, the current source can operate down to the 1.2-1.5V range. Table 9.10 shows a summary of the power reduction techniques that were applied to the DAC design.

Table 9.10 Summary of power reduction techniques applied to the DAC

Approach	Power Reduction	Comments
Reduction of Load Resistance	x16	The output resistance was chosen to be 1.2KΩ since the load into the LCD is capacitive. This is much higher than the normal 75Ω resistance
Single Ended Architecture	x2	Due to the relaxed speed requirements, a single-ended architecture was sufficient and statistically the power was reduced by a factor of 2
Voltage Reduction	x3.3	A linear reduction in power consumption from 5 to 1.5V

Figure 9.30 shows the die photo of the color space converter and triple DAC.

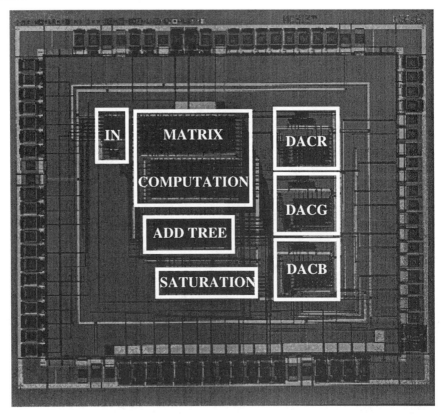

Figure 9.30: Die photo of the color space converter and digital to analog converter.

9.4.3.6 Summary of Video Decompression Chipset

Table 9.11 shows a summary of the video decompression chipset which is implemented using a standard Mosis 1.2µm CMOS process. The minimum operating supply voltage for each chip is also shown along with the power consumption at the worst case voltage for this system implementation which is 1.5V. At the worst case supply voltage, the entire chipset consumes less than 2mW to perform the video decompression for the 4" active matrix SHARP display.

Table 9.11 Summary of the video decompression chipset implemented in 1.2µm CMOS.

Chip Description	Area (mmxmm)	Number of Transistors	Minimum Supply Voltage	Power at 1.5V
Video Controller	6.7 x 6.4	31400	1.1V	150µW
Luminance Decompression	8.5 x 6.7	~250000	1.1V	115 µW
Chrominance Decompression	8.5 x 9.0	~400000	1.1V	100µW
Color Space Conversion and Triple DAC	4.1 x 4.7	12500	1.3V	1.1 mW

9.5 System Implementation

9.5.1 First Generation System: Pen, Speech I/O and Text/graphics

A PCB board containing the protocol chip, 6 SRAM chips, a speech codec, and pen interface logic has been fabricated and tested. Figure 9.31 shows a block diagram of the first generation InfoPad terminal called the IPGraphics. Various power supply voltages needed for the design including -17V (adjustable using a trim-pot) for display drive, 12V for DC to AC inverter for the backlight, 1.1 V for the custom chips, and -5V for the speech codec have been realized using commercial chips from Maxim.

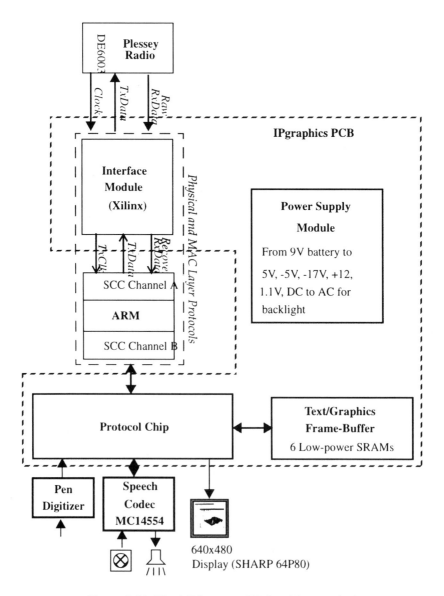

Figure 9.31: Block Diagram of IPGraphics terminal.

This board is integrated with a commercial radio modem, a half-duplex, 1 Mbps RF modem: the DE6003, from GEC Plessey, to realize a complete I/O terminal with a 1 Mbit/sec wireless channel. The DE6003 provides a demodulated signal to the baseband receiver, implemented in a Xilinx 4000 FPGA, without any explicit timing information or attached protocol. This "raw" received data is passed to the clock recovery module (the "Interface Module"), which extracts timing information from transitions in the data stream. Once it has attained bit-level synchronization -- achieved by sending a training sequence of alternating 1's and 0's at the beginning of a packet -- the interface board attempts to reconstruct the received bits from the raw bitstream, and passes the recovered clock and data to the ARM board. Early on in the design of the wireless link, the decision was made to use a general-purpose microprocessor or microcontroller as part of the protocol support hardware, providing flexibility in the design of the protocol. The ARM610, a 32-bit RISC device, was the microprocessor chosen because it offered a relatively low-power solution for our needs. Media access control, flow control, and tracking error statistics for link management are the most important duties performed by the ARM, while physical layer functionality, such as clock and data recovery, line coding, and channel sensing, is implemented in the Xilinx FPGA. Figure 9.32 shows the photograph of the first generation InfoPad terminal.

Figure 9.32: IPGraphics (Pen, Speech I/O and Text/graphics) Terminal.

9.5.2 Video Decompression Test Board

A test board using the video decompression chipset has been designed and tested. Figure 9.33 shows a block diagram of the video test board. There are 5 custom PGAs (a video controller chip, three luminance decompression chip - one for Y, I and Q, and a color space and triple-DAC chip), a crystal oscillator which generates the master clock (2.5MHz) from which all other timing signals are derived, and a power supply chip to generate the supply voltage for the custom chips (1.3V-1.5V). The video controller takes a serial compressed video stream (for test purposes from an Areial board on SPARC station) and drives the Y, I, and Q decompression chips. The decompressed busses (Y,I, and Q) output from the decompression chips go to the color space converter which generates the analog signals that drive the LCD display. The video controller generates the composite sync for the LCD display. The LCD display requires 5V and -8V power supplies. Figure 9.34 shows the output of the video decompression chips operating at a supply voltage of 1.3V.

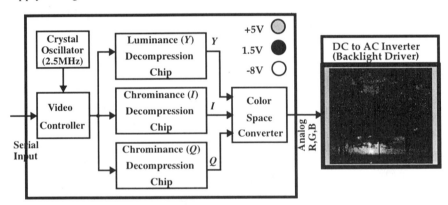

Figure 9.33: Block diagram of the video decompression test board.

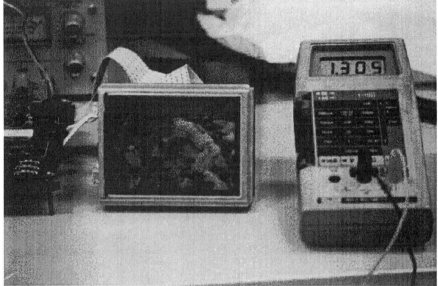

Figure 9.34: Output of video decompression chips running at $V_{dd} = 1.3$V.

9.6 Summary

Recently, the concept of "personal communications" has come to the forefront of communications research, in which individual users will have portable, private access to fixed computing facilities. The ultimate goal is to provide a personal communications system, which will move information of all kinds to and from people in all locations, through an advanced wireless network supporting a wide range of services. As a step beyond today's portable computers, a high-speed wireless link allows the advent of small, lightweight, multimedia graphics terminals, whose primary function would be to connect the user instantaneously and transparently into powerful fixed processing units and data storage. It would be capable of providing simplified user interface through pen input and speech I/O, data transfer and retrieval, computing services, and high-quality, full-motion video. The design of a low-power chipset has been described for a portable multimedia terminal which supports pen input, speech I/O, text/graphics output and one-way full-motion video.

A system approach to power reduction was used which involved optimizing the physical design, the circuitry, the logic design, the architectures, the algorithms, and system partitioning.

At the system level, the power consumption is reduced by optimizing the partitioning of computation between the terminal and the backbone network. By exploiting wireless communications capability, all general-purpose user-programmable computation is *removed* from the terminal and performed by compute servers on the backbone network. This significantly reduces the power requirements of the terminal.

At the algorithmic level, the number of operations to implement a given function was minimized. For example, for the video decompression, vector quantization algorithm was chosen since the computational complexity was significantly lower than conventional DCT based algorithms. In the matrix multiplication, *YIQ->RGB*, coefficient optimization and bitwidth reduction resulted in significant reduction of the computational complexity over an implementation that uses true multipliers. In addition to minimizing the number of operations, the data representation was optimized for reducing the switching activity for memory access and in the logic of the matrix multiplication.

At the architecture level, the biggest power savings came from using the architecture driven voltage scaling strategy which involved parallelism and pipelining and allowed supply voltages as low as 1.1V. The memory architecture for the decompression and frame-buffer was optimized, eliminating the need for an output line buffer used to convert block scan to raster scan therefore reducing the number of memory operations. Time-multiplexing was avoided since it can destroy correlations and increase the switching activity; that is, uncorrelated data was processed in different hardware units.

At the logic level, gated clocks were used extensively to minimize the effective clock load and the switched capacitance in logic circuits. Various circuit level optimizations were performed to minimize the voltage swings and switched capacitance. In the memory circuits (SRAM and FIFO), reduced swing logic (using NMOS precharge) was used to limit the swing on the bit lines to 300mV at a supply voltage of 1.1V. The memory modules also used self-timed circuits to eliminate glitching on the data busses, hence minimizing the effective capacitance switched. The various logic modules were implemented using a low-power cell-library that featured minimum sized devices, optimized layout, and single-phase clocking. The DAC conserves power by using a larger output load resistance to reduce load currents, a single ended architecture to reduce the average power by a factor of 2, and voltage scaling down to 1.5V.

At the layout level, an activity driven place and route strategy was used to route high activity nets using short wires while allowing low activity nets to have long wires.

The various low-power techniques result in the power consumption of the entire chipset to be less than 5mW at 1.5V, which is more than three orders of magnitude lower than existing commercial solutions performing the same functionality. It is clear from this design example that there is no single magical technique that can be used to achieve orders of magnitude reduction of power consumption; instead, a system level solution is required that optimizes the algorithms, the architectures, the logic and the circuits.

REFERENCES:

[Benson90] D. Benson *et al.*, "Silicon Multichip Modules," *Hot Chips Symposium III*, Santa Clara, CA, August 1990.

[Burd94] T. Burd, Low-power Cell Library, M.S. Thesis, U.C. Berkeley, June 1994.

[Burstein93] A. Burstein, Low-power Cell-library Documentation, U.C. Berkeley, 1993.

[Chandraksan94] A. P. Chandrakasan, A. Burstein, and R. W. Brodersen, "A Low-power Chipset for Multimedia Applications," *IEEE International Solid-state Circuits Conference*, pp. 82-83, February 1994.

[Clarke85] R.J. Clarke, *Transform Coding of Images*, Academic Press, NY, 1985.

[Eager91] J. Eager, "Advances in Rechargeable Batteries Pace Portable Computer Growth," *Silicon Valley Personal Computer Conference*, pp. 693-697, 1991.

[Feig90] E. Feig, "On the multiplicative complexity of the Discrete Cosine transform," 1990 *SPIE/SPSE symposium of Electronic Imaging Science and Technology*, Santa Clara, CA, February, 1990.

[Fournier91] J.M. Fournier and P. Senn, "A 130MHz 8-b CMOS Video DAC for HDTV Applications," *IEEE Journal of Solid-State Circuits*, pp. 1073-1077, July 1991.

[Gersho92] A. Gersho and R. Gray, *Vector Quantization and Signal Compression*, Kluwer Academic Publishers, 1992.

[Gray84] P. Gray and R. Meyer, *Analysis and Design of Analog Integrated Circuits*, Second edition, John Wiley & Sons, 1984.

[Miki86] T. Miki, Y. Nakamura, M. Nakaya, S. Asai, Y. Akasaka, and Y. Horiba, "An 80-MHz 8-bit CMOS D/A Converter," *IEEE Journal of Solid-state Circuits*, pp. 983-988, Dec. 1986.

[Miki92] T. Miki, Y. Nakamura, Y. Nishikawa, K. Okada, and Y. Horiba, "A 10bit 50MS/s CMOS D/A Converter with 2.7V Power Supply," *1992 Symposium on VLSI Circuits*, pp. 92-93.

[Rao90] K.R. Rao and P.Yip, *Discrete Cosine Transform*, Academic Press, 1990.

[Sheng92] S. Sheng, A. P. Chandrakasan, and R. W. Brodersen, "A Portable Multimedia Terminal," *IEEE Communications Magazine*, pp. 64-75, December 1992.

10

Low Power Programmable Computation

Coauthored with Mani B. Srivastava
AT&T Bell Laboratories, Murray Hill, NJ

Figure 10.1 shows a block diagram of the InfoPad I/O terminal, that was described in the previous chapter, emphasizing the I/O functionality that is suitable for implementation on a programmable processor.

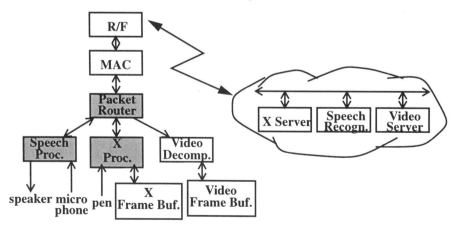

Figure 10.1: Energy efficient programmable computation required in portables.

Even with the minimalist approach to computation adopted by InfoPad, there are parts of its functionality that are best implemented using general-purpose microprocessors, programmable DSPs, and programmable processor cores embedded in ASICs. Examples include media access and data link layer protocol processing for wireless RF communication between the portable terminal and a basestation, the DSP processing for speech coding, and processing for parts of the graphics display server (X server for example) that may reside on the portable terminal.

The need for the use of software programmable components is even greater in other portable devices - such as most PDA type devices - that do not adopt as extreme a partitioning of computation between the network servers and the portable device as is adopted in InfoPad. For example, the InfoPad system requires that the X-server computation be completely performed on the backbone network. Another partitioning scheme could be to move the X-server computation, or some part of it, to the terminal [Weiser93] (while the client applications still run on the backbone network), which then will require a programmable processor; however, as described earlier this approach will require extra bandwidth for protecting error sensitive control data of X-protocol. Implementation using software may be needed because: (i) the algorithmic and logical (control) complexity of the application may preclude dedicated hardware and make software the only practical choice, or (ii) the application may not operate continuously or different functionality may be needed at different times so that a time-multiplexed software implementation is more cost effective.

10.1 Architectural Approaches to Low-power

The primary focus of this chapter is on architectural techniques to improve the energy efficiency for parts of the portable devices that need to be implemented using software programmable components such as microprocessors, DSPs, and embedded processor cores. So far the main approaches for lower power consumption and increased battery life for general purpose computers have been restricted to using a limited amount of voltage scaling in the form of using 3.3V parts instead of 5V parts (often coupled with lower clock speed), and straightforward shutdown of the system power supply and/or clock - various notebook and laptop computers on the market illustrate these techniques. However, much higher reductions in power consumption are possible by using more sophisticated architectural and implementation strategies than straightforward shutdown and the use of 3.3V components. For example, a proper addressing of the problems of *when* to shutdown, and *how* to scale the voltages can result in substantial improvement in energy efficiency with little or no loss in performance.

As discussed in Chapter 3, the average power consumption of a CMOS gate is dominated by the switching component and is given by $\alpha\ CV_{dd}^2 f$. The expression for power consumption suggests several strategies for increasing the energy efficiency (reducing the power consumption while maintaining the computation speed):

Activity based system shutdown: Many computations are "event-drive" in nature with intermittent computation activity triggered by external events and separated by periods of inactivity - examples include X server, communication interfaces, etc. An obvious way to reduce average power consumption in such computations would be to shut the system down during periods of inactivity. Shutting the system down, which would make power consumption zero or negligible, can be accomplished either by shutting off the clock (f=0) or in certain cases by shutting off the power supply (V_{dd} = 0). However, the shutdown mechanism has to be such that there is no or little degradation in speed - both latency and throughput are usually important in event-driven computations.

Supply voltage reduction: Not all computation implemented as software are "event-driven" in nature - data-flow functions such as DSP are "continuous" in nature. Obviously, shutdown is not an effective mechanism for these systems. An alternative strategy is to operate at the lowest possible supply voltage, as is suggested by the quadratic dependence of power on supply voltage V_{dd}. Unfortunately, operation at reduced supply voltage comes at the expense of reduced circuit speed. Fortunately, however, if only throughput and not latency is the metric of speed - as is true for many "continuous" applications such as speech coding - the reduction in circuit speed can be compensated for by architectural techniques like pipelining and parallelism that increase throughput so that the net result is a more energy efficient system operating at a lower voltage but same throughput. Compiler techniques for effective parallelization and pipelining play an important role in the success of this strategy to energy efficiency.

Switching activity reduction: In addition to the above strategies that are specific to event-driven and continuous computations, there is a range of architectural strategies that are applicable to both types of computation. Such strategies in general try to make the system more energy efficient by reducing the switching activity α. As described in Chapter 7, reduction in α can be accomplished in a variety of ways - computation restructuring, communication restructuring, optimizing the memory storage architecture and hierarchy, changing the data encoding, etc.

10.2 Shutdown Techniques

An obvious mechanism for saving energy is to shut down parts of the system hardware that are idle because they are waiting for I/O from outside the system or from other parts of the system. While it is shutdown, the system consumes near zero power. Shutdown is a particularly relevant mechanism for event-driven interactive computation, such as is found in the graphics server and other user-interface related functions of InfoPad like portable devices. As shown in Figure 10.2, such computations are in one of two states: either blocked while waiting for an I/O event or performing computation. When running on a dedicated CPU, the application will alternate between a *blocked* state where it stalls while waiting for external events such as a key press or a mouse click and a *running* state where it will execute instructions to perform computation. If $T_{blocked}$ and $T_{running}$ are the average time spent in the *blocked* and the *running* states respectively, then one can improve the energy efficiency by as much as a factor of $1 + T_{blocked} / T_{running}$ provided the system is shutdown whenever it is in the *blocked* state.

There are two main problems in shutdown - *how* to shutdown, and *when* to shutdown. The first problem is addressed by mechanisms for stopping and restarting the clock ($f=0$) or for turning on and off the power supply ($V_{dd}=0$). The second problem is addressed by policies such as "shut the system down if the user has been idle for 5 minutes". Although the two problems are not really independent because the decision about when to shutdown depends on the cost (in time and power) of shutting down and restarting the system, the focus here is primarily on the problem of deciding when to shutdown while being cognizant of the available shutdown mechanisms.

Simple shutdown techniques, for example shutting down after a few seconds of no keyboard or mouse activity, are already used to reduce power consumption in current notebook computers. However, the event-driven nature of modern window systems, together with efficient hardware shutdown mechanisms provided by newer microprocessors and system controllers, suggests the possibility of a more aggressive shutdown strategy where parts of the system may be

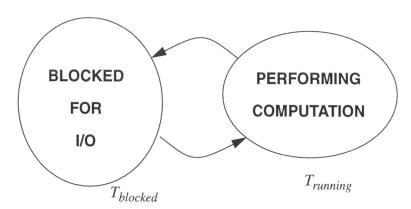

Figure 10.2: Event-driven applications alternate between *blocked* and *running* states.

shutdown for much smaller but more frequent intervals of time while waiting for I/O events. Such a shutdown mechanism is presented here for use on a portable X-terminal type device. A simple algorithm derived from analysis of actual traces is used to predict the length of the time spent in *blocked* state in order to minimize the impact on interactive speed, while the more frequent system shutdowns result in large reductions in effective computation energy.

10.2.1 Conventional Shutdown Approaches

The portable computers available now use various shutdown techniques that are all variants of the following basic scheme: "Go to Reduced Power Mode after the user has been idle for a few seconds/minutes". Figure 10.3 illustrates the philosophy underlying the conventional approaches to shutdown. A drawback of this straightforward policy is apparent - the system continues to waste energy while it idly waits to check for lack of user activity for a few seconds/minutes. Experimental traces obtained from an X server (Section 10.2.2) showed that while the X server spends 96-98% of its time in the *blocked* state, the average time spent on each visit to the *blocked* state is short (<< a second). The conventional shutdown schemes will therefore fail to exploit the large reduction in energy that is otherwise possible.

Typical of the conventional approaches to shutdown are the schemes used in Apple's popular Macintosh PowerBook series of portable computers, which have three different types of reduced power modes based on shutting down parts of the system [Apple92]. A Power-Management IC controls the process of enter-

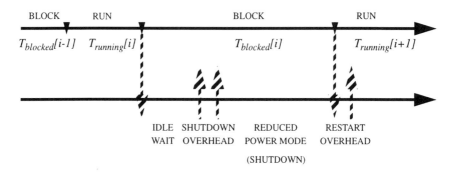

Figure 10.3: Conventional shutdown approaches.

ing and exiting these shutdown modes by monitoring the input devices and the battery voltage, controlling the contrast of the LCD display.

10.2.1.1 Rest Mode

A technique termed *Power Cycling* is used in the rest mode on most models of PowerBooks. The computer enters the rest mode after 2 seconds of idle time upon which the processor registers are saved and the processor is powered down. However, the I/O devices remain on, the screen cursor continues to blink, and the keyboard continues to be scanned. After 1/60 second (16.7 ms) the power to the main processor is restored. If there has been no I/O activity, the processor is again shutdown for another 1/60 second.

The power consumption is reduced by up to 90%, i.e. by as much as x10, *while in the rest mode*. However, because the rest mode is entered only after 2 seconds of idle time, the effective reduction in power consumption is much less dramatic. Analysis of experimental battery life data reported in [Berkoff93] for various usage scenarios of a PowerBook suggests that enabling the rest mode reduces the power consumption of the processor and logic section alone by x6 whereas the power consumption of the entire system, including the disk and display backlight, is improved by x2. Experiments, described later, suggest improvements in the processor power consumption in the neighborhood of x1.5-x2 if the power cycling strategy is used for a processor running an X server. While even this x1.5-x2 improvement in power consumption is desirable, later sections show that a more aggressive shutdown technique yields a much larger improvement.

The impact of power cycling on the performance of interactive applications is negligible.

10.2.1.2 Sleep Mode

The sleep mode is entered after the computer has been idle for a user selected period of time, typically in the range of a few minutes. The computer exits the sleep mode and reverts back to the normal mode on the occurrence of an external event, such as a key press or a modem ring-detect. A decrease in computers responsiveness compared to the rest mode is traded-off against an increase in energy conservation by shutting down the peripheral functions as well: the I/O ports, the disk ports, the display, the sound circuits. Power is retained only to the RAM and the Power-Manager IC.

Since the disk goes to sleep too, the effect on interactive applications is substantial - spinning the disk down and back up takes a substantial time. Further, the break-even point when it is worth spinning-down the disk and spinning it back up (because spinning-up the disk takes much increased power) is about 15 seconds [Berkoff93], which implies that sleep mode is effective only for moderately long periods of idle time.

10.2.1.3 Shutdown Mode

This mode is typically entered on an explicit command from the user. It has the worst impact on the responsiveness of the computer but saves the maximum power - on most PowerBook models the entire computer, including the Power Manager IC, is turned off - only a tiny amount of power is drawn for a parameter RAM. The computer does not respond to external events at all while in this mode, and one needs to restart it.

10.2.2 Predictive Shutdown Approaches

The straightforward shutdown schemes described above either show only a moderate overall improvement in energy consumption with negligible loss of computer responsiveness as in the rest mode or show a higher degree of improvement but at the cost of much decreased computer responsiveness, as in the case of sleep mode and shutdown mode. In this section a shutdown mechanism that uses the computation history to predict the length of idle time is explored. The processor is shut down if the predicted length of idle time justifies the cost - in terms of both power and responsiveness - of shutting down. The basic philosophy behind the predictive approach can be summarized as follows: "On entering the blocked state, use computation history to predict whether $T_{blocked}$ will be large enough $(T_{blocked} > T_{cost})$ to justify a shutdown". Analysis done using real-life traces suggests that the predictive shutdown approach leads to a much higher reduction in effective processor power, than is obtained with straightforward non-predictive shutdown schemes, with only a small loss in responsiveness.

10.2.2.1 Helpful Trends in Computing & Communications

Two developments in computing and communication motivate more sophisticated shutdown mechanisms. First, newer microprocessors, such as Intel 486SL, and AT&T Hobbit, provide power management support integrated with the system control, thus enabling implementations of shutdown that are efficient both in terms of time and hardware cost. Power management features typically include the ability to shut down the clock and/or power supply to various parts of the system and efficient mechanisms to store and restore the processor state for power down.

Second, the integration of computing and communication is resulting in a paradigm where increasingly the computer will become a device to access remote computation and information servers across a network. This is particularly the case with the proposed wireless devices such as the InfoPad. With the actual application-specific computation being done primarily on network servers, the computation that is left on the personal device is increasingly related to the graphic user interface and windowing system functionality. A good example of this would be a wireless multimedia device which, besides audio and video services, provides the capability to run X system based remote client applications. In such a case the general purpose programmable computer in the portable device would primarily be running the X display server (or a part of the server), and maybe some important local clients such as the window manager - similar to conventional X terminals. Such software is even-driven in nature, where the events arise because of user interaction. Except for extremely graphics intensive applications, a vast majority of time such software just idles while waiting for user input. Shutting down the hardware while the software is waiting for external events can give rise to potentially huge savings in computation energy.

10.2.2.2 Potential for Reduction in Computation Energy by Shutdown

A study of the potential energy reduction was done for an X display server that adopts the "X Terminal" model of computation described above. The results of our analysis below, based on experimentally obtained traces of X server state, suggest that potential energy savings as high as x30 to x60 are possible.

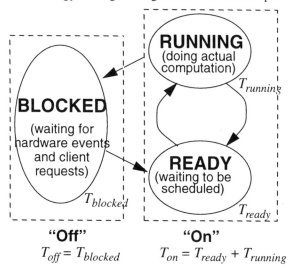

i. X Server running under multi-tasking OS

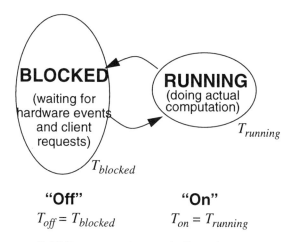

ii. X Server running on dedicated processor

Figure 10.4: States of X Display Server Process.

Figure 10.4 shows that the X server process running under a multitasking operating system, such as UNIX, is in one of the three states:

a. It may be *blocked* or stalled while waiting for either an hardware event (key press or mouse activity) or requests from one of the existing client applications or a connection request from a new client.

b. It may be *running* on the processor doing some computation.

c. It may be *ready* to run but is waiting for the scheduler of a multi-tasking time-shared operating system to schedule it to run.

Let $T_{blocked}$, $T_{running}$, and T_{ready} be the time spent in the three states. If the X server were to run on a dedicated processor, as would be the case in the X Terminal model of computation, the server will never be in the *ready* state - it would always be either *blocked* or *running* - i.e. $T_{ready} = 0$. The hardware does nothing while the server is in the *blocked* state and can therefore be shut down by stopping the clock or by saving the state and then power down. The fraction $T_{blocked} / (T_{blocked} + T_{running})$ of the total time (in this case $T_{ready} = 0$) that the server spends in the *blocked* state corresponds to the maximum possible reduction in computation energy.

In order to estimate $T_{blocked} / (T_{blocked} + T_{running})$ under the condition that $T_{ready} = 0$, and thus the potential energy reduction in the X Terminal model of computation, the X server on an UNIX workstation was instrumented to measure $T_{blocked}$ and $T_{running} + T_{ready}$. Unfortunately, T_{ready} is non-zero due to the multi-tasking time-shared nature of UNIX, and there is no easy way to measure $T_{running}$ and T_{ready} separately without modifying the kernel. Even though the experiments were run on an unloaded workstation with no local X clients except for the console window and the window manager, there are background daemon processes over which one does not have any control. However, it is important to realize that the fraction $T_{blocked} / (T_{blocked} + T_{running} + T_{ready})$, which can easily be measured, is a lower bound on the fraction of energy that can be saved under ideal conditions. So the experimental results err on the safe side - the estimate of the potential energy reduction that will result from using $T_{blocked} / (T_{blocked} + T_{running} + T_{ready})$ will thus be pessimistic.

From now on the *blocked* state will be referred to as the *off* state, and the *running* and *ready* states will jointly be referred to as the *on* state. Further, $T_{off} = T_{blocked}$ and $T_{on} = T_{running} + T_{ready}$. The instrumented X server measures T_{off} and T_{on}, and the ratio $T_{off}/(T_{off} + T_{on})$ is used as a lower-bound on the maximum possible reduction in energy under ideal shut down. Figure 10.5 shows a sample trace

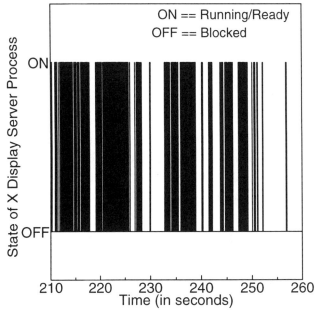

Figure 10.5: Sample trace of X server state under UNIX.

of the time spent by the X server in the *on* and the *off* states over a small stretch of time, and Table 10.1 shows the results obtained by analyzing several traces

	Trace 1	Trace 2	Trace 3
Trace Length (sec)	5182.48	26859.9	995.16
T_{off} (sec)	5047.47	26427.4	960.82
T_{on} (sec)	135.01	432.5	34.34
$T_{off}/(T_{off}+T_{on})$	0.9739	0.9839	0.9655
Maximum Energy Reduction (conservative estimate)	x 38.4	x 62.1	x 29.0

Table 10.1 Analysis of X Server Traces for Potential Energy Reduction

obtained from real X sessions. These X sessions consisted of running the instrumented X server together with the window manager *olvwm* and console window *contool* running locally, and several typical X clients, such as *xterms, Frame-Maker, xclocks, mailtool, cm* (calendar manager) etc., running remotely. As is evident from these traces the X server spends most of its time, ranging from 96.5% to 98.4%, in the *off* state suggesting that energy reductions range from x29 to x62 under ideal shutdown conditions.

10.2.2.3 Shutdown Overhead

Although the X server trace analysis in the previous section suggests that there is a tremendous potential reduction in processor energy consumption, in practice it is much harder to realize this reduction. The chief reason for this is that the process of shutting down and restarting has a cost associated with it.

Enough processor state needs to be stored before shutting down so that the computation can be restarted and the state needs to be restored to restart the computation. This process requires additional compute time and power although the precise numbers vary depending on the hardware and software organization. Some newer microprocessors, such as ARM, AT&T Hobbit and some versions of 80386 and 80486, use static CMOS logic which gives them the ability to shutdown by stopping the clock, thus reducing the power consumption to the microwatts range. Very little state needs to be saved for this as the processor registers retain their values even when the clock is stopped. This type of shutdown can be accomplished in a few microseconds. However, in other cases the entire processor state may need to be stored in the memory - for example if the processor uses dynamic logic (as most processors do) or if one wants to conserve even more energy by doing a power down instead of just stopping the clock. The overhead now increases substantially to hundreds of microseconds, in some cases even to several milliseconds, as work similar to a context switch needs to be performed. At the penalty of more overhead, even more energy can be saved by storing the state on the disk as opposed to the main memory as it is no longer necessary to refresh the main memory which is typically DRAM. The overhead now increases to hundreds of milliseconds to several seconds.

The overhead due to shutdown creates problems. If the overhead is large enough, then there is problem of deciding *when* to shutdown. The time spent in the blocked state must be long enough to justify the overhead. If, after deciding to shut down, the blocking interval turns out to be too small, then one has to pay not only a power penalty, but more importantly, an effective slowing down of the computation speed because now the computer has to block for an interval higher than necessary. This slowing down translates into increased latency which, once it increases beyond a certain point, has an adverse impact on the interactive behavior of applications like X server.

10.2.2.4 Energy Reduction by Predictive Shutdown

The ideal situation, of course, would be to make the overhead zero or very small - but as the previous discussion suggested, overhead really is a function of the hardware and type of shutdown. The next best thing would be an a priori knowledge of the length of the blocking interval right at the beginning. Unfortunately, this is physically not possible.

One therefore has to resort to a heuristic to decide when to shut down. The approach used is based on the recent computation history and a prediction is made whether the idle time would be long enough to break-even with the shutdown overhead. Results demonstrate that for reasonable values of shutdown overhead, the predictive approach allows much higher energy savings to be achieved compared to the straightforward non-predictive approach, while the degradation in interactive performance is negligible.

Restricting the analysis to X server running on a dedicated processor, one can see that the server process starts in the *on* state, and makes alternate transitions from *on* to *off*, and from *off* to *on* state.

Let $T_{on}[i]$ and $T_{off}[i]$ be the time spent by the X server in the *i-th* visit to the *on* and the *off* state respectively, for $i = 1, 2, 3, ...$ Relating to the previous terminology, T_{on} is the sum of $T_{on}[i]$ over all i, and similarly T_{off} is the sum of $T_{off}[i]$ over all i.

Further, let T_{cost} be the time overhead associated with the process of shutting down. This is interpreted to mean that once shutdown, it takes time at least equal to T_{cost} before the computation can restart. Thus, if it turns out that the event that wakes up the X server and hence restarts the computation occurs before the expiry of this T_{cost}, a penalty is paid in terms of increased latency over the case with no shutdown.

Conventional idle-time based shut-down mechanisms, such as Apple's *Power Cycling*, are non-predictive in nature. Such non-predictive schemes decide to shutdown the computation once the time spent in the *off* state after the most recent entry (say, the *i-th* entry) exceeds a certain idle-time threshold, which is

2 seconds in Apple's scheme. In other words, shutdown is done on *i-th* entry to the *off* state once $T_{off}[i]$ has exceeded 2 seconds. Of course, once shutdown the computation cannot be restarted for at least T_{cost} time, so that if $T_{off}[i]$ would have been less than $2 + T_{cost}$, a penalty is paid in increased latency. The intuition behind such a scheme is that if the idle time has exceeded 2 seconds then most likely the user is going to remain idle for a long time. However, this scheme has two disadvantages. First, no shutdown is done for the first 2 seconds (or whatever is the idle time threshold that is chosen), and power is wasted during that period. Second, this scheme is able to take advantage of only relatively long idle periods - as the analysis presented earlier showed, the average $T_{off}[i]$ is less than a second, and thus relatively short idle periods are more common.

Prediction of $T_{off}[i]$

To address the deficiencies of the conventional idle-time based shutdown scheme, a predictive scheme for deciding when to shut down was designed and studied by us. The idea is that a simple heuristic rule is used that, on the *i-th* entry into the *off* state, predicts whether $T_{off}[i]$ is going to be long enough to justify shutting down. Specifically, the heuristic rule predicts whether $T_{off}[i] \geq T_{cost}$, and if so it is decided to shut the processor down.

The heuristic rule uses the computation history to make the prediction. The previous values of $T_{on}[i]$ and $T_{off}[i]$ form the obvious choice of computation history in our model. In particular, $T_{on}[1]$ ... $T_{on}[i]$, and $T_{off}[1]$ ... $T_{off}[i-1]$ can be used to predict $T_{off}[i]$. Further, the prediction rule itself needs to be simple to evaluate and not require too much state information. This intuition led to two approaches and the corresponding prediction rules:

- The first approach was to use regression analysis to arrive at a model for predicting $T_{off}[i]$. An analysis of the traces obtained from the instrumented X server resulted in the following model for $T_{off}[i]$ in terms of $T_{off}[i-1]$ and $T_{on}[i]$:

  $T_{off}[i] = 0.0740018 + 0.553733\, T_{off}[i-1] - 0.00947348\, T_{off}[i-1]^2 + 1.42233\, T_{on}[i] + 1.13883\, T_{off}[i-1]\, T_{on}[i] - 1.49143\, T_{on}[i]^2$

  A quadratic model with cross terms was used because the scatter plot of $T_{off}[i]$ versus $T_{on}[i]$ in Figure 10.6, and a similar plot of $T_{off}[i]$ versus $T_{off}[i-1]$ shown in Figure 10.7, both suggested a hyperbolic behavior. If the value of $T_{off}[i]$ predicted by the above model is $\geq T_{cost}$, a decision is made to shut the processor down.

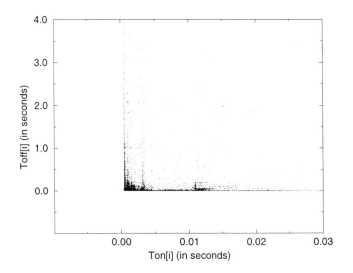

Figure 10.6: L-shaped $T_{off}[i]$ versus $T_{on}[i]$ Scatter Plot.

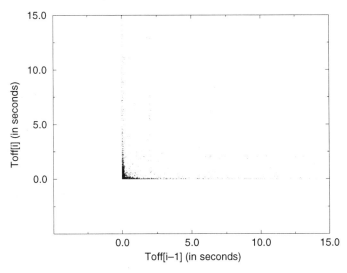

Figure 10.7: $T_{off}[i]$ versus $T_{off}[i-1]$ Scatter Plot.

The second approach is based on an even simpler intuition - $T_{on}[i]$ corresponds to the most recent history of $T_{off}[i]$, and the scatter plot of $T_{off}[i]$ versus $T_{on}[i]$ in Figure 10.6 is a L-shaped plot with the points concentrated along the two axes. A large value of $T_{on}[i]$ is followed by a small $T_{off}[i]$ with a very high probability, and the $T_{off}[i]$ following a small value of $T_{on}[i]$ is fairly evenly distributed. This suggests the following simple filtering (or thresholding) rule as the prediction heuristic:

$$T_{off}[i] \geq T_{cost} \quad \Leftrightarrow \quad T_{on}[i] \leq T_{on\_threshold}$$

and, Figure 10.6 suggests that for $T_{cost} = 10$ ms, a reasonable value of $T_{on\_threshold}$ is in the range of 10 ms to 15 ms. Note that $T_{cost} = 10$ ms is a very safe upper bound on the shutdown cost if the state is being saved in the main memory - in reality it is more likely to be x10-x100 smaller.

Hit Ratios for Prediction Schemes

The above heuristic rules to predict whether $T_{off}[i] \geq T_{cost}$ are similar in nature to the replacement rules in caches. A good idea of their efficacy can therefore be obtained by measuring the probability with which the prediction is correct, i.e. the *hit ratio*. The experimentally obtained traces were used to simulate the X server running with predictive shut down. Figure 10.8 shows the hit ratio for three schemes: the *model-based* prediction, the *on-threshold-based* prediction with $T_{on\_threshold} = 15$ ms, and the *on-threshold-based* prediction with $T_{on\_threshold} = \infty$. The last case corresponds to a scheme where one always decides to shut down. The hit-ratio curve corresponding to the *model-based* prediction shows a sudden jump at around $T_{cost} = 160$ ms. This is an artifact of the quadratic model. Finally, note that the hit ratio for the *on-threshold-based* prediction scheme is relatively insensitive to $T_{on\_threshold}$ changing from 15 ms to ∞ (the two curves almost overlap), whereas it does degrade with increase in T_{cost}.

Impact on Responsiveness - *Slowdown*

The time overhead T_{cost} associated with shutdown can have a negative impact on the responsiveness of the computer due to increased latency. This *slowdown* is difficult to quantify as it involves ill-defined psychological and biological metrics - in fact some studies even suggest second order effects in interactive behavior such as an increase in the user think time as the computer responsiveness degrades [Brady86]. In the absence of any well-defined metric, a simple and intuitive measure of slowdown was used: slowdown is the factor by which the total length of the X server session is increased due to shutdown being used. This increase occurs because many of the $T_{off}[i]$'s are longer than they would have

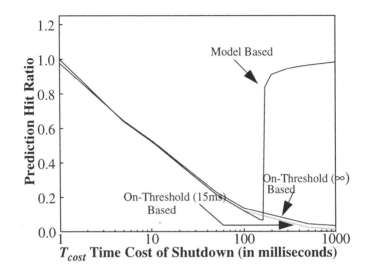

Figure 10.8: Hit ratio curves.

been without shutdown because of the overhead T_{cost} associated with shutdown. Figure 10.9 plots the slowdown resulting from the various schemes for different values of T_{cost}. For comparison the slowdown resulting from using a conventional *idle-time-threshold* based scheme with a threshold of 2 seconds, similar to Apple's Power Cycling scheme, has also been plotted.

As the curves demonstrate, the conventional schemes have almost no slow-down for all values of T_{cost}. However, the *on-threshold-based* prediction scheme performs reasonably well for values of T_{cost} smaller than 10-15 ms - the slow-down is less that 3 % to 4 % which is not noticeable at all. This was verified by creating a special version of the X server where an artificial delay, corresponding to T_{cost}, was added on entry to the *off* state. In fact, even a T_{cost} of 100 ms, which corresponds to a slowdown of 50% to 60% for *on-threshold-based* prediction scheme, resulted in a very usable, though noticeably sluggish, X server. As mentioned earlier, T_{cost} corresponding to saving state in the main memory and restoring state from the main memory is typically going to range from just a few microseconds for processors with ability to stop the clock, to a few hundreds of microseconds or a few milliseconds for other processors - and for these range of values of T_{cost} the slowdown is negligible for the predictive schemes.

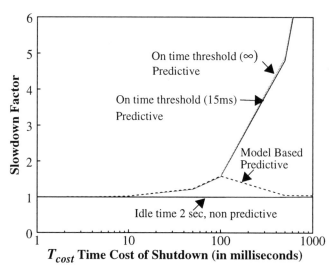

Figure 10.9: *Slowdown* as a function of T_{cost}.

Reduction in Computational Energy

Finally, various predictive, non-predictive, and ideal shutdown schemes were evaluated from the point of view of reduction in computational energy. The results are summarized in Table 10.2 for three values of T_{cost} - 0 ms, 10 ms and 100 ms - the former being the ideal case, and the latter two being safe upper bounds for storing the processor state in main memory and on disk respectively. The results demonstrate that for T_{cost} = 10 ms, a conservative upper bound for most processors, the predictive schemes have much superior energy savings at negligible slowdown. However, for T_{cost} = 100 ms, which may correspond to the case where the processor state is saved on the disk, the predictive schemes have noticeable degradation of interactive performance.

10.3 Architecture-Driven Voltage Reduction

Not all programmable computation, even on user-interaction dominated portable devices, is event-oriented - many functions that are not event-driven and instead execute continuously also require implementation as software running on microprocessors or embedded core processors on an ASIC. In fact embedded software for DSP core processors on an ASIC has emerged as the dominant method of implementing speech coding, speech compression, text-to-speech synthesis, sim-

		Energy Reduction	% Slowdown
$T_{cost} = 0(\text{ideal})$	Known $T_{off}[i]$	x 38.4	0 %
	Non-Predictive Idle-Time Threshold (2 seconds)	x 1.7	0 %
	On-Threshold-Based Predictive Scheme ($T_{on\_threshold} = 15$ ms)	x 20.1	0 %
	On-Threshold-Based Predictive Scheme ($T_{on\_threshold} = \infty$)	x 38.4	0 %
	Model-Based Predictive Scheme	x 38.4	0 %
$T_{cost} = 10$ ms	Known $T_{off}[i]$	x 25.7	0 %
	Non-Predictive Idle-Time Threshold (2 seconds)	x 1.7	0 %
	On-Threshold-Based Predictive Scheme ($T_{on\_threshold} = 15$ ms)	x 20.1	2.7 %
	On-Threshold-Based Predictive Scheme ($T_{on\_threshold} = \infty$)	x 38.4	2.8 %
	Model-Based Predictive Scheme	x 38.4	2.8 %
$T_{cost} = 100$ ms	Known $T_{off}[i]$	x 5.1	0 %
	Non-Predictive Idle-Time Threshold (2 seconds)	x 1.7	.025 %
	On-Threshold-Based Predictive Scheme ($T_{on\_threshold} = 1$ ms)	x 2.1	21.1 %
	On-Threshold-Based Predictive Scheme ($T_{on\_threshold} = \infty$)	x 38.4	59 %
	Model-Based Predictive Scheme	x 38.4	59 %

Table 10.2 Summary of Energy Reduction

ple speech recognition, modem functionality, and many other signal processing functions in portable devices such as cellular phones, PDAs, etc. Given the "continuous" nature of signal processing functions, shutdown strategies such as the

predictive shutdown technique of the previous section are not effective. The
architecture driven voltage reduction techniques presented in Chapter 4 is however very effective for such applications. Unfortunately, as described earlier and
as shown in Figure 10.10a, operation at reduced supply voltage comes at the
expense of increased gate delays, and consequently slower clock frequency for a
given circuit. However, DSP algorithms possess some nice attributes which allow
the effect of increased gate delays to be overcome. First, throughput is the sole
metric of computation speed for most DSP algorithms - latency is not an issue.
Second, DSP algorithms have ample inherent concurrency. Together, these two
attributes often make it possible for one to use architectural techniques such as
parallelism and pipelining to increase the computation throughput which can then
be traded-off against the loss in speed due to voltage reduction for a net gain in
energy efficiency for unchanged throughput.

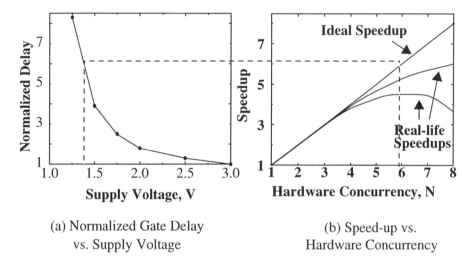

(a) Normalized Gate Delay   (b) Speed-up vs.
    vs. Supply Voltage   Hardware Concurrency

Figure 10.10: Trade-off between voltage and hardware-concurrency.

Previous chapters have described techniques for energy efficient computation in ASICs that use dedicated architectures for implementing DSP algorithms
through aggressive voltage scaling (down to 1-1.5 Volts) and concurrency exploitation; here architecture-driven voltage reduction will be applied to software programmable computation. While the supply voltages of many microprocessors

have indeed come down to 3.3V or 3V from 5V, the choice of 3.3V has been driven by the fact that for this level of voltage scaling the degradation in system performance (clock rate) is not very significant. Architecture-driven voltage reduction, on the other hand, reduces voltage even further into the realm where there is a significant reduction in clock rate, but uses architectural concurrency to keep the system throughput at its original level.

10.3.1 Voltage-Concurrency Trade-Off and Architectural Bottlenecks

As described in Chapter 4, the basic idea in architecture-driven voltage reduction is to compensate for the loss of speed due to increased gate delay when operating at a lower voltage, with the increase in speed due to increased hardware concurrency. The power consumption is reduced at a fixed computation speed (throughput). Hardware concurrency can be due to parallelism or pipelining or a mix of the two. For example, hardware that performs N operations in parallel has concurrency of N. Similarly, a hardware with N pipeline stages so that N successive operations are overlapped also has a concurrency of N. Compiler transformations such as parallelization and software pipelining play an important role in restructuring the computation so that the hardware concurrency is fully exploited.

Figure 10.10 illustrates this trade-off between voltage and hardware concurrency by plotting the curve for increase in gate delay (normalized to gate delay at 3.0V) next to hypothetical curves for increase in throughput (speedup) due to hardware concurrency, and using the same scale for normalized gate delay and speedup. With a hardware concurrency of 6 - for example by using a six processor parallel computer - an ideal speedup of x6 can be obtained as shown by the plot in Figure 10.10b. If the voltage is now reduced from 3.0V to a level where the gate delay increases by x6, the clock frequency will have to be reduced by x6 as well. Each of the processors will therefore slow down by x6 so that the net speedup with the six processor machine operating at the reduced voltage will be 1 compared to the uniprocessor machine operating at 3.0V. In general, the strategy would be to reduce the voltage to a level where the normalized gate delay increases by the same factor as the speedup. The throughput of the concurrent hardware operating at the reduced voltage level will then be the same as the throughput of the non-concurrent hardware operating at the original voltage of 3.0V. For our example the reduced voltage level at which the gate delay increases by x6 is 1.3V, as the dotted lines in Figure 10.10 show. Assuming that the switched capacitance for the six processor machine was x6 higher than for the uniprocessor machine, the power is reduced by a factor of $(1/6)*(3/1.3)^2*(6) = 5.3$.

Consider a more detailed analysis. Let $S(N)$ be the speedup for a hardware concurrency of N from Figure 10.10b. Obviously, for non-concurrent hardware $N=1$, and $S(1)=1$. Let $V(d)$ be the voltage for a normalized gate delay in Figure 10.10a, with $V(1)=3.0V$ being the reference point. The initial hardware has $N=1$, $S(1)=1$, and $V(1)=3.0V$. Let $f(1)$ and $f(N)$ be the clock frequencies for the initial and the final hardware, let $C(1)$ and $C(N)$ be the switched capacitances, and let $P(1)$ and $P(N)$ be the power consumptions.

If the hardware concurrency is due to parallelism, then one can operate each of the N parallel hardware units at a frequency $f(N)=f(1)/S(N)$ so that the effective throughput is unchanged. Further, parallel hardware units will require some overhead circuit so that the total capacitance can be expressed as $C(N)=(N+\varepsilon(N))*C(1)$ where $\varepsilon(N)$ is the overhead capacitance. Assuming unchanged switching activity, it follows that:

$$\frac{P(1)}{P(N)} = \frac{C(1)\,V(1)^2 f(1)}{C(N)\,V(S(N))^2 f(N)} = \frac{S(N)}{N+\varepsilon(N)} \times \left(\frac{V(1)}{V(S(N))}\right)^2 \quad (265)$$

Similarly, if the hardware concurrency is due to N-level pipelining, then the hardware can be operated at a reduced voltage of $V(S(N))$ so that the clock frequency will be $f(N)=N*f(1)/S(N)$, which in turn gives an effective throughput that is unchanged. Let the overhead capacitance due to pipelining be $\varepsilon(N)*C(1)/N$, so that the total capacitance is $(1+\varepsilon(N)/N)*C(1)$. Assuming unchanged switching activity, the ratio of the power consumption is again given by an expression identical to that in Equation 265.

An increase in energy efficiency is obtained whenever $P(1)/P(N)$ is > 1. In the ideal case of linear speedup $(S(N)=N)$ and no capacitive overhead $(\varepsilon(N)=0)$, it is clear from Equation 265 that $P(1)/P(N)$ is indeed > 1 because $V(1) > V(S(N))$. In the ideal case one can get arbitrarily large improvements in energy efficiency by continuing to increase hardware concurrency and decrease voltage until the devices stop working.

However, Equation 265 points to two fundamental architectural bottlenecks that prohibit an arbitrary increase in energy efficiency by reducing voltage. First, the speedup is not linear in most cases - as shown in Figure 10.10b, real examples typically show a speedup that starts saturating, and even decrease, as N is increased. In parallel hardware this may be due to lack of enough parallelism in the computation, or due to effects like bus contention. In pipelined hardware the speedup may not be linear because of granularity of pipelining, or because of the existence of pipeline interlocks and feedback cycles in the computation. The second architectural bottleneck occurs because $\varepsilon(N)$ is not zero in real world. In parallel hardware, capacitive overhead is contributed by data multiplexing/demultiplexing, bus arbitration etc., while in pipelined hardware the capacitive overhead is contributed by pipeline registers.

Together these two architectural bottlenecks - nonlinear speedup $S(N)$ and capacitive overhead $\varepsilon(N)$ - place a restriction on the increase in energy efficiency that can be obtained.

10.3.2 Exploiting Concurrency in Programmable Computation

From the preceding discussion it is clear that the energy efficiency of a computation can often be increased by trading voltage reduction with increased concurrency. There are two independent issues in this trade-off: the concurrency that is available in the hardware, and the concurrency that is inherent in the computation algorithm. A key role is played by compiler optimizations and transformations in exposing and mapping the algorithmic concurrency to the hardware concurrency. This problem has been an important area of research in computer science for many years, but with a different goal - increasing the computation speed. Nevertheless, the results are equally relevant to the goal of energy efficiency.

Unlike computation implemented on ASICs, the computation done in software is relatively general-purpose, and efficient compiler transformations and parallelization techniques to exploit the inherent concurrency are not as easily available except in special cases such as DSP and numerical computation. Applications like windowing systems, database access etc. are control intensive and logically more complex than the dataflow oriented DSP and numerical applications. Besides the obvious potential of pipelining execution of successive instructions, here are three common types of concurrency that are available in software computation:

- **Instruction Level Parallelism**
  The fine-grained instruction level parallelism inherent in a code is discovered by a dataflow analysis, either dynamically in hardware, or stati-

cally in software. Research [Butler91] has shown that instruction level parallelism in excess of 17 per cycle exists in the SPEC suite, a set of applications that are the *de facto* standard compute-bound benchmarks. Further, with properly balanced hardware, an execution rate of 2 to 5.8 can be sustained. Compiler optimization techniques, such as software pipelining, loop unrolling, and branch prediction, help in enhancing the available instruction level parallelism.

- **Thread Level Parallelism**
  Many applications have a natural coarse-grained parallelism that is suitable for mapping as separate thread of executions. A good example of this would be a multi-threaded X server that handles different clients via multiple threads. Even on a uniprocessor a multi-threaded organization can improve speed by reducing unnecessary blocking. When multiple processors are available then the thread level parallelism can be exploited by mapping different threads to different processors. The thread level parallelism is usually specified by the user - compilers that can restructure and/or partition an algorithm into multiple threads do not exist, except in special domains such as DSP [Hoang92]. Unfortunately, very few existing applications are multi-threaded, although this has started to change with the availability of threads in various operating systems.

- **Process Level Parallelism**
  Process or task level parallelism is inherent in the assumption behind multi-tasking operating systems like UNIX. Multiple processes, usually corresponding to different applications and often different users, time-share the CPU. Since processes often have to block for I/O, time-sharing results in improved CPU utilization (computation throughput) at the cost of increased latency for compute-bound tasks.

Complementary to the algorithmic concurrency is the concurrency available in the hardware. There are three main ways of increasing the concurrency in a software programmable computer:

- **Increase the number of processors**
  This is the approach taken by MIMD computers such as Sequent and Sun SPARC-10s. The overhead comes in the form of more complex hardware for sharing data between processors as well as more complicated operating system. This approach can exploit process and thread level parallelism, although the latter will require a more sophisticated OS.

- **Increase the number of functional units**
  This is the approach taken by SuperScalar microprocessors and VLIW processors. Of course this comes at an overhead in the form of a more

complex instruction issue unit. This approach is meant to exploit instruction level parallelism.

- **Increase the levels of pipelining**
  This is the approach taken by superpipelined microprocessors. This too comes at an overhead in the form of an increased pipeline bubble. This approach is also meant at exploiting instruction level parallelism.

The success of a compiler in utilizing the intrinsic concurrency of an algorithm to exploit the available hardware concurrency is a major determinant in achieving energy efficiency through architecture-driven voltage reduction.

10.3.3 A Power Consumption Model for MIMD

To evaluate the effect of hardware parallelism on power, a model to estimate the power consumption needs to be developed. For this one needs to define the following quantities:

Let $S(N)$ = computation speed when the MIMD system has N processors. The speed metric $S()$ could represent either throughput or latency. Also, let $V(N)$ = lowest supply voltage at which one can run the N processors while maintaining the same throughput as with the uniprocessor system. Assume that the uniprocessor is running at a reference voltage of 3V (i.e., $V(1) = 3V$). Given the speedup factor (based on the number of processors and algorithmic constraints), the supply voltage $V(N)$ can be determined from the function shown in Figure 10.10a which is obtained by characterizing the semiconductor process. For example if $N = 4$, and if $S(4)=4$, then from Figure 10.10a, $V(4)$ is approximately 1.5V.

Ignoring static power consumption due to leakage currents, the power consumed by CMOS circuits is given by $C\,V^2 f$. Therefore the power consumption of a uniprocessor is:

$$P(1) = \left(C_{processor} + C_{interconnect} + C_{memory} \right) V(1)^2 f \quad (266)$$

where $C_{processor}$, $C_{interconnect}$ and C_{memory} are determined for a specified set of hardware modules (e.g. DSP32C or 386SL) and technology (PCB interconnect => 3-4 pF/inch routing capacitance). For example, for a DSP32C processor, the average capacitance switched is given by $1.2W / ((5.25)^2\ 25MHz) = 1.75nF$ (the power of 1.2W, obtained from data sheets, does not include the I/O power). For interconnect, considering both the bus and pin components, and assuming an average switching activity, the capacitance was determined to be 0.15nF.

Now consider extending the model presented in Equation 266 to multiple processors. Ideally, speedup grows linearly with the number of processors; however, due to interprocess communication overhead, the speedup is typically not linear. Similarly, parallelism will introduce capacitance overhead. The overhead is often attributed to an increase in interconnect capacitance (longer wires on the PC board/MCM), loading capacitance (more processors on the shared bus), control capacitance, interprocess communications overhead (e.g. more memory I/O transactions), etc. Taking these factors into account, the power consumption as a function of the number of processors is given by:

$$P(N) = [NC_{processor} + NC_{interconnect} (N + \%C_{over}) + \\ C_{memory} (N + \%C_{over})] f V(N)^2/S(N) \qquad (267)$$

where $\%C_{over}$ is the interprocessor communications overhead. Note that the physical interconnect capacitance, the second term, increases linearly with the number of processors. This is a reasonable assumption since for PCB interconnect technology, the processors can be placed right next to each other (in a linear fashion) and routing can be performed on multiple layers (unlike ASICs where the routing is confined typically to 2 or 3 metal layers).

Note that a minimum bound on the supply voltage, V, can be set by one of the following:

- Noise immunity requirements.
- $S(N)$ stops increasing with N. Using the curve in Figure 10.10a, a maximum bound on N can be translated into a minimum bound on V.
- There may be a maximum bound on N due to operating system and other software considerations. This too can be translated into a corresponding minimum bound on V.

Intuitively, a model of speedup in MIMD will have the following characteristics:

- For small values of N the speedup will be close to linear.
- At higher values of N, the speedup will start saturating and reach a peak.
- Finally, for even larger values of N, the speedup may actually start falling off. This may happen because, for example, the arbitration overhead for access to a shared bus may just grind the system to a halt.

Such a "hump" like speedup curve is characteristic of shared bus centralized memory systems, and there is no rational reason to parallelize beyond the point of saturation. In other words, one is only interested in the initial part of the curve. One such model of speedup, using analytical results from [Stone87], is given by the following:

$$S(N) = \frac{N}{1 + \dfrac{N-1}{r}} \qquad (268)$$

where r is a constant. As N gets large, the speedup will saturate to r. The value of r depends on, among other things, the ratio of computation to communication, the granularity and number of interacting tasks, and the communication network.

10.3.4 Example: CORDIC on a MIMD using Thread Level Parallelism

This example is a signal processing application running on programmable processors. For such applications, it is typically throughput (and sometimes latency) which is the design constraint rather than trying to compute as fast as possible. For example, a speech codec has to compress and decompress speech at a sample rate of 8Khz; however, once this throughput requirement is met, there is NO advantage in making the computation faster. As described earlier, by making the computation faster (using more parallelism for example), the supply voltage can be dropped and hence the power can be reduced while still meeting the functional throughput. Typical applications include filters, speech processing, robotics and image processing.

Most examples in this class exhibit a substantial amount of concurrency. For example, all DSP applications are executed in an infinite time loop, giving rise to temporal concurrency. In addition, several exhibit spatial concurrency. For low-power operation it is important to detect and exploit concurrency since one can run at reduced supply voltages. For example, temporal concurrency can often be exploited by pipelining, resulting in dramatic speedup and hence lower power for a fixed throughput. For application that have a lot of recursion (feedback loops), it often necessary to apply a series of transformations (like loop unrolling) to achieve speedup. Compilers have been developed to effectively exploit concurrency in DSP applications [Hoang92].

To study the power trade-offs of DSP multi-processor implementations, a cordic algorithm (obtained from [Hoang92]) is studied that converts cartesian to polar coordinates iteratively in 20 steps. It takes as input an (X,Y) coordinate, as well as an array of correction angles. The loop iteratively calculates the corresponding amplitude and phase. Since each iteration is dependent on the results of the previous iteration, the computation is sequential in nature.

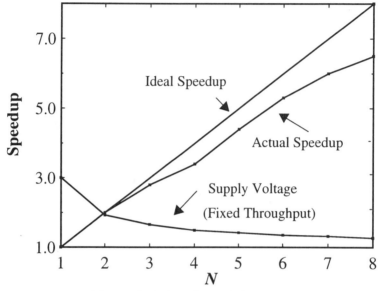

Figure 10.11: Speedup and V_{dd} vs. N

A scheduling algorithm which only exploits spatial concurrency would perform poorly on this example. However, optimizing using transformations like pipelining the loop and assigning successive loop iterations to successive processors, significant speedup can be achieved. The analysis is based on the DSP32C processor power numbers. Figure 10.11 shows the ideal speedup (linear with processors) and the actual speedup obtained. Figure 10.11 also shows the lowest supply voltage ($V(N)$) at which the various multiprocessor implementations can run while meeting the same functional throughput as the uniprocessor version.

Figure 10.12 shows the power consumption as a function of N. It is seen from the figure that a factor of 3.3 reduction in power can be accomplished by reducing the supply voltage from 3V to 1.5V. The power starts to increase with more than 6 processors since the overhead circuitry (interconnect capacitance) starts to dominate, resulting an optimum number of processors for minimal power.

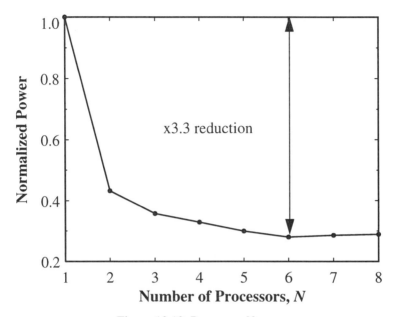

Figure 10.12: Power vs. N.

10.3.5 Low Power Computers - Multiple Slow CPUs More Energy Efficient than a Single Fast CPU?

The discussion in this section shows that if energy efficiency is the consideration, it is better to use concurrent (parallel or pipelined) slower hardware operating at a lower voltage than to use non-concurrent faster hardware at a higher voltage. To put it differently, multiple slower CPUs are more energy efficient than a single fast CPU for applications with enough algorithmic concurrency. This suggests that from a power perspective it may be better for future microprocessor chips to consist of multiple slow and simple CPUs operating at a slow clock frequency and a low voltage - a conclusion which is just the opposite of the current trend towards complex CPUs operating at extremely high frequencies (many 100s of MHz).

10.4 Summary

Two architectural approaches for energy efficient programmable computation were described: predictive shutdown and concurrency driven supply voltage reduction. A significant reduction in power consumption can be obtained by employing these architectural techniques. For example, parallel processors operating at a reduced voltage can substantially improve the energy efficiency of a fixed throughput computation (by a factor of 4 in the CORDIC example that was described). For applications where continuous computation is not being performed (like X-server, etc.), an aggressive shut down strategy based on a predictive technique has been presented which can reduce the power consumption by a large factor compared to the straightforward conventional schemes where the power down decision in based solely on a predetermined idle time threshold.

REFERENCES:

[Apple92] Apple Computer Inc., "Power Manager IC", and "Reduced Power Modes," in Chapter 20 of *Technical Introduction to the Macintosh Family*, 2nd Edition, Addison Wesley, October 1992.

[Berkoff93] B. I. Berkoff, "Taking Charge: PowerBook-Battery Management," *MacUser*, February 1993.

[Brady86] J. T. Brady, "A Theory of Productivity in the Creative Process," *IEEE CG&A*, May 1986.

[Butler91] M. Butler, T. Yeh, Y. Patt, *et al.*, "Single Stream Parallelism is Greater than Two," *18th International Symposium on Computer Architecture*, May, 1991.

[Hoang92] P. D. Hoang, "Compiling Real-Time Digital Signal Processing Applications onto Multiprocessor Systems," Ph. D. Thesis, EECS Department, University of California, Berkeley, June 1992.

[Stone87] Harold S. Stone, "Multiprocessor Performance," in Chapter 6 of *High-Performance Computer Architecture*, Addison Wesley, 1987.

[Weiser93] M. Weiser, "Some Computer Science Issues in Ubiquitous Computing," *Communications of the ACM*, Vol. 36, No. 7, pp. 75-85, July 1993.

11

Conclusions

Portable operation is becoming increasingly important for many of the most exciting new electronic products. The strict limitation on power dissipation which this imposes must be met by the designer while still meeting ever higher computational requirements. To meet this need, a comprehensive strategy is required at all levels of the system design, ranging from the algorithms and architectures to the logic styles, circuits, and underlying technology. An integrated methodology and the associated CAD tools for low power CMOS design has been presented here which incorporates optimization all levels of system design.

For a properly designed CMOS gate, power is consumed primarily in charging and discharging parasitic capacitors and is given by $\alpha C_L V_{dd}^2 f$ where α is node transition activity, C_L is physical load capacitance, V_{dd} is supply voltage, and f is operating frequency.

The choice of supply voltage clearly has the greatest impact on the power-delay product, which is the amount of energy required to perform a given function. It is only necessary to reduce the supply voltage to quadratically improve the power-delay product. Conventional low-power approaches have exploited the velocity saturated nature of sub-micron devices to apply a limited amount of voltage scaling from 5V down to 3.3V - the "critical" voltage at which there is little degradation in circuit speed relative to 5V. Unfortunately, further reduction in supply voltage is associated with a reduction in the circuit speed, with the gate delays increasing drastically as the voltage approaches the threshold voltages of the devices. However, if the goal is to reduce the power consumption for realizing a fixed functionality (or fixed throughput), then an architectural voltage scaling strategy, that involves compensating the increased gate delays at reduced voltages with increased level of parallelism, should be used to scale the power supply voltage down to 1-1.5V (for a standard CMOS technology) while maintaining system throughput. Not only can such an approach that trades silicon area for lower power consumption be applied for signal processing applications and other dedicated applications that have a fixed computation rate, but also for those general purpose applications where the goal is to increase the MIPS/Watt for a fixed MIPS level. Therefore, future power conscious microprocessor chips should use multiple slow and simple CPUs operating at a slow clock frequency and a low voltage - a conclusion which is just the opposite of the current trend towards complex CPUs operating at extremely high frequencies.

To scale power supply voltages beyond the 1-1.5V scaling presented in this work, it is necessary to increase the leakage component of power by scaling the device threshold voltages to levels that are much lower than the values currently used (0.7V to 1V). Device threshold voltages should be scaled to less than 300mV, so as to make the subthreshold leakage power and switching power equal. The use of low V_t devices will increase the overall power consumption for circuitry that is powered down using gated clocks since sub-threshold current will continue to flow even though no switching power is consumed. Special circuitry will be required to dynamically adjust the leakage power for such applications - that is, increase the leakage when active and decrease it when idle. One approach is to dynamically adjust the substrate bias to modulate the threshold voltages. Another approach is to use a process with a high V_t and a low V_t and use the high V_t devices to turn off devices when idle to eliminate leakage power.

High efficiency DC-DC conversion is an integral part of low-power system design. The freedom to scale the power supply voltages to arbitrary levels depends strongly on the ability to provide high efficiency power supply conversion at arbitrary levels under varying loads. Adaptive power supply circuits can be used to minimize power by dynamically changing the voltage levels for applications that require processing of bursty input data. Also critical to the design of multi-voltage systems, in which different modules are optimized to operate at their own "optimal" supply voltage, is the use of level conversion circuits.

Besides architecture driven voltage scaling, another approach to low power design is to reduce the switching activity to the minimal level required to implement a given function. To reduce the average number of transitions per operation to the minimal possible level, it is necessary to exploit the temporal and spatial correlations present in the data being processed while optimizing at all levels of the system design. When minimizing transition activity, it is not enough to just reduce the number of transitions due to the static behavior of the circuit but also those transitions that occur due to the dynamic nature of the circuit.

The choice of algorithms has the greatest impact on minimizing total switched capacitance. Just by redefining the approach to solve a problem, the computational complexity can be significantly altered (for example, using VQ decoding vs. DCT decoding). At the algorithmic level, it also important to optimize the data representation and data coding (in which controlled correlation is introduced to reduce the total number of bit transitions). Next to algorithmic optimization, architecture optimization has the greatest impact on the switching activity. A key conclusion on the architectures that minimize transition activity is that time-multiplexing of hardware can destroy naturally occurring temporal correlations that exist in data. Therefore, resource sharing should be avoided as much as possible. When the amount of computational resources is limited, an optimized assignment and schedule must be used which keeps uncorrelated data on different hardware units. The sequencing of operations should be optimized to reduce the switching activity by altering the dynamic range. Race free topologies should be used to minimize spurious transitions that occur due to timing skew. Self-timed circuits are useful only when the overhead is minimal, such as is the case for memory circuits.

At the logic and circuit level, gated clocks should be carefully used to eliminate wasteful logic and bus transitions. At the physical design level, the place and route should be optimized such that signals that have high switching activity can be assigned short wires while signals that have low switching activity can be allowed to have long wires. This is different from current approaches which typically use placement schemes to minimize area.

Clearly, design tools which encapsulate the above methodologies are needed. The techniques presented to reduce power consumption are by no means orthogonal, making manual optimization very difficult. For example, an optimization that reduces the number of operations and therefore switched capacitance can increase the critical path dictating a higher voltage. Therefore, high-level design tools are required to automatically find computational structures that result in the lowest power consumption. Such design tools will require fast estimation of power consumption for a given topology from a high-level of abstraction, which allows the exploration of several structures without having to map high-level descriptions to low-level implementations. The only viable approach to estimate interconnect and control without mapping a high-level design to layout, is to develop statistical models based on a representative benchmark set. The efficient application of high-level optimization techniques requires an "intelligent" ordering of optimizing transformations and it was shown that a combination of heuristic and probabilistic strategy produces efficient solutions.

The various low-power techniques developed have been demonstrated in the design of chipset for a portable multimedia terminal that allows users to have untethered access to multimedia information servers. Six chips were designed to provide the interface between a high speed digital radio modem and a commercial speech codec, pen input circuitry, and LCD panels for text/graphics and full-motion video display. The chips provide protocol conversion, synchronization, error correction, packetization, buffering, video decompression, and D/A conversion at a total power consumption of less than 5 mW - which is three orders of magnitude lower than existing systems. This power reduction was possible due to the system level approach used that involved system partitioning, minimizing the number of operation, choice of data representation, architecture driven voltage scaling, activity driven hardware assignment, use of gated clock, optimized cell-library, and activity driven place and route.

We have demonstrated circuitry for multimedia applications that works at power supply voltages as low as 1.1V using a standard process that has a worst case threshold voltage of 1V. This is by no means a lower bound on the operating power supply voltage. In fact, inverter circuits have been demonstrated in 1974 that operate from a 200mV power supply voltage (Chapter 2). The lower bound on power supply voltage needs to be investigated for the implementation of large and high throughput digital systems by scaling the threshold voltages of the devices. Threshold voltage scaling will also require new circuit and power management techniques. DSP circuits that operate continuously, can tolerate a fixed low threshold voltage, but circuitry that is frequently powered down such as the multiply unit of a microprocessor cannot tolerate a fixed low threshold voltage. Adaptive bias circuitry to turn off devices when idle need to be developed. Other techniques for power reduction such as adiabatic switching techniques, which reduce the energy per switching event at a fixed voltage, should be further explored.

Index

Active-matrix displays, 313
Adaptive dead-time control, 172–174
Adaptive filters, 268
Adder-comparators, 132–137
Adiabatic amplification, 186–195
Adiabatic buffers
  conventional buffers compared with,
    192–194
  one-stage, 189–190, 195
  two-stage, 190–191, 195
Adiabatic charging, 184–186
Adiabatic logic gates, 195–203
Adiabatic switching, 6, 181–216
  amplification in, 186–195
  charging in, 184–186
  logic gates in, 195–203
  pulsed-power supplies in, 186, 210–214
  stepwise charging in, 203–210
Air-core coils, 163–164
Algebraic transformations, 272, 273
Algorithmic optimization, 219–234, 257,
    see also Coding; Vector quantization
Algorithms
  heuristic, 297–299
  probabilistic, 297–299
  for video decompression, 339–342
AND gates, 84
  inter-signal correlations and, 79
  signal statistics and, 77
  technology mapping and, 246
Architectural bottlenecks, 387–389
Architectural power estimation, 264–265
Architecture
  memory, 334–336
  overview of approaches in, 368–370
  of portable multimedia terminals,
    317–319, 364
  switched capacitance minimization and,
    235–245, 257
Architecture-driven voltage scaling,
    117–138, 384–395
  hardware duplication in, 118–125
  optimal supply voltage for, 125–129
  parallel, *see* Parallelism
  pipelining, *see* Pipelining
  voltage-concurrency trade-off in,
    387–389, 395–396
Area capacitance, 63–64
Arithmetic computation, 235–239
ARM, 378
Associativity, 279, 298–299

Asynchronous designs, 253–254
AT&T Hobbit, 374, 378

Batteries
  in DC-DC conversion, 142–145
  for portable multimedia terminals, 312
BiCMOS, 30
Binary coding, 225–228
Bitmaps, 315–317, 318, 326
Blocked state, 370, 371, 373–374, 375, 376
Body diode reverse recovery, 167
Body effects, 188
Boltzmann's constant, 19
Boltzmann's probability distribution
    function, 20
Boost converters, 153–155, 157
Buck-boost converters, 153–156, 157
Buck converters, 152–153, 154–155
  conversion ratio and, 156
  converter miniaturization and, 157–158
  power transistor sizing and, 174
  synchronous rectification and, 168, 169
  zero-voltage switching and, 170, 171–172,
    176, 177
Bus inversion coding, 228–230

Capacitance estimate, 281–294
  of control logic, 287–294
  of execution units, 282–284
  of interconnect, 284–287
  registers in, 284
Capacitive switching loss, 167, 281–282
Carry-lookahead topology, 246–247
Carry-select topology, 246–247
Carry-skip topology, 247
Chain implementation, 84–85
Channel hot carriers (CHC), 107
Charge pumps, *see* Switched capacitor
    converters
Charge sharing, 58–59
Chip performance index (*CPI*), 13, 47–48,
    49
Chips, 44–48
Chrominance video decompression,
    347–348
Circuit limits, 4, 12, 29–34
Circuit optimization, 166–177, 249–254
  adaptive dead-time control, 172–174
  power transistor sizing, 174–175